T0227684

Algorithmic and Geometric Aspects of Robotics

First published in 1987, the seven chapters that comprise this book review contemporary work on the geometric side of robotics. The first chapter defines the fundamental goal of robotics in very broad terms and outlines a research agenda each of whose items constitutes a substantial area for further research. The second chapter presents recently developed techniques that have begun to address the geometric side of this research agenda and the third reviews several applied geometric ideas central to contemporary work on the problem of motion planning. The use of Voronoi diagrams, a theme opened in these chapters, is explored further later in the book. The fourth chapter develops a theme in computational geometry having obvious significance for the simplification of practical robotics problems — the approximation or decomposition of complex geometric objects into simple ones. The final chapters treat two examples of a class of geometric 'reconstruction' problem that have immediate application to computer-aided geometric design systems.

Algorithmic and Geometric Aspects of Robotics

Edited by
Jacob T. Schwartz
Chee-Keng Yap

First published in 1987
by Lawrence Erlbaum Associates, Inc.

This edition first published in 2016 by Routledge
2 Park Square, Milton Park, Abingdon, Oxon, OX14 4RN
and by Routledge
711 Third Avenue, New York, NY 10017

Routledge is an imprint of the Taylor & Francis Group, an informa business

ISBN 13: 978-1-138-20347-1 (hbk)
ISBN 13: 978-1-315-47145-7 (ebk)

Algorithmic and Geometric Aspects of Robotics
Volume 1

Edited by

JACOB T. SCHWARTZ
CHEE-KENG YAP
Courant Institute of Mathematical Science
New York University

LEA LAWRENCE ERLBAUM ASSOCIATES, PUBLISHERS
1987   Hillsdale, New Jersey   London

Lawrence Erlbaum Associates, Inc., Publishers
365 Broadway
Hillsdale, New Jersey 07642

Library of Congress Cataloging in Publication Data

Algorithmic and geometric aspects of robotics.

(Advances in robotics ; vol. 1)
Includes bibliographies and indexes.
1. Robotics. 2. Robot vision. I. Schwartz, Jacob T.
II. Yap, Chee-Keng. III. Series: Advances in robotics
(Hillsdale, N.J.) ; vol. 1.
TJ211.A44 1987   629.8′92   86-23963
ISBN 0-89859-554-1

Printed in the United States of America
10 9 8 7 6 5 4 3 2 1

Contents

Preface   ix

Introduction   1
Chee-Keng Yap

1. Why Robotics?   *1*
2. An Overview of the Contents of the Volume   *3*
3. Brief Review of Standard Terminology   *4*

1. **The Challenge of Robotics for Computer Science**   7
    John E. Hopcroft and Dean B. Krafft

    1. What is Robotics?   *7*
    2. A New Science   *9*
    3. Representations for Physical Objects
        and Processes   *10*
    4. Manipulating Object Representations   *22*
    5. Reasoning   *28*
    6. Conclusions   *40*
    References   *40*

2. **Computational Geometry—A User's Guide**   43
    David P. Dobkin and Diane L. Souvaine

    1. Introduction   *43*
    2. Hierarchical Search   *44*
    3. Hierarchical Computation   *66*
    4. Geometric Transformations   *77*
    5. Conclusion   *90*
    References   *91*

3. Algorithmic Motion Planning   **95**
Chee-Keng Yap

   1. History   *97*
   2. Basic Concepts   *99*
   3. A Taxonomy of Motion Planning Problems   *103*
   4. Issues Related to Algorithmic Motion
      Planning   *106*
   5. Moving a Disc   *108*
   6. Moving a Ladder   *112*
   7. Two Approaches to Motion Planning   *119*
   8. Solution to the General Motion Planning
      Problem   *123*
   9. Retraction Via Cell Complexes   *129*
  10. Lower Bounds   *132*
  11. Summary: New Directions and Open
      Problems   *137*
  12. Conclusion   *140*
  References   *141*

4. Approximation and Decomposition of Shapes   **145**
Bernard Chazelle

   1. Introduction   *145*
   2. Approximation of Shapes   *146*
   3. Decomposition of Shapes   *165*
   4. Epilogue   *180*
  References   *181*

**5. Intersection and Proximity Problems
   and Voronoi Diagrams**   **187**
Daniel Leven and Micha Sharir
   1. Introduction   *187*
   2. A Simple Intersection Detection
      Algorithm in Two-Dimensional Space   *190*
   3. Generalized Planar Voronoi Diagrams   *192*
   4. Applications of Voronoi Diagrams   *197*
   5. Efficient Construction of Voronoi
      Diagrams   *202*
   6. Dynamic Intersection and Proximity
      Problems   *216*
   7. The Three-Dimensional Case   *216*
  References   *226*

6. **Fleshing Out Wire Frames: Reconstruction of Objects, Part I**   **229**

George Markowsky and Michael A. Wesley

1. Introduction   *229*
2. Basic Concepts   *231*
3. The Wire Frame Algorithm   *237*
4. Examples   *252*
Appendix A: Topological Concepts   *256*
References   *258*

7. **Fleshing Out Projections: Reconstruction of Objects, Part II**   **259**

Michael A. Wesley and George Markowsky

1. Introduction   *259*
2. Basic Concepts and Results   *261*
3. Fleshing Out Unlabeled Projections   *269*
4. Additional Information from Drawing Conventions   *280*
5. Examples   *282*
6. Summary   *292*
References   *295*

AUTHOR INDEX   297
SUBJECT INDEX   303

Preface

Although robotics has by now been a recognized area of computer science for several decades, it is only during the last few years that the level of theoretical work in it has begun to expand rapidly. Nevertheless, it seems certain that research in this area is destined to affect computer science profoundly. Till now, computer science has been largely combinatorial and symbolic, having the manipulation of patterns and tables of data as its principal content. In robotics, however, computer science makes contact with real-world geometric and physical phenomena such as the compliance of elastic bodies, the frictional phenomena which occur when bodies come in contact, errors in modeling which are inevitable in the real world, the sudden changes of state which occur when bodies collide unexpectedly, and so forth. It is to be expected that much interesting new computer science will emerge from contact with these rich conceptual domains and, in particular, that computer science will become more traditionally mathematical and "continuous."

Fields of science have their own internal rhythms, in which periods of slow progress conditioned by a lack of ideas or by the exhaustion of old ideas alternate with the excitement of rapid advance triggered by conceptual breakthroughs or by maturation of supporting technology. Sustaining technological development and systematic conceptual advance often go together and reinforce one another. After its slow start during the past several decades, robotics stands at the start of a period of rapid advance, current theoretical and pragmatic developments foreshadowing major progress in many of its subfields. The massive computational power created by VLSI technology is a significant driving force: By making computing cycles available in whatever quantity required, the work of VLSI designers is rapidly creating most of the purely "electronic" side of the technological base which robotics will require. Armed with this technology, robotics researchers have begun to perceive ways of radically strengthening the basic capabilities of robots, e.g., their ability to see, to manipulate, and to plan. As these capabilities are improved, additional work integrating them into composite software environments facilitating robot use must also follow.

In recognition of the increasing importance and pace of work in this area, the present volume initiates an annual series of publications which will track and reflect the progress of robotics over the coming years. We have chosen to begin with a relatively theoretical subarea, namely automatic planning of robot motions, which has attracted considerable attention of late and which serves well to illustrate the depth and a variety of the mathematical and geometric issues arising out of robotics. The problem here is to develop algorithms that will allow a robot which knows the geometry of the environment in which it must move to plan the details of its motion automatically. Moreover, if the robot is grasping a body (of known geometry) it must allow for this in the motion it plans. As the chapters in the present volume show, this motion planning problem has begun to yield to the efforts of theorists and algorithm designers who have found it possible to apply methods developed by topologists and algebraic geometers to this practical area.

However, this is only one of many theoretical and pragmatic problems in a very broad field. The general aim of robotics can be defined as the mechanization of that elementary *operative* intelligence which people use unthinkingly in locating and handling ordinary objects. Such research has two principal aspects: sensory and manipulative. Sensory studies aim to develop techniques which make it possible to organize the raw data gathered by sensors such as video cameras and ultrasonic rangefinders into perceptually meaningful gestalts. Studies of manipulation deal with all the tactics and strategy needed to control bodies moving slowly or rapidly through three dimensional space, both when the controlled (robot) bodies must move avoiding contact with other bodies or obstacles in their environment, and also when the controlled robot bodies need to make contact with portions of their environment or with other robots, e.g., to grip an object which is to be moved, to insert a peg into a hole, etc. Control of highly dynamic motions is another intriguing area of the *manipulative* or *dynamic* side of robotics; control of delicate and dextrous motions is yet a third. Work in this latter area aims at robots that can adjust smoothly and simply to the shapes and physical behavior of delicate 3-D bodies. For example, one wishes to be able to grasp an egg, either to draw some figure on it with a stiff pen, or to carry it to the edge of a cup, crack it, and then (more dynamically) pour its contents into the cup. Here a variety of problems arise. Sophisticated multidimensional feedback control methods are needed and are rapidly being defined. Work in computational geometry is elucidating the interesting geometric issues involved in management of mechanisms with many degrees of freedom. Theoretical attention is being directed to one of the most neglected areas of classical physics, the analysis of the frictional motions of rigid and flexible bodies. Interesting robot hands, which will provide appropriate levels of experimental challenge to control theorists, computational geometers, and robotic software designers, are being developed by several engineering groups.

The enormous field of robot vision illustrates the sensory side of robotics. The problem here is to find techniques which make it possible to organize the raw data gathered by a video camera into perceptually meaningful gestalts. A wide variety of approaches to this deep problem are available. For example, in *model-based* vision studies, attention is confined to scenes containing only known objects or objects belonging to known parametric classes (e.g., cylinders with spherical caps and cylindrical holes bored in them, but of heights and radii not known a priori). This contrasts with *general* vision studies, which aim to impose helpful perceptual groupings on entirely general scenes, e.g., landscapes containing shrubbery. The great advantage of the first problem is that it is entirely objective: Its aim is simply to reduce a scene known to contain objects drawn from a fixed finite set to a table giving the identities and orientations of all objects actually present. In contrast, the deeper *general* vision problem has inherently psychological aspects: Here one aims to devise (the image-analysis portions of) a robot "eye" whose perceptual groupings are close enough to those formed by the human eye for easy communication and mutual understanding to be possible. Even the narrower *model-based* vision problem can be put at various levels of difficulty, to which theoretical and experimental work can be expected to advance progressively over the next few years. Specifically, one can consider:

(i) images of either 2D or 3D bodies, which can either been seen in isolation or as parts of compound scenes.

(ii) the bodies with which one deals can either be wholly visible or partially obscured, and can be present either in constrained or in perfectly general orientations.

(iii) the bodies seen can either be stationary or can be allowed to move.

(iv) the bodies seen can either conform exactly to their models, or can be affected by extra error features such as "burrs," "dents," "flash," etc.

Recent work makes successful treatment of all these problems appear feasible.

Vision studies are also conditioned by the form of input they assume, and a variety of schemes have been developed for acquiring information-rich images when use of simpler images complicates the object identification problem. The images from which a model-based approach works can either be:

(a) ordinary intensity images

(b) high quality silhouettes, obtained, e.g., by "backlighting" a scene

(c) "depth images," in which each pixel records the true geometric distance of an observed point P on a body surface from the camera, or, equivalently, P's true geometric position in 3-dimensional space.

It appears likely that effective analysis of all of these kinds of images will be possible provided that only finitely many objects of shapes known a priori can be present in the scene being viewed.

The topics briefly noted in the preceding paragraphs, plus many others that can be expected to arise as new aspects of the rapidly expanding field of robotics research, will be subjected to detailed scrutiny in the series of volumes hereby initiated. Early volumes dedicated to such subareas as robot vision, control-theoretic aspects of robotics, robot kinematics, robot software systems and programming languages, as well as the theoretical aspects of such apparently pragmatic subjects as parts mating and other manufacturing-science related issues will be discussed. It is hoped that these volumes succeed in conveying the interest and excitement of this relatively new field to the widest possible audience.

J. T. Schwartz

Algorithmic and Geometric
Aspects of Robotics

Introduction

Chee-Keng Yap

1. WHY ROBOTICS?

Robotics is the discipline concerned with the science and technology of computer controlled autonomous mechanisms. It includes both an *effectuator* side (which focuses on such matters as the geometry and control of manipulator arms, grippers, robot locomotion and balance, etc.) and a *sensory* side (concerned with robot vision, tactile sensation, object identification using sensory inputs in visual, tactile, or other modalities such as ultra sound). As this remark indicates, robotics is highly interdisciplinary in character, combining computer science (e.g., real-time software design, computer graphics, computational geometry, design of special-purpose computational devices), engineering (design and control of mechanisms and of sensors), and applied physics (materials science, theory of elasticity and of friction.) The *Advances* series which this volume initiates will aim to serve the many academic and industrial communities whose skills must play a role in its healthy development, and will cover all parts of this extensive field. We plan to address issues in computer science, theoretical mechanics and mechanism design, control theory, sensor technology and analysis of sensory data, and automated assembly technology, insofar as these contribute to robotics proper.

This first volume concerns itself with the algorithmic and geometric side of robotics, an area of the subject whose technical sophistication has increased very greatly in the last few years. The techniques reviewed reflect three fundamental appreciations that have grown out of 2 decades of work on *algorithm design* and *complexity theory*:

(a) Suitably ingenious design of key computational procedures can increase their efficiency far beyond the levels which naive "first cuts" at these same procedures might lead one to think possible;

(b) In the design of high-quality algorithms, formal analysis of procedure efficiency is required to quantify success (or to pinpoint remaining opportunities for efficiency improvement):

(c) Certain computational tasks are inherently intractable (and therefore should never by attempted in full generality if they are to be used in larger computations that aim to be practical).

During the past decade these fundamental insights have inspired rapid growth of many fields of algorithm design, including the very new field of *computa-*

tional geometry. Most recently the methods developed by computational geometers have been applied to robotics (whose extensive geometric content is obvious), and conversely robotics has begun to enrich the range of problems that challenge the geometer.

Each of the review articles collected in this initial volume focuses on some aspect of the fruitful interaction between robotics and computational geometry. Many of the methods presented are applicable not only to robotics, but also to such other related areas as geometric modeling and simulation to computer aided design and manufacturing, and (to a more limited degree) to computer vision and pattern recognition. Moreover, though necessarily specialized, the techniques described in these articles exemplify the beautiful interaction between data structures and geometric properties that are often key to attaining efficiency in computational geometry, and will reward their careful readers by suggesting general efficiency paradigms applicable to many geometric problems other than those specific to the present volume.

From a deeper point-of-view, the algorithms presented by our authors highlight the fundamental difficulties of robotics by pinpointing some of the very challenging problems that arise whenever one attempts to make a computer duplicate even the most elementary human manipulative or sensory activities. Consider, for example, the problem of using a robot hand to move an object from a given starting position (in a cluttered environment) to a specified target position. Though trivial for humans, even this elementary task raises many challenging algorithmic subproblems: If the position of the object or the geometry of the environment is not fully known, we may have to utilize some sort of sensory device and find some appropriate way of analyzing its output to supply the missing information. Once this information is available, we need to generate a plan for grasping the object, a task which involves analysis of the geometry of the object in relationship to the geometry of the moving manipulator and its grasping hand, and beyond this requires us to analyze the state and frictional reactions of an object to which forces are applied at several points. After deciding on a grasp, we need to plan a motion from start to finish position of the hand-object composite which results. All this must be done with acceptable efficiency, generality, and robustness.

Some of this list of problems belong to the domain of computational geometry to which this volume is devoted. Others raise entirely different issues, relating, e.g., to computer vision, statics, dynamics, theory of frictional movements, etc.; subject areas to be treated in subsequent volumes in this series. Finally, practical application of the algorithms and analyses generated by detailed study of the special problems raised by such practical tasks as object transport by robots creates extensive software engineering problems, an aspect of robotics which also will be reviewed in subsequent volumes in this series.

The foregoing reflections justify the claim that robotics has a truly rich scientific content and is likely to exert a major influence on computer science during coming years.

2. AN OVERVIEW OF THE CONTENTS OF THE VOLUME

The seven chapters comprising this volume review new work on the geometric side of robotics. It is hoped that these articles will serve the needs of the growing community of students and researchers wishing to acquire systematic knowledge of recent developments in this field.

Hopcroft and Krafft (*The Challenge of Robotics for Computer Science*) define the fundamental goal of robotics in very broad terms: to develop techniques for the *representation of, and formal reasoning about, physical objects and physical processes*. In accordance with this view of the field, they propose the name *stereo-phenomenology* as a better name for it than the too application-oriented term *robotics*. They emphasize the view that the concerns of this new field are likely to involve all properties of systems of solid objects amenable to mathematical formalization (and thus ultimately amenable to automated reasoning), including not only geometric properties, but also physical properties such as force, mass, velocity, energy, friction, vibration, etc., as well as more special materials-related properties. They outline a staggering research agenda, each of whose items may well come to occupy a generation of researchers.

Dobkin and Souvaine (*Computational Geometry—A User's Guide*) present some of the recently developed techniques that have begun to address the geometric side of the research agenda propounded by Hopcroft and Krafft. Their article illustrates some of the most basic paradigms of efficient algorithm design in the geometric area: divide-and-conquer, hierarchical search, duality transformation, ordered scanning, etc. Although some of these ideas, and the data structures which carry them, may be familiar from other fields of algorithm design, their applications in computational geometry are often remarkable, unexpected, and of great elegance.

Yap (*Algorithmic Motion Planning*) reviews several applied geometric ideas that have been particularly central to recent work on the important problem of motion planning. He describes the two main approaches, the so-called *retraction* and *decomposition* methods, on which systematic algorithmic treatments of the motion-planning problem have been based, and discusses some of the technical problems involved in developing efficient algorithms within the general frameworks defined by these approaches. The manner in which complexity theory techniques can be used to bound the difficulty of some robotic problems from below is also described and illustrated. The chapter closes with a list of open problems.

Chazelle (*Approximation and Decomposition of Shapes*) develops a theme in computational geometry having obvious significance for the simplification of practical robotics problems, namely approximation or decomposition of complex geometric objects by (or into) simple ones. For example, given a polygon P we can consider the number of sides of P as a measure of P's complexity. Then, for any P, one can ask for that other polygon, of a specified smaller number of sides, which best approximates P. Such a reduction can be very useful in applications, since the simplified objects may be substantially easier to manipulate. Another possibility is to decompose a complex object into several simple objects, e.g., to decompose a nonconvex polygon or polyhedron into a minimal number of convex figures, which may or may not be allowed to overlap. This chapter is one of the first expositions to provide a systematic account of such *decomposition* and *covering* problems, and of the related *enclosure* problem namely that of including a specified figure optimally in some larger figure drawn from a specified class.

Sharir and Leven (*Intersection and Proximity Problems and Voronoi Diagrams*) deepen a theme opened in Chapters 2 and 3, namely the use of Voronoi diagrams (which are defined to be the subspaces consisting of all those points, lying within larger abstract spaces of "positions," which represent "critical positions" in one or another sense, e.g., positions simultaneously nearest to several of a given collection of points or other geometric elements). Such diagrams have turned out to be among the most versatile and effective structures in computational geometry, having application in such diverse areas as motion planning, geometric searching, pattern recognition, computational fluid dynamics, solid state physics, chemistry, and statistics. Sharir and Leven review some of the most important applications of this technique to robotics and describe many of the most efficient algorithmic techniques known for calculating and for applying Voronoi diagrams.

Markowsky and Wesley (*Fleshing Out Wire Frames and Projections*) treat an intriguing class of geometric "reconstruction" problems having immediate application to computer-aided geometric design systems. Given any polyhedral object, we can easily derive a "wire frame" representation for it by dropping all its faces and retaining only its edges and vertices. (When one needs to enter the object into a mechanical design system, it may be significantly easier to supply just its edges than to define the object's faces explicitly.) The first problem considered by Markowsky and Wesley is that of reconstructing the object's surfaces, given only its wire frame. Since there can exist intriguing ambiguities of reconstruction, it is best to require the algorithm to compute all possible reconstructions. A second related problem is to compute the same representation given only several two-dimensional projections of an object's wire frame. The two Markowsky–Wesley chapters give elegant, general, and complete solutions for these problems; the algorithms that they describe have already begun to find practical application.

A Brief Review of Standard Terminology

Readers may find this brief review of concepts of complexity theory helpful. By a *complexity function* we mean a real function f of a real variable, where the range of f may include ∞.

Definition.   (*Big-oh notation*) If f, g are two complexity functions, we say that f *dominates* g if for some n_0, for all $x \geq n_0$, $g(x) \leq f(x)$. We write $O(f)$ for the set of all complexity functions g such that for some constant $C > 0$, g is dominated by $C \cdot f$.

'$O(f)$' is read 'big-oh of f' or 'order of f'. A typical idiom using this notation is '$O(1)$' which denotes the set of functions that are bounded by some constant, e.g., $f(n) = 1/n$ is in $O(1)$. Two conventions customarily govern the use of this notation:

(i) The sum $F + G$ of two sets F, G of complexity functions is defined as the set of functions of the form $f + g$ where $f \in F$, $g \in G$. Similarly for the expressions $F - G$, $F \cdot G$, F^G and $F \circ G = \{f \circ g : f \in F, g \circ G\}$ where $f \circ g(n) = f(g(n))$ denotes functional composition. If F consists of a single function f then we use 'f' rather than the correct '$\{f\}$' in such expressions; for instance, $O(n^2) + O(n \log n)$. Another example is $n^{O(1)}$, denoting the set of functions g that is dominated by the function n^k for some $k \geq 1$.

(ii) One often writes equations in which the O-notation appears on one side or on both sides of the equality symbol. Examples are $n = O(n^2)$ and $O(n \log_2 n) = O(n^2)$. The expressions on *both* sides of the equation should be interpreted as denoting sets of functions and the 'equality' sign should actually be interpreted as set inclusion '$\subseteq$'. This so-called 'one-way equality' is clearly transitive but not reflexive. For example, $n = O(n \log_2 n)$ is true even though $n \log_2 n = O(n)$ is not.

If $f = O(g)$ and $g = O(f)$ then we write $f = \Theta(g)$ and read "f is big-omega of g". Clearly the big-omega relation is an equivalence relation. We also say f and g are Θ-*equivalent* if $f = \Theta(g)$. In algorithmic analysis, it is usual to distinguish complexity functions only up to Θ-equivalence. It is this convention that allows us, for instance, to use logarithms without regard for the base of the logarithm. The rationale for this convention is that the complexity of a program required to realize a given algorithm can often be changed up to some small constant factor by rather trivial modifications in the programming language used; however such the changes remain within the Θ-equivalence class of the original function. An example of a trivial modification is where we allow one instruction in the new language to stand for a fixed sequence of instructions of the original language. Thus, up to Θ-equivalence, the complexity of realizing a given algorithm is typically independent of the details of the machine on which is run or the actual programming language involved.

We say a programming language L can O-*simulate* another programming language L' if for any program with complexity f in language L' there is another equivalent program in language L' with complexity $O(f)$. Say L and L' are Θ-*equivalent* if they can O-simulate each other. Even if two languages are not Θ-equivalent, it often turns out that they can simulate each other up to a polynomial difference: for each program of complexity of f in one language we can find a program in the other language of complexity $f^{O(1)}$ with the same input/output behavior. We say that the two language are *polynomial-equivalent* in this case. In fact, it turns out that all reasonable[1] general-purpose programming languages that have been proposed or used in algorithmic descriptions are polynomial-equivalent: this may be called the *polynomial simulation thesis*.

A function is called *polynomial* if it is in the family $n^{O(1)}$; otherwise it is *super-polynomial*. A fundamental phenomenon in complexity theory is the gap between problems that can be solved in polynomial time and those that require super-polynomial time. Super-polynomial functions grow so rapidly that even for problems of moderate sizes, algorithms with such complexity cannot be solved within a lifetime (on a human or astronomical scale), assuming the fastest imaginable computer. This reality makes for the following crucial distinction which goes back to Alan Cobham and Jack Edmonds: A problem is *tractable* if it can be solved in polynomial time; otherwise, it is *intractable*. One consequence of the polynomial simulation thesis is that *the tractable-intractable dichotomy is invariant under change of programming languages*. We can call this assertion the "complexity analogue" of Church's thesis in computability theory. Couched in a corresponding framework, the original Church's thesis states that the computable-uncomputable dichotomy is invariant under change of programming languages.

The reader is cautioned that "tractability" of a problem does not necessarily mean that it can be solved in practice, since its polynomial degree might be impractically large. Indeed, in some applications only problems that are $O(n^2)$ are considered practical. On the other hand, "intractability" does not necessarily imply that a problem cannot be solved in practice, provided we are willing to go to "computational modes" that are nonsequential (parallel) and/or are not deterministic (e.g., nondeterministic or probabilistic).

In case we suspect that a problem P_0 is intractable, how can we show, or perhaps just give evidence of, intractability of P_0? Several technical tools in complexity theory serve this end. One method goes as follows: choose a class C of problems for which there are strong evidence that it contains intractable problems. Then we must establish two facts. (a) Every problem in C can be 'easily reduced' to P_0. (b) If P_0 is tractable, then every problem that is easily reducible to P_0 is also tractable. Observe that the desired evidence is thereby obtained. One such class C is known as *NP* which stands for the class of problems that can be solved in "nondeterministic" polynomial time. We do not know whether *NP* is intractable although it is widely conjectured not to be tractable. (Section 10 in the chapter by Yap has examples of such problem reduction.)

---

[1]Actually we must also add further qualifications on the class of programs: they must be deterministic and sequential.

1 The Challenge of Robotics for Computer Science

John E. Hopcroft
Dean B. Krafft
Cornell University

ABSTRACT

Computer scientists have the opportunity to make contributions to the field of Robotics in a way that traditional engineering disciplines cannot. This chapter describes some of the opportunities, problems, and challenges that Robotics poses for computer science. Fundamentally, these fall into the area of representing and reasoning about physical objects, tasks, and processes—an area that we call stereo-phenomenology.

1. WHAT IS ROBOTICS?

The traditional view is that robotics is the study of designing, controlling, and instrumenting robots. These topics form an essential part of the field of robotics. They have been, and are continuing to be, successfully investigated by researchers in engineering disciplines.

Robots manipulate physical objects in accordance with a set of predefined instructions and, possibly, sensory input from the workspace. General-purpose programmable robots differ from computer-controlled hard automation, which has long been the province of the mechanical, electrical and industrial engineer, in that they are not specialized for a single problem. Instead, they can potentially perform arbitrary physical manipulations.

Robots can be used simply as a replacement for hard automation. They can be programmed, with a teach pendant perhaps, for a specific task. When the task changes, rather than scrapping the production line, the robots need only to be

7

reprogrammed for the new task. However, this is a rather limited use for a very powerful tool.

By making use of off-line programming, a robot system can provide more powerful capabilities and deal with much more complex situations. For example, information from a computer-aided-design database can be used to control machining or assembly. The robot can make use of sensory feedback to guide its actions. If it detects errors, it can take alternative actions to correct the problem and proceed with the task. Finally, off-line programming allows the system to deal with variations in the objects or processes it is handling, without having had all the cases preprogrammed by hand.

However, off-line programming brings with it new requirements. Programmers can no longer make use of the actual robot and physical prototypes to answer all their questions about position, action and movement. Instead, there need to be computer models of the workspace and the objects being manipulated. The robot task can then be described in terms of these models.

The next step in sophistication immediately suggests itself. If the robot system has models of the physical objects with which it is dealing, why not make use of these models to simulate and reason about the objects? This would allow robot programmers to work at a much higher level, saving time and reducing the chance for error. Such reasoning might allow the robot system: to figure out a grip, to plan motion, to determine the stability of an assembly (and thus how to move it from one location to another), to understand the behavior of nonrigid objects (when picking up a pair of pliers, for example), and so forth.

This leads to a much broader view of robotics. Areas involving high-level programming, reasoning about objects, and perception are qualitatively different from the issues involved in designing, controlling and instrumenting robots. The problems and goals are much less clear, and the techniques required to elucidate and solve them are very different. It is in dealing with these areas that computer science can make great contributions.

The study of robotics and physical systems is a natural step for computer science. Computer science was mostly begun by numerical analysts, physical scientists and engineers. Moving from scientific applications, the discipline drew into itself to develop a set of fundamental intellectual tools. Having specialized in developing representations and languages and in reasoning about abstractions, computer scientists are now well prepared to attack complex applications in other disciplines. The field is turning outward and applying its tools and knowledge in a broad range of areas, of which robotics is one of the most important.

The process of applying computer science to robotics has already begun. Researchers are studying problems in robot language design, motion planning, fine motion strategies, and many others. The AI community has been studying certain aspects such as vision and navigation for years. However, a much wider application of computer science tools to the problems of robotics is needed.

In this paper we sketch out some ideas, suggestions and challenges for the field of computer science. Although these were inspired by the problems of robotics, as we shall see in the next section, they are applicable to a much wider range of problems and disciplines. This paper is intended to provoke thought and discussion, not to provide answers. If the paper gets existing robotics researchers thinking in new ways and new researchers interested in the problems of robotics, then the authors will have achieved their goal. We *are* convinced that, whatever the effect of this paper, the area of research we are outlining is difficult, exciting, and critically important. Over the next decade, it will become a vital area of computer science research.

2. A NEW SCIENCE

The problems and potential benefits of representing and reasoning about physical objects and processes extend far beyond the realm of robot arms. At the simplest level, representation techniques clearly extend to controlling and reasoning about other automated production equipment. Thus, numerically controlled milling machines, lathes, and drill presses, automated parts handling systems, and even typical hard automation machinery could all be described and reasoned about in a general system of physical object representations.

Consider a computer-integrated manufacturing system. Such a system could involve graphics-based computer-aided design for specifying individual parts and simple assemblies, an off-line programming/simulation system for controlling individual workcells, and finally, a production control system to manage parts flow throughout the factory. Currently, integrated systems of this type are constructed on an *ad hoc* basis for a single factory application (when they are possible at all). The existence of general systems that could represent physical objects and processes and use that information to control production machinery would greatly simplify the design and implementation of truly integrated manufacturing facilities.

Moving on to engineering analysis, consider the problem of prototyping. Currently, verifying the behavior of a mechanical assembly almost always requires building a physical prototype. Not only is this a time-consuming and expensive process, but later design changes frequently require scrapping the prototype and starting again. Worse yet, there are situations where a physical prototype *cannot* answer necessary questions about the design. It is easy to imagine a satellite having an antenna system that, when fully deployed, cannot support its own weight under normal gravity. It seems extremely difficult to verify the correctness of the deployment process for such an antenna without using computer-based object representation and simulation techniques. Accurate electronic prototyping to determine correct operation, stability, behavior under mechanical and other loads, and so forth would be quicker, less expensive, more flexible, and potentially more informative than current physical prototyping techniques. What is needed is the scientific knowledge about representing physical objects that guarantees that the electronic prototype is as accurate as the physical one.

More sophisticated and dramatic examples are easy to come by. The design and construction of medical implants provides an entire class of them. Currently, an artificial knee joint is designed by a surgeon for each individual patient using a set of *ad hoc* rules. A sufficiently powerful object representation system could take into account the physical characteristics of a given patient and model the forces acting on the joint, the shape of the joint surfaces, and all the other important factors. The action of and stresses on the joint under such activities as sitting down or walking upstairs could be simulated. Thus, the system would allow verifying the specific design for an individual patient *before* the operation to install the knee is performed.

It is easy to imagine still more complex and speculative examples in such areas as modeling plastic flow in molds, realistically animating objects from a representation database, and many others. Some of these may be beyond the capability of a general object representation and reasoning system, but some will not. Until the field has been developed, this cannot be determined.

A new science of representing physical objects and of manipulating and reasoning about those representations is needed. This new science draws on material from a host of existing fields, including computer science, physics, and many branches of engineering. It is a *science* because researchers will be establishing and systematizing the facts, methods and principles underlying the representation of physical objects. Drawing on the Greek, we take the word for "solid," *stereo,* and combine it with the word for "the study of representations," *phenomenology,* to christen this new science—*stereophenomenology.*

The importance of this new discipline is not, of course, in its name. It is instead in the concept that there is a collected set of principles, tools, and methodologies, forming the science of stereophenomenology, that can be applied to a broad range of important problems. We believe this to be true, but only time, and the collected efforts of many researchers, will provide the final answer.

3. REPRESENTATIONS FOR PHYSICAL OBJECTS AND PROCESSES

If we are going to reason about physical objects and simulate physical processes with computer systems, then the systems must have representations for the objects and the physical processes that modify and manipulate them. This section examines both the kinds of information that must be represented and techniques for doing so. As the individual sections make clear, on at least some aspects of representation much excellent research has already been done. However, there are large areas that are almost untouched, and no work has been done that truly ties together all the various aspects of physical object representation.

The real challenge seems to be developing general representation models that are:

- *abstract* enough so as not to require computationally infeasible deductions;
- *formal,* so that reasoning about objects is possible;
- *structured,* so that complex physical objects and processes can be understandably represented;
- *complete,* or at least strong enough that an interesting set of objects and tasks can be modeled.

3.1. Surfaces and Solids

Many representations have been developed for modeling objects in computers. One popular representation is the wire frame model (Foley & Van Dam, 1983). Its main attribute is its computational simplicity, which allows modeling complex objects such as aircraft and illustrating motion with modest computational resources. However, the wire frame model suffers from ambiguity as to actual shape of the object (see the chapter by Wesley and Markowsky in this volume), does not support the ability to calculate mass properties, and provides insufficient detail to allow visualizing complex objects.

Another representation, the octree (Meagher, 1982), is used because of computational expediency in producing images and combining objects by operations such as union and intersection. In the octree representation an object is enclosed in a cube. The cube is divided into eight smaller cubes and each subcube that contains a piece of the object is recursively subdivided, unless it is completely full or empty, until the resulting subcubes are of sufficiently small granularity. The object is represented by a tree, each node of which corresponds to a cube and the piece of the object enclosed in the cube. The children of a given node correspond to the subset of the eight smaller cubes of the parent cube that are nonempty. Unfortunately, the octree representation is not convenient for reasoning about objects, and it is not amenable to supporting finite element methods.

Researchers in computer graphics frequently build models from bicubic patches (Ferguson, 1964). The current state of the art allows highly realistic pictures to be generated by ray tracing methods from these models, including reflections and shadows. However, these representations are primarily surface representations rather than object representations. As a consequence they are not conducive to answering questions about the internal structure of objects or determining, for example, an object's response to external forces (see Section 3.4 for further discussion and examples).

Most general-purpose solid modelers (Requicha & Voelcker, 1982) use one of two representation methods. The first method, called constructive solid geometry (CSG), represents a solid by combining primitive solids using Boolean operators. The second method, called boundary representation, represents a solid by the topological structure of its vertices, edges and surfaces, along with their geometrical description.

In the CSG approach an object is represented by a tree. With each interior node there is associated an operator that either combines two subtrees together or translates or rotates the object represented by a single subtree. With each leaf is associated a primitive object from a class of simple objects that typically includes spheres, cylinders, halfspaces, and so forth. The root node then represents the complete object.

The fact that objects are usually defined to be connected, closed sets creates a minor difficulty. The intersection of two unit cubes that touch by sharing a common face is a two dimensional surface rather than a three dimensional object. Similarly, the difference of two such cubes is not a closed set, since a face is missing. To ensure that the operations applied to legal objects produce legal objects, CSG uses regularized Boolean operations. A regularized Boolean operator produces the closure of the interior of the object produced by the ordinary operator. It is these regularized operations that are at the interior nodes of a CSG tree. This method is guaranteed to produce only topologically correct objects.

A major advantage of the CSG approach is that it needs only simple recursive algorithms to determine if a point is inside an object, to find the first point of intersection with a ray, and to solve many similar problems. Each of these problems can be easily solved for the primitive objects at the leaves of the tree, and the results propagated up the tree by combining or modifying them in accordance with the operations at interior nodes of the tree. An operation such as intersection is even simpler. It is performed by combining the trees for the two original objects into a single tree by creating a new root labeled with the regularized intersection operator and having the roots of the two original trees as children. Thus the combining operations themselves can be performed very rapidly.

It is generally recognized that the CSG approach is very convenient for input. In fact, some modelers that use boundary representations as their underlying data structure still use the constructive solid geometry approach for the user interface. Another advantage of the CSG approach is that it provides some internal structure to the object. For example, if a cylinder is subtracted from two arms of a bracket, the resulting two holes are guaranteed to have their axis aligned. Moving one hole will automatically cause the other to move.

The CSG approach does have disadvantages. Consider the question of determining if an object is null. Other than computing the volume there is no simple way to use the CSG tree to answer this question. Thus the problem is usually solved by constructing the boundary representation. The object is the null object if and only if the boundary representation is empty.

In motion planning it is often convenient to use the notion of the swept volume of an object (Lozano-Perez & Wesley, 1979). The swept volume is the region of space swept out as the solid object is moved along a trajectory.

FIG. 1.1. CSG tree and boundary representation for an object consisting of the difference of two rectangles.

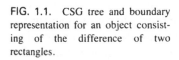

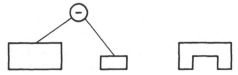

Unfortunately, sweeping does not commute with an operation such as difference. Thus, one cannot use CSG techniques to compute the swept volume by appropriately combining the swept volumes of the primitive pieces. To illustrate the problem, consider the solid displayed in Fig. 1.1 formed by taking the difference of two blocks.

Figure 1.2 shows the volume resulting from taking the difference and sweeping. Figure 1.3 shows the effect of sweeping the two components and then taking the difference of the results.

The boundary representation of an object consists of two components. The first component gives the topological structure of the object. The second component describes the geometrical structure of the vertices, edges and faces that make up the object. The surface of the object is assumed to be a manifold, that is, each point on the surface has a neighborhood that is topologically equivalent to a plane. The reason for requiring that the surface be a manifold is to provide for a natural ordering of edges about vertices. However, this condition rules out contact between two parts of the surface where the contact consists of only a point or a line. Such a contact can be handled by replacing it with a sphere or cylinder of infinitesimally small radius ϵ. The sphere or cylinder serves to keep the topological structure correct. Algorithms operating on the representation can then rely on this structure for their termination conditions.

The topological structure consists of:

1. for each vertex, an ordered list of edges about the vertex
2. for each edge, the head, tail and face on either side
3. for each face, the list of edges forming its boundary

The above information is actually redundant. The list of edges forming the boundary of a face can be constructed from the order of edges about each vertex.

The choice of an appropriate surface representation raises many interesting issues. Using a low degree surface has the advantage of allowing the use of

FIG. 1.2. Computing the difference of components and then sweeping.

FIG. 1.3. Sweeping components and then computing the difference.

algebraic techniques as opposed to numerical methods in reasoning about objects. Mechanical parts, for example, can be described almost completely by quadric surfaces. Thus quadrics are a particularly inviting choice for representing surfaces. The following discussion illustrates some of the algebraic techniques that can be used to handle such surfaces. (For a somewhat different treatment, the interested reader should consult Ocken, Schwartz, & Sharir, 1985.)

The general equation of a quadric surface is of the form $x^T A x = 0$, where x is a column vector. The intersection of two quadric surfaces in 3-space is a curve. This curve may lie in a plane, in which case it is a quadratic and has a rational parameterization. This allows a convenient and efficient representation, as well as the ability to sort points along the curve or its projection. However, if the curve is a 4th degree space curve, then it has no rational parameterization. This forces us to use a nonparametric form for curves, which complicates ordering specific points along the curve, or to use more general parametric forms involving radicals.

That there is no rational parameterization can be seen as follows: Project the intersection curve from a point on the curve onto a plane. The result is a cubic plane curve with no singular points. Since a nonsingular cubic plane curve has no rational parameterization (Walker, 1950), and since a rational parameterization for the original curve would imply one for the cubic, the intersection curve cannot have a rational parameterization.

Although the intersection of two quadrics does not always have a rational parameterization, it does have a parameterization of a more general type involving square roots. If $x^T S_1 x = 0$ and $x^T S_2 x = 0$ are the equations of two quadratic surfaces, then the pencil of S_1 and S_2, $x^T S(\lambda) x = 0$, where $S(\lambda) = (1 - \lambda) S_1 + \lambda S_2$, is a parameterized one dimensional family of quadric surfaces. The important property of the family is that any two members of the family intersect in the same curve. The roots of the discriminant of $S(\lambda)$ correspond to values of λ for which $S(\lambda)$ is a singular quadratic. One can show that $S(\lambda)$ always contains a ruled surface, such as a cylinder, cone or hyperbolic paraboloid (Levin, 1979). These ruled surfaces can be generated by a line moving along the directrix, which is a quadratic curve. Thus, they can be parameterized.

Let $x_1 = f_1(s, t)$, $x_2 = f_2(s, t)$ and $x_3 = f_3(s, t)$ be a rational parameterization of the ruled surface. If parameter t defines the point on the directrix and s defines the distance along a line on the surface through the directrix, then the x_i are linear in s. Substituting the parametric equations for x_1, x_2 and x_3 of the ruled surface into the implicit equation for the other quadric surface results in a quadratic equation in s. Solving for s by the quadratic formula yields an equation for s in terms of t but involves a square root of a rational expression in t. Thus, in the parameter space of s and t we have a parametric form for the curve of parameter values corresponding to the intersection. Substituting the parametric form of s into f_1, f_2 and f_3, yields a parameterization in t of the curve of intersection.

Of course, quadrics are not sufficient for all mechanical parts. A helical spring is not even an algebraic surface, since a line can intersect the spring infinitely often. One approach is to approximate the surface with a reasonably low degree algebraic surface. A possible candidate is the Steiner surface (Sommerville, 1934), a surface with a quadratic polynomial parameterization in homogeneous coordinates. Another alternative is to use some form of patch, of which bicubics are perhaps the most common example. Unfortunately, using bicubic patches probably rules out algebraic techniques, as an analysis of Sederberg et al. indicates that the bicubic patch is a degree 18 surface (Sederberg, Anderson, & Goldman, 1985).

Sederberg's analysis proceeds as follows. Let $x_1 = p_1(s, t)$, $x_2 = p_2(s, t)$ and $x_3 = p_3(s, t)$ be the parametric equations of a bicubic patch. If $p_i(s, t) - x_i = 0$, $1 \leq i \leq 3$, then

$$D(s,t,\alpha,\beta) = \begin{vmatrix} p_1(s,t) - x_1 & p_2(s,t) - x_2 & p_3(s,t) - x_3 \\ p_1(\alpha,t) - x_1 & p_2(\alpha,t) - x_2 & p_3(\alpha,t) - x_3 \\ p_1(\alpha,\beta) - x_1 & p_2(\alpha,\beta) - x_2 & p_3(\alpha,\beta) - x_3 \end{vmatrix}$$

must equal 0, since the top row of the determinant is identically zero. Now when $s = \alpha$ or $t = \beta$, then $D = 0$, since the top two rows or the bottom two rows are identical. Thus D is divisible by $s - \alpha$ and $t - \beta$. Let $\delta(s, t, \alpha, \beta) = D(s, t, \alpha, \beta)/(s - \alpha)(t - \beta)$. If the polynomials $p_i(s, t)$ are of degree n in both s and t, then δ is of degree $n - 1$ in s and β and of degree $2n - 1$ in t and α. Thus we can write

$$\delta = \sum_{i=0}^{2n-1} \sum_{j=0}^{n-1} f_{ij}(s,t)\alpha^i\beta^j$$

Since δ must vanish for all α and β if s and t satisfy $p_i(s, t) - x_i = 0$, the $f_{ij}(s, t)$ must be 0 for all i and j. Therefore we get a system of $2n^2$ polynomials $f_{ij}(s,t)$ equal to zero each with $2n^2$ terms (s appears in degree $n - 1$ and t in degree $2n - 1$).

$$M[1, s, s^2, ..., s^{n-1}, t, ts, ts^2, ..., t^{n-1}s^{n-1}]^T = [0, 0, ...]^T$$

Each coefficient of the system of equations is linear in x_1, x_2 and x_3 or is constant. For a solution to exist $det(M)$ must be zero. This gives a polynomial of degree $2n^2$ in the variables x_1, x_2 and x_3 for the surface. [One should note that $det(M) = 0$ is not a sufficient condition for a solution to exist, since the column vector does not consist of independent variables. Also note that a system of parametric equations may actually give rise to a semialgebraic rather than an algebraic set, as in the example $x - t^2 = 0$, $y = 0$, $z = 0$. A more detailed analysis of the semialgebraic set would seem to require a procedure such as Collins' cylindrical algebraic decomposition (Collins, 1975).] For the bicubic patch, $n = 3$, and the above analysis indicates a surface of degree 18. A general degree n surface in 3-space has $\binom{n+3}{3}$ terms. Thus a degree 18 surface has 1330 terms. The intersection of two degree 18 surfaces is an $18^2 = 324$ degree curve. This complexity surely precludes using algebraic methods for dealing with surfaces made up of bicubic patches.

Although we have only provided the briefest of surveys of what is known about surfaces, solids and their representations, we have tried to convey some of the depth of current knowledge. One thing that has not been researched is the choices and trade-offs among the various possible representation techniques. There is no comprehensive theory to guide the builder of an object representation system in choosing those approaches most appropriate to his needs. Such a theory would be a major contribution to the science of stereophenomenology.

3.2. Abstract Models

In existing solid modeling systems, an object is modeled simply by its physical surface. This surface may be determined by CSG operations, boundary representations or patches, but in any of these, the solid and its surfaces are directly modeled. This approach is simultaneously too detailed and insufficiently informative. For example, (Wesley et al., 1980) illustrates a procedure to create a representation of a machine screw. The threaded shaft of the screw is modeled by a very large number of faces. It is thus quite detailed. However, given a similar model of a threaded nut, determining if the screw and nut mate would be a very difficult problem. Thus, the representation lacks the information needed to quickly answer this obviously important question.

Clearly, neither increasing the detail of the representation nor decreasing it will help. The solution to this dilemma is to use an *abstract model* for the important information about the object being represented. Rather than merely modeling the physical surface, the threaded surface must be represented abstractly, for example, by indicating that it is a 1 inch 8-32 thread.

Abstract models of physical objects have been previously considered by the AI community (see, for example, Winston, Binford, Katz, & Lowry, 1983). However, their normal approach is to define objects by a set of initially uninterpreted attributes. The interpretation is then supplied through examples and heuristics. This results in a rather unstructured representation for a highly structured physical object. We believe that a more formal and structured abstract model will result in more straightforward and useful systems.

Let us analyze the requirements for a formal abstract model. Consider a physical object. Normally, it consists of large expanses of easily represented surfaces (forming the structure of the object) and a few important *features*. Informally, we define a *feature* as the site of some potential physical activity, such as an attachment point in an assembly or a grip area for a robot. Standard solid modeling systems provide the same representation mechanisms for the vast expanses as for the key points of interest. If the features of an object are modeled directly, through abstract rather than physical surface models, then deductions about such problems as the mating of threads become immediate.

The notion of *feature* that we present here is somewhat similar to that described in Popplestone, Ambler, & Bellos (1982), in that they also view features as the focus for action and specification, but it is different in that they assume that geometric descriptions for features are available, while the rest of the object is not described geometrically (and hence may be arbitrarily complex). Although having features that can be fully modeled with simple geometric descriptions does allow one to reason formally about them, it results in a system that is not sufficiently powerful to handle real mechanical assembly. For example, the types of features that seem particularly susceptible to abstract modeling include points of attachment, such as a screw and a nut, grip points for a robot assembly system, and objects in constrained contact, as in the meshing of gears. Working from a standard solid model or geometric description, it would be very difficult to determine that the gears mesh, that the nut and bolt mate, and that the particular point is a good one for a robot grip. Research is needed to devise a concept of complex features that supports both formal reasoning and interesting physical objects.

3.3. Hierarchical Representations

The concept of hierarchical descriptions for physical objects is a common one. When describing a car, we talk about it at a number of levels. There is the car itself, the major componets (engine, transmission, body, etc.), subassemblies (carburetor, steering column, starter, heater), and so forth. Similarly, in designing or specifying a physical object or assembly, it is very valuable to be able to consider a group of parts as forming a new entity, whose internal structure can, to some extent, be ignored.

This notion of hierarchical specification is very similar to the programming notions of encapsulation and data abstraction. The *mental complexity* of dealing with a complex object (or program) can be greatly reduced by hiding the internal details of its implementation and considering only its inputs and outputs. In designing a car, at some point the carburetor becomes a black box. It is fed throttle and choke signals, given access to air and gasoline, and the result will be a known mix of fuel and air. The designer is no longer interested in the details of idle ports and float valves. Indeed, any other box that provides the same outputs for the same inputs (and meets some other physical constraints) would do just as well.

The above example illustrates another advantage of encapsulation. By forcing an accurate definition of the interface between the subassembly and the rest of the structure, it becomes simple to substitute different subassemblies that meet the design requirements. Conversely, the carburetor can be used in any assembly for which it meets the requirements.

Existing solid modeling systems have some notion of hierarchical description. In PADL II (see Brown, 1982; Brown & Voelcker, 1979; Voelcker et al., 1974) these notions are very primitive. The only structures are solid primitives, parts, and assemblies. GDP (Fitzgerald, Gracer, & Wolfe, 1981; Wesley et al., 1980) has more powerful constructors and allows specification of named subassemblies, however there is no notion of functional encapsulation. A problem with all existing systems is that they partition objects by physical attachment rather than by function. The two partitionings are not necessarily the same, and it may be that the functional approach is more appropriate.

The notions of hierarchical levels of description and encapsulation lead naturally to an important related concept. This is the notion of generic or parameterized objects, or *object classes*. The concept of an object class was described and implemented in the context of solid modeling for computer vision by the ACRONYM system (Brooks, 1981; Brooks, Greiner, & Binford, 1979). GDP supports a weaker mechanism of implementing object classes by having procedures that produce instances of the class.

Rather than representing just a single carburetor, a carburetor class could define a continuum of carburetors, parameterized by the acceptable ranges of inputs and outputs. The concept of parameterized objects, and the closely related notion of functional specification, is discussed further in Section 4.1 of this paper.

There is a completely different aspect of hierarchical representation that is of equal importance in dealing with physical objects. This is the concept of *hierarchy of detail*. Essentially, this notion involves attempting to solve problems concerning physical objects by starting with very simple representations of the properties involved, and introducing more complex representations only as they are required to solve the problem.

An example of a possible use of detail hierarchies is in motion planning. Since motions involving spheres are relatively easy to calculate, a system could initially approximate all objects by interior and exterior enclosing spheres. When the exterior spheres did not intersect at all during a planned motion, the motion would be known to be safe. When the interior spheres intersected during such a motion, the motion would be impossible. If neither of these conditions were met, then the system could proceed to a finer level of detail, possibly approximating the objects with simple polyhedra. Ultimately, at a high computational cost, exact surface representations could be used. (For more on motion planning, see Section 5.2 and Yap's chapter in this volume.)

This notion of hierarchy of detail applies to other object properties. For example, the center of mass will suffice for most applications, but detailed information could be available on the masses and extent of regions of uniform or continuously varying density. Similarly, a structural member could have an overall breaking strength associated with it, but a function over the length of the part could provide greater detail when required.

To summarize, the concepts of encapsulation, levels of refinement, and hierarchy of detail are very important ones for any object representation system. Work on developing systems that use these notions would be extremely valuable.

3.4. Internal Structure

One area of object representation that current modeling systems fail to address adequately is that of internal structure. Given any sort of moderately complex object or assembly, the structure of the object, its degrees of mechanical freedom, internal states, and the nature of the attachments of subassemblies and components, is at least as important to understanding its nature as the shapes of its parts. If we are to develop systems that fully represent physical objects, the nature of objects' internal structure must be clarified, and methodologies for describing and reasoning about it must be developed.

For the purposes of this discussion, we identify two distinct but related types of internal structure. The first is object states, and the second is functional interdependency. This is not intended as a complete classification, but it serves as a starting point for exploring internal structure.

Let us first consider object states. Informally, we define an object to have a set of potential states whenever there is some degree of freedom in the configuration or properties of the object. A mechanical joint, as in a pair of pliers, is the most obvious example, but there are many others. A spring has a single resting state (if it is sufficiently stiff), but under compressive forces it can take on a continuous, but limited, range of states. A flexible piece of wire or string is another example. Here, the range of potential states is limited by the length of the piece and its maximum curvature. More esoteric examples of states would include such properties as temperature and magnetism.

It should be clear that we would like to represent both the (probably infinite) range of states that an object can attain, and the specific state in which it currently rests. One difficulty lies in deciding where the state information resides. Consider the assembly of a box and top plate (see Fig. 1.4). Initially, all the pieces are rigid and separate and, taken individually, have only one state each. Regarded as an entire system, of course, they have a large number of degrees of freedom. Now we attach the top plate with a single screw. The partial assembly has one rotational degree of freedom as the plate swings around the screw. When a second screw is attached, the assembly is suddenly a rigid single-state object. Should the previous state information be maintained, and, if so, how?

One might ask why we would wish to represent object states directly. The most important reason is so that we can reason about the behavior of objects that

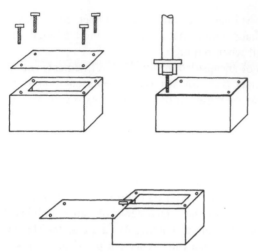

FIG. 1.4.   Fastening a lid on a box.

can assume a range of states. To determine what will happen when we pick up a pair of pliers by one handle, we clearly cannot model the joint as a rigid connection. In assembling the box of Fig. 1.4, reasoning about object states allows the system to determine when it no longer needs to worry about the lid moving relative to the box. Finally, trying to decide if a wire cable will fit when the front panel is fastened to an enclosure clearly requires a sophisticated model of object states.

There are a few current systems that allow some form of representation for object states. The work of Forbus (1981), for example, allows the qualitative representation of certain state information, although it does not deal with mechanical structures. ACRONYM (Brooks, 1981; Brooks, Greiner, & Binford, 1979) supports a simple notion of object states, but does not distinguish them from generic or parameterized objects. Both concepts are lumped under the heading of *object classes*. GDP/GRIN (Fitzgerald, Gracer, & Wolfe, 1981) supports a low level of simulation of object states through procedures, but does not allow expression of the range of states in a manner amenable to reasoning. Existing kinematic simulation systems (e.g., the robot arm simulation system of Leu and Mahajan, 1984) can represent a range of object states, but they tend to be extremely specific in their application, and they do not really support general representation and reasoning at all. A general system for representing and reasoning about object states would be a significant advance.

The second type of internal structure we wish to discuss is functional interdependency. Clearly, the various parts of an assembly are dependent upon each other. The gears in a transmission must successfully mesh, a bolt must fit its corresponding nut, and so forth. Currently, modeling systems allow only for the implicit representation of this information. The gears were designed to mesh by the individual who built the model, and so they do. We feel that it is important that these dependencies be represented explicitly.

The most significant advantage of directly representing functional dependency occurs in the design, modification, and redesign of physical systems. Both the notion of functional editing, discussed in Section 4.1, and the concept of reasoning about design, discussed in Section 5.3, require that functional interdependencies be explicitly represented. This information might also be used in direct behavioral reasoning, for example in predicting the result of turning one of two meshed gears. Finally, together with functional encapsulation, discussed in section 3.3, explicit functional dependency specification would allow the designer easily to replace one subassembly with a functionally equivalent one.

While we have briefly discussed two small categories of internal structure, what is really needed is a wide-ranging investigation into all aspects of structural reasoning and representation. Dealing with the internal structure of physical objects is an essential step toward the overall goals of stereophenomenology.

3.5. Transformations and Processes

To describe a physical process or transformation, a system must have representations for both the objects involved and the task itself, even if the objects are merely represented as spatial coordinates and the task as a series of grip and move commands. Currently, robot systems represent activities by programs built up from physical action primitives and control structures. While such programs provide a powerful representation of tasks, they make it very difficult to build systems that can manipulate and reason about descriptions of physical activities.

An analogy drawn from the area of computer programming and programming language design may be instructive. Initially, designers of programming languages were only concerned with developing clear, higher-level representations for computer programs. The program itself was the only representation of the task, and the inherent undecidability of whether an arbitrary program achieved some specific goal was not viewed as a problem.

As the field developed and programs grew more and more complex, people began to wonder about the correctness of programs. Informal proof methods (at the level of normal mathematical proofs) were developed by Dijkstra (1976), Floyd (1967), Hoare (1969) and others, for reasoning about the behavior and correctness of programs. These led to improvements in the design of the languages themselves.

Eventually, complete formal systems for reasoning about program behavior were developed, for example, PL/CV (Constable & O'Donnell, 1978) and PRL (Bates & Constable, 1985). These systems provide a formal intensional representation for essentially arbitrary procedures. Thus, the behavior of programs expressed in these systems is explicit, and reasoning about them is straightforward.

In developing languages for describing physical processes, we would hope to learn from and build on the programming language experience. If we are to develop systems capable of any sort of automatic task synthesis, we need explicit, intensional representations of physical activities.

There are a number of other important and challenging problems in expressing transformations and processes. We briefly describe a few of these below:

1. Some method is needed for describing *compliant motion* in specifications of physical tasks and processes. Currently, robot arms are constructed to be as rigid as possible. It has been proposed that inherently compliant arms might be much easier to build and at least equally useful for a variety of tasks. Current systems, however, provide no good way of expressing compliance.

2. *Sensory feedback,* from vision systems, force sensors, range sensors, and so forth, is an essential component of robot systems. Somehow, task representations must include control based on sensory input. Representing and reasoning about such inputs seems to pose unique problems.

3. Unlike computer programming systems, real physical systems must deal with the problems of *error and tolerance.* The representation of tolerance information is needed both for objects and tasks. Some work has already been done on object tolerances (Requicha, 1983), but more is needed. The area of task errors and tolerances is almost untouched.

4. A system is needed that supports *generic* transformations. A procedure written to screw together a ¼ -20 hex head bolt and corresponding nut would presumably be very similar to one written for a 10-32 machine screw and nut. Somehow, the generic nature of these tasks must be recognized and expressed in their representations.

The problem of representing transformations and processes is a major issue in implementing both high-level robot programming systems and more general physical reasoning systems. Part of the challenge lies in trying to represent physical tasks in an explicit, intensional fashion rather than an implicit, procedural one. Another part of the challenge arises from the wide variety of tasks that must be described. These range from simple placement problems, to attachment processes, to destructive processes such as machining and drilling, to tasks requiring compliant motion, such as erasing a blackboard. Any progress towards a general language for describing this wide range of activities, inputs and consequences will be a major achievement.

4.   MANIPULATING OBJECT REPRESENTATIONS

Together with the basic computer representations for physical objects, we need algorithms and systems for manipulating these representations. In particular, there must be systems for the object representations that support high-level creation and modification, fast, realistic display, Boolean combining operations, and so forth. This section discusses some of the algorithmic and systems work that has been and still remains to be done in three of these areas.

4.1. Editing

Representing solid objects in a computer will require user-efficient object editors with which to design and modify object descriptions. A designer needs the ability to manipulate object descriptions in high-level terms appropriate to a particular engineering discipline.

A major problem with existing solid modeling systems is that they fail to capture the relationships that exist among the various surfaces of the object and the internal structure from which these relationships flow. In any physical object certain relationships must be maintained. With today's systems a user modifies one geometrical feature at a time and tediously edits a large number of dimensional entities. If representations move in the direction of capturing functionality, objects could be parameterized so that dimensioning data is derived from a small set of parameters. Varying a single parameter would result in several aspects of the object varying simultaneously. Other aspects of the object that are related to the varying features would experience appropriate changes as well.

The nature of such a system is perhaps best illustrated by an example. One might want, for instance, to define a class of "elbow joints" for pipes. The design would require that two cylinders of equal radius be attached so that their center lines meet at a right angle. Further, it might specify thickness of the pipes by having two nested cylinders in each direction from the joint, representing respectively outside and inside radii. The design might also require that the joint created meet some smoothing criterion. Finally, all of the interrelated geometric parameters might be summarized by an equation relating flow through the joint to the inner and outer radii, and to geometric (or better, functional) parameters of the mode of attachment of the elbow pipe to other pipes. Ultimately, the pipe dimensions might be dependent on some parameter of an entire system.

In a complete three-dimensional editing system, users would create composite objects from previously defined (or in some cases primitive) subcomponents. Composite objects would be parameterizable in both geometric and functional terms, and the designer would specify how the subcomponents would be spatially and functionally related. Having done that, the designer (or someone else who uses the design) would have the flexibility to modify the composite object either by varying functional parameters, or by geometrically manipulating the figure representing the object. In a manner analogous to the Prolog (Clocksin & Mellish, 1981) style of programming, the computer could then manipulate the geometric and functional equations describing the object to produce a new object that incorporates the user's changes.

This constraint and dependency approach was pioneered in two-dimensional graphic editing by researchers interested in typefont design and phototypesetting of drawings (Nelson, 1983; Van Wyk, 1981) and simulation (Borning, 1981). They produced systems that defined two-dimensional drawings in terms of geometric constraints. This work provides a base for future efforts in editing representations of realizable three-dimensional objects.

Systems for editing and manipulating representations can provide a fruitful source of interesting theoretical and algorithmic problems. To give some understanding of this, we sketch out below one such problem and a possible solution.

In designing an object it is often convenient to construct a quick graphical sketch of the object to be used in defining the precise surfaces and dimensions. Only after the parameters are defined are relations developed that determine precise shapes and locations. These relations may well be nonlinear, and hence have more than one solution. It is important that, where possible, the system automatically selects the correct solution based on the approximate locations indicated by the user. The alternative would be to require the user to be aware of the multiple solutions and to force specification of the correct one, an annoying task.

One approach to the problem is to create a homotopy between the initial approximation and the desired constraints. For example, Fig. 1.5 illustrates a previously created unit square. The user wishes to place a point B to the right of the square at a distance 1.5 from a point A at coordinates $(1, 1)$ and at a distance 2.5 from the origin. The user places B' approximately where B should be and specifies the AB and OB distances. The system then creates the desired constraints:

$$x^2 + y^2 = (2.5)^2$$
$$(x - 1)^2 + (y - 1)^2 = (1.5)^2$$

and the set of constraints currently satisfied by B; say $x = 4$, $y = 0$. The homotopy

$$(4(x - 1)^2 + 4(y - 1)^2 - 9)t + (1 - t)(x - 4) = 0$$
$$(4x^2 + 4y^2 - 25)t + (1 - t)y = 0$$

is created. This homotopy produces a path from $x = 4$, $y = 0$ to the desired solution at $(2.43, 0.57)$. The second path runs from infinity to the undesired solution. Thus the system modifies the approximate position to the correct one based on the desired constraints.

There are a number of theorems that can be proved concerning the region of

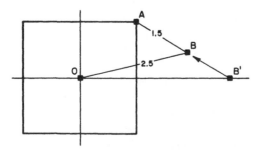

FIG. 1.5.   Positioning a point.

initial approximations that will give the desired result. However, the real test of such techniques occurs when they are built into an actual editing system.

We have tried to give a brief outline of some of the problems involved in creating and manipulating object representations. Systems issues, user interface design, data representation, and so forth, tend to predominate; however, there are significant algorithmic and theoretical problems as well. Good editing systems form an essential component of general object representation systems.

4.2. Display

The algorithmic problems involved in graphical display have been thoroughly covered elsewhere. In particular, Foley and Van Dam (1983) give a general overview of graphic display techniques. There is also extensive literature in more specialized areas. For example, there is the work of Blinn (1982), Levin (1976), and Hanrahan (1983) on the subject of quadric surface display. A description of the Cook-Torrance model for highly realistic surface rendering can be found in Cook and Torrance (1982). In view of the available literature, we will simply give a brief description of our view of the state of display as it relates to object representation and then present a few areas for future algorithmic research.

There are currently two general classes of graphic display systems. Those based on polygonal approximations are reasonably fast, and can even approach real-time display speed on moderately priced hardware. The other class of graphic displays make use of raytracing methods to produce highly realistic images. Shadows, reflections and transparency can all be accurately rendered. Unfortunately, these techniques are extremely slow, and can take a number of hours on fairly powerful machines to produce a single image.

As we discussed in Section 3.1, we expect that surface representations for physical objects will be algebraic surfaces. If the surfaces are restricted to planar polygons, one can make use of the existing fast graphics display systems. If higher-order surfaces are used, rapid surface display poses a problem with several possible solutions.

One way to get rapid display of mathematical surfaces is to approximate them with polygonal panels. Gouraud shading can be used to obtain fairly realistic displays. This requires an algorithm to do the necessary polygonal approximation of the mathematical surface. In this process there is a tradeoff between speed (the fewer the number of panels, the faster) and accuracy (the greater the number of panels, the more accurate). Thus, no approximating solution is completely satisfactory.

To obtain fast displays of algebraic surfaces without approximating will require faster algorithms. On a raster display system this involves:

1. identifying the partition of screen pixels among the displayed surfaces,
2. using information about the light source, shadows, depth, surface com-

position and surface orientation to determine the exact color and illumination of the surface.

Both of these are interesting problems. The first involves processing the solid description (perhaps based on a specific viewpoint) to determine which faces occlude other faces and to account for the edges and vertices of the solid. With polygonal faces, only the edges need to be taken into account in determining which faces are in front. With higher-order surfaces, convex and concave bulges in the face must also be considered. For efficient rendering, the second problem involves identifying regions of the display where simple functions can be used to describe the color and intensity. Once this is done, rapid surface display can proceed.

To summarize: A great deal of work has been done in the area of graphic display, but there are some specific problems that still need to be solved. Faster algorithms for realistic algebraic surface display are needed. These methods should be general enough to support reasonably high-order surfaces. The goal is a system that adequately supports interactive solid editing, for example, of the type described in Section 4.1.

4.3. Boolean Combining Operators

A major component of all object modeling systems is a set of algorithms for manipulating representations. The nature of these algorithms is, of course, dependent on the form of the underlying representations. As we discussed in Section 3.1, the main representations used by current modeling systems are constructive solid geometry and boundary representations.

CSG representations have essentially trivial combining operations. For example, representing the intersection of two objects involves creating a new tree with a regularized intersection operator at the root. The two original trees are made children of this root.

Determining if a CSG object is null is slightly more complex. Either its volume is computed by ray casting and compared to zero, or, more commonly, the CSG representation is converted to a boundary representation. After discarding tiny fragments of essentially zero volume that arise due to the finite precision of numerical computations, the null boundary representation object will have a null representation.

Thus, for all current solid modeling systems, combining boundary representations is an important problem. At the heart of any such system is a set of algorithms for computing intersections, unions and differences of objects. As the general structure of these algorithms is the same, we illustrate them by giving an overview of an algorithm for constructing the boundary representation of $A \cap B$.

Consider two objects A and B. To construct a boundary representation for their intersection, traverse the edges of A starting from some vertex. Each time an edge intersects a face of B, mark the point. This will partition the edges of A into segments that are inside and outside of B. When an edge of A enters B, it can do so by crossing the interior of a face or by crossing an edge of a face. In the latter case three situations can occur. The edge can go into the interior of B, it may go along the boundary of B or it may leave B. The appropriate case must be determined. From this information we can construct $A - B$, $B - A$, $A \cup B$ and $A \cap B$.

It may be the case that no edge of B intersects A and no edge of A intersects B, yet B is neither entirely inside A nor entirely outside A. This can occur when the interior of a face of B intersects the interior of a face of A. For example, imagine two faces that are portions of a sphere. Thus, in addition to computing the intersection of edges with faces, every face of A must be intersected with every face of B. The intersection of two faces is calculated by intersecting the two surfaces defining the two faces. If the two surfaces intersect, then one must determine if any point of the intersection is inside both faces. Assuming that the two surfaces are distinct, the intersection is a curve. It is sufficient to test a single point on this curve. If a point on this curve lies on both faces, then the faces intersect.

For simplicity, assume that the objects are polyhedral. For nonpolyhedral objects the basic structure of the algorithm is the same although the computations involved in computing the geometrical intersection of faces is more complex and a modification to detect certain additional intersections of faces must be made.

After the edges of A have been partitioned into segments that are inside, outside, or on the boundary of B, $A \cap B$ is constructed by traversing edge segments of A that are inside B or on the boundary of B. When an edge of A leaves B, new edges are created on the face of B. Let F_1 and F_2 be the faces of A whose intersection determine an edge E and let F_3 be the face of B through which E leaves B. New edges are constructed by computing the intersection of the faces F_1 and F_3 and the intersection of the faces F_2 and F_3. This provides all edges needed for $A \cap B$. However, in constructing the boundary representation of $A \cap B$, extraneous one and two dimensional components must be discarded. Thus at each vertex a check is made to ensure that at least three edges are present. If not, then the surface or edge is removed. Algorithms for $A - B$, $B - A$, and $A \cup B$ are analogous.

The above techniques are reasonably well understood. However, there are still problems to be solved. One, in particular, is choosing among various representations for the edges of nonplanar faces in attempting to design algorithms. Worse yet, given two algorithms based on different edge representations, there is no good way to compare their computational complexity. The only well understood measures, such as worst-case asymptotic time analysis, do not give realistic results. We need to develop a measure of complexity that better reflects the notion of *expected* or *average-case* performance. For example, the worst-case asymptotic time complexity of the Simplex method is exponential, but its actual behavior in normal cases is far better than this would indicate. Therefore, we feel that to categorize realistic algorithms an entirely new approach to measuring complexity is required.

Thus, we see that even while looking at fairly simple algorithms we can run up against serious, fundamental problems. Both by bringing forth new problems and by forcing researchers to reevaluate their current techniques, stereophenomenology will challenge computer science.

5. REASONING

So far, we have discussed the problems of representing physical objects and processes and the algorithms and systems needed to manipulate these representations. The most important part of a science of representation still remains to be addressed. This is the capability for reasoning about physical systems. A system that does not allow determination of the future states of objects, that does not permit posing and answering questions about object behavior, and that does not support the accurate simulation of physical processes is a very weak system indeed.

In the following sections, we examine three general areas of reasoning about physical systems. The first, reasoning about behavior, deals with determining the response of a physical object (or assembly) to predetermined forces or actions upon it. Determining stability under gravity for an assembly is one example of this type of reasoning. The second, reasoning about tasks, concerns the more difficult problem of trying to achieve some desired state. Spatial planning falls into this category. The final and most speculative area, reasoning about design, deals with trying to determine shape and composition based on desired function. An example would be deducing the structure of a screwdriver from the specification of a screw.

We intend the discussion in this section to be illustrative of the types and methods of reasoning that might be suitable to a general system of object representation. It is certainly not intended to be exhaustive. In particular, we do not address any of the problems of perceptive reasoning and computer vision, which have been discussed extensively elsewhere (see, for example, Ballard and Brown, 1982). We hope that this examination of reasoning about object representations will serve as a useful starting point for discussion and analysis in this area.

5.1. Reasoning About Behavior

There are currently three main modeling techniques used in reasoning about the behavior of physical objects. To some extent, activities can be partitioned into categories based on the modeling technique needed to support them. The first category consists of those activities that can be supported by an abstract model or by a simulation driven by an abstract model. The second category of activities needs a solid model to support the level of reasoning required. The third category consists of problems that require the solution of partial differential equations. Currently, finite element methods are used to solve these problems.

Frequently, for a given problem there is a choice of modeling techniques. In particular, the tradeoffs between solid and abstract models must be carefully evaluated. For example, solid modeling is usually used for simple design verification. In the design of a complex assembly, such as a computer mainframe, a relatively simple application of solid modeling techniques to constructing a static electronic prototype suffices to verify that no two components occupy the same physical location.

More complex situations arise when we must reason about physical properties of an object or an assembly; for example, in what orientations can a partial assembly be moved so as not to fall apart. Although the reasoning is more sophisticated, the solid model is often not necessary and, in fact, merely adds complexity. These activities are best addressed by means of abstract models. To illustrate their use, we present a sequence of simple examples requiring more and more complex reasoning.

Consider a stack of blocks as shown in Fig. 1.6 and the result of removing the lower right hand block. Each block is represented by a point mass and moments of inertia about three orthogonal axes. Assuming that the objects are cuboids, we can determine the points of contact between the objects without a solid model. To simulate general objects, on the other hand, requires the solid model to determine the points of contact. Assuming a force of gravity, a force of reaction at each point of contact, and a friction force equal to the product of the coefficient of friction with the normal force, we can construct the equations stating that the summation of forces in each direction equals the mass times acceleration and that the summation of torques about each axis equals the product of the moment of inertia with the rotational acceleration.

These assumptions allow simulation of the blocks until new contacts occur. At that point an assumption about the nature of collisions is needed. Either the collisions are perfectly elastic and both the kinetic energy and the momentum are preserved, or only the momentum is preserved and some appropriate assumption

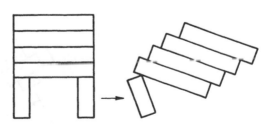

FIG. 1.6. Simulating a stack of blocks.

g Gravity
f Applied Force

FIG. 1.7.   Object with two possible motions.

concerning the elasticity of the collision is needed to continue the simulation. At the level of detail being modeled, the actual physics of the collision are summarized in the assumption concerning the nature of the elasticity. The model is obviously not fully realistic. For example, the effects of air resistance are ignored, and at the moment of collision the objects are modeled as having infinite accelerations.

The fact that mechanics with friction is not deterministic might force a more complex model. For example, the object in Fig. 1.7 might rotate and move upward to the right or might slide to the right depending on whether or not there is an assumed friction force at the point of contact. Either assumption as to its motion is completely consistent with the laws of physics at this level of modeling. To determine what actually happens requires a tribology model of the contact region. Another example requiring a more complete model of the contact arises when a drawer jams. Here a model of stored energy is needed to explain why the force needed to release the drawer is as large as the force that jammed it (Burridge, Rajan, & Schwartz, 1984).

Returning to the collapsing stack of blocks, when a new set of points of contact occurs, a new set of equations is produced and the simulation continued. This simple example requires only an abstract model with point masses, features consisting of possible points of contact, friction, and the ability to support simulation.

A more sophisticated model is needed for the device shown in Fig. 1.8. The mechanism is a simple swing consisting of a rigid rod with a pivot at the top. Attached to the rod is a movable mass. The swing is pumped by moving the mass up the rod at the extreme of the swing and down the rod at the center of the swing. Reversing the pumping motion brings the swing to a halt. The mechanism

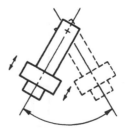

FIG. 1.8.   A swing.

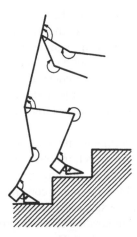

FIG. 1.9.   A stick person climbing stairs.

has the added complexity of an active control that senses the position and acceleration of the swing and issues commands to move the mass. Capturing this added complexity requires incorporating control programs and sensory data to the model.

Once the model supports control based on sensory feedback, the system can deal with the more complex modeling required to simulate a stick person walking up a flight of stairs (see Fig. 1.9). By ignoring situations such as that illustrated in Fig. 1.10, where the foot of the stick figure is moving through the stair, we can continue to use an abstract model. Incorporating collision avoidance appears to require integrating a volume model with the abstract model—a more complex task.

In order to carry out the simulation of a walking stick figure, control programs are needed for each of the three joints of the figure. In the case of locomotion,

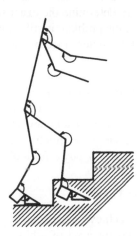

FIG. 1.10.   Situation requiring a solid model.

control models such as those of Raibert (1984) or McMahon (1984) could be used. Using one of these models, the simulation could calculate the forces produced at the various joints. If the simple abstract model of the knee joint were replaced by a more sophisticated solid model capable of representing the surfaces in contact, an actual artificial knee joint could be modeled. This could potentially allow testing the behavior of such a joint designed for a specific person under a wide range of human activities, for example, walking up stairs or sitting in a chair.

A third category of problems involves the solution of partial differential equations. This class includes such problems as air flow over airplane wings or car bodies, heat flow in injection molding processes, and determining stresses in beams. Solving these problems requires constructing a finite element mesh over some region bounded by the surface of a solid model.

There are a number of interesting research problems related to finite element method (FEM) models and meshes. The automatic generation of meshes from a solid model is one such problem. Work is needed on achieving well-conditioned meshes with minimum user input. At the same time, the system must retain the flexibility to allow the user to refine meshes in any region of special interest.

An even more challenging problem arises when the shape of the object, and hence the mesh, changes with time. In studying the operation of a cylinder in an internal combustion engine, the FEM mesh must be adjusted as the piston moves in the cylinder. In this problem the situation is simplified by prior knowledge of the exact change in shape of the volume with time. However, in more general situations the change in shape is determined only by solving the differential equations at the previous time unit.

Modeling injection molding of plastic requires a very general mesh update strategy. In this process, molten plastic is forced into a mold. As the plastic is injected, the surface of the flowing plastic is determined by the rate of injection, the shape of the mold, and the pressure and temperature distribution in the plastic. If we wish to try to determine the exact surface shape, rather than just a set of grid points across it, the problem is still more complicated, since the shape of the surface of the plastic is determined by the differential equations and is not a simple mathematical surface.

Finding computationally efficient solutions to problems such as these, and indeed to all the problems of reasoning about physical objects, presents a strenuous challenge to computer science. Both new approaches to reasoning about the old modeling techniques and completely new modeling strategies must be investigated. The ability to accurately and quickly determine the behavior of complex physical objects would be a very powerful tool indeed.

5.2. Reasoning About Tasks

Reasoning about how to achieve a desired physical state or configuration is inherently difficult. In fact, for a set of arbitrary ideal mechanical linkages the question of whether a given configuration can be reached has been shown to be PSPACE-hard (Hopcroft, Joseph, & Whitesides (1984). Motion planning for robots has been intensively researched and has proved to be extremely difficult (see Lozano-Perez, 1981; Schwartz & Sharir, 1983, and also the chapter by Yap in this volume).

In this section we examine both motion planning and fine motion strategies as two major examples of reasoning about tasks. This area is one of the most challenging and difficult in the entire science of stereophenomenology.

5.2.1. Motion Planning.   Configuration space is an important tool in motion planning (Lozano-Perez, 1983). An object or a configuration of objects can be represented as a point in a space of configurations. For example, triangles in 2-dimensional space can be represented as points in a 6-dimensional space where the first two coordinates give the location of the first vertex of a triangle, the next two coordinates give the location of the second vertex, etc. Similarly the location and orientation of an object can be represented by a point in a six-dimensional configuration space (three dimensions give position and the other three dimensions give orientation).

Certain points in configuration space correspond to configurations in which the object does not overlap any obstacle. Other points are illegal, in that the object overlaps one or more obstacles. The sets of illegal points are called configuration space obstacles.

One difficulty in planning motion for physical objects is the need to check if an object overlaps an obstacle at any point along its trajectory. Using configuration space to plan motion for a single rigid object reduces the problem to planning motion for a mathematical point. This latter problem is trivial since a point can be moved without restriction in any connected region. The difficult part of the technique comes in the construction of the configuration space obstacles.

Let A be an object and B be a fixed obstacle. In the case where A is free to move under translation but not rotation, configuration space is three-dimensional. A point of A, say a_0, is designated as the origin of A. A point in configuration space represents the location of a_0. Let A_x denote the set of points in 3-space covered by A when A is located with a_0 at the point x. A_0 denotes this set of points when a_0 is at the origin. Let B denote the set of points in 3-space occupied by the obstacle B. We define $A - B$ to be the set of points $[p \mid p = a - b,\ a \epsilon A,\ b \epsilon B]$ in 3-space, where $a - b$ is the vector difference of a and b. The configuration space obstacle corresponding to B is the set of configuration space points $[x \mid b \epsilon B]$ in 3-space, where $a - b$ is the vector difference of a and b. The configuration space obstacle corresponding to B is the set of configuration space points $[x \mid A_x \cap B \neq \varnothing]$.

1. The configuration space obstacle is given by $A_0 - B$.
2. If A and B are convex polyhedral objects then

$$A_0 - B = \text{convexhull(vertices}(A) - \text{vertices }(B))$$

If A and B are not convex polyhedra, then part (2) can be applied by first decomposing A and B into convex components.

In the two-dimensional version of the problem it can be shown that if A and all obstacles B_j are convex polygons, then the shortest path from start to goal in configuration space is piecewise linear, and the endpoints of the linear pieces of the path are vertices of the configuration space obstacles. For three-dimensional objects this is no longer true. The shortest path may graze an edge of a configuration space object. If rotation is allowed, then configuration space obstacles are no longer polyhedra, and no efficient algorithms for their construction are known. For this case Brooks and Lozano-Perez (1982) have developed techniques to approximate the configuration space obstacles by slicing the configuration space obstacles and enclosing each slice by a rectangular solid.

It can be shown that motion planning is, in principle, a computable task. Although the natural formulation of the decision problem for moving a set of objects from one configuration to another appears to be outside the theory of reals, Schwartz and Sharir have developed a decision procedure based on Collins' cylindrical decomposition decision technique for the theory of reals (Collins, 1975) to solve this problem.

The theory of reals is concerned with quantified formulas, where quantification is over the reals, made up of Boolean expressions of algebraic equalities and inequalities. Let's look at an example of a formula in the theory of reals: Let A be the triangle with vertices $(0, 0)$, $(0, 1)$ and $(1, 0)$ and let B be a square with vertices (x, y), $(x + 1, y)$, $(x + 1, y + 1)$ and $(x, y + 1)$. The statement that A and B do not overlap is given by

$$\forall r, s \ (0 \leq r \leq 1) \ and \ (0 \leq s \leq 1 - r) \Rightarrow$$
$$(r < x) \ or \ (r > x + 1) \ or \ (s < y) \ or \ (s > y + 1).$$

To express a motion planning problem one appears to need quantification over continuous functions. The obvious way to express that there exists a path in the xy plane from a to b is by the statement

$$\exists \ continuous \ functions \ f_1(t) \ and \ f_2(t) \ \forall t \ (0 \leq t \leq 1) \Rightarrow$$
$$x = f_1(t) \ and \ y = f_2(t) \ are \ not \ in \ any \ obstacle,$$
$$a = (f_1(0), f_2(0)) \ and \ b = (f_1(1), f_2(1)).$$

However, adding quantification over functions makes the theory undecidable. The Schwartz-Sharir technique overcomes this difficulty by applying the cylindrical decomposition method to the formula

$$\forall x \forall y \ (x, y) \ \text{is not in any configuration space obstacle.}$$

The space is partitioned into cells such that for all points (x, y) in a cell, the predicate is always true or always false. Planning motion is equivalent to finding a sequence of adjacent cells in which the predicate is true starting from a cell containing a to a cell containing b. The complexity of the method is double exponential in the size of the problem. However, for problems with a finite number of degrees of freedom, such as moving a single rigid object (six degrees of freedom) or a hinged object with one or more internal degrees of freedom, the time complexity is polynomial, although the degree of the polynomial grows exponentially with the degrees of freedom.

Other complexity results, such as Hopcroft, Joseph, and Whitesides (1984), Hopcroft, Schwartz, and Sharir (1984), show that motion planning for linkages or even rectangles is PSPACE-hard. Thus procedures guaranteed to solve the problem will of necessity have prohibitive worst case complexity. However, it is our belief that efficient algorithms *can* be developed for handling those cases likely to arise in practice.

In planning coordinated motion of independent objects, not only must paths be determined, but also the relative timing of objects as they move on their individual trajectories. A path in the appropriate configuration space defined for multiple moving objects contains this information about relative timing. Thus searching for valid paths in configuration space conceptually simplifies the problem (Hopcroft and Wilfong, 1984). However, in general the dimension of configuration space is very large. One method of reducing the dimension of the space to be searched is to push the objects together in both their initial and final configurations. Once the objects come in contact we continue to push them causing them to rotate and make higher and higher dimensional contact with each other until no higher degree of contact is possible. For example, when two cubes are pushed together, first a vertex touches a face, then the cubes rotate so that an edge of one lies on the face of the other, and finally two cube faces become adjacent.

A configuration space object can be defined consisting of all configurations in which two or more objects overlap. The surface of this object is defined to be the set of configurations in which the objects form a single connected object, but no two objects overlap. The motion involved in pushing the physical objects together can be viewed as moving a point in configuration space towards the surface of this configuration space object. The k-dimensional surface is composed of k-dimensional faces that intersect in (k-1)-dimensional faces that in turn intersect in (k-2)-dimensional faces and so forth. On reaching the surface of the configuration space object, the motion in configuration space continues through a sequence of lower and lower dimensional faces of the surface until the initial and final configurations have been converted to configurations corresponding to vertices of the configuration space object.

Under suitable restrictions, the vertices of the surface of contact in configuration space will be edge connected. In this case, the problem of searching configuration space is reduced to a much simpler graph searching problem. In general the number of vertices of the graph will be astronomical. However, the entire graph need not be constructed; vertices and edges can be generated as the search progresses. With a suitable heuristic it may be possible, in practical situations, to find the desired path having only generated a tiny fraction of the graph. The knowledge of such a path could be used in constructing a trajectory where the objects do not touch one another. In practical cases, where objects are being moved in a work space, an effective algorithm might be possible even though the worst-case complexity would be prohibitive. The search process envisioned is somewhat analogous to linear programming, where only a small number of vertices of a polytope are examined.

The above technique rests on a theorem the simplest form of which states that if it is possible to move two objects from one configuration where they are in contact to a second configuration where they are in contact, then it is possible to do so by a motion that always maintains contact.

The theorem can be established by basic techniques from algebraic topology. Algebraic topology describes the structure of a space by associating with it a sequence of groups called homology groups. The groups associated with a space reflect the topological structure of the space. The $0th$ homology group H_0 provides information about path connectivity of the space. For a space with k path connected components, H_0 is isomorphic to k copies of the integers. The first homology group H_1 provides information concerning holes in the space and is isomorphic to a number of copies of the integers equal to the number of holes. If a space is contractible to a point, i.e. has no holes, H_1 is isomorphic to the trivial group consisting only of the identity element. If a space is isomorphic to a torus, then H_1 is isomorphic to Z, the group of integers under addition.

To prove the above theorem: Let A be the configuration space object and let B be the closure of the complement of A. A and B are both closed sets, $A \cup B$ is the totality of configuration space and $A \cap B$ is the boundary of the configuration space object. The Mayer-Vietoris theorem (Massey, 1967; Vick, 1973) relates the homology groups of the spaces A, B, $A \cap B$ and $A \cup B$. In particular, it states that the sequence

$$H_1(A \cup B) \xrightarrow{h_1} H_0(A \cap B) \xrightarrow{h_2} H_0(A) \oplus H_0(B) \xrightarrow{h_3} H_0(A \cup B) \xrightarrow{h_4} \emptyset$$

along with the natural sequence of homomorphisms is an exact sequence.

An exact sequence of homomorphisms is one in which the image of one homomorphism is the kernel of the next. Assuming that $A \cup B$ is path connected, contractible to a point and that the configuration space object A is path connected

$$H_1 (A \cup B) = \emptyset, H_0 (A \cup B) = Z \text{ and } H_0 (A) = Z$$

Thus,

$$\emptyset \rightarrow H_0 (A \cap B) \rightarrow Z \oplus H_0 (B) \rightarrow Z \rightarrow \emptyset$$

From the fact that the sequence of homomorphisms is exact one can deduce that the $0th$ homology group of the boundary of the configuration space object,

H_0 $(A \cap B)$, is isomorphic to the $0th$ homology group of the legal regions of configuration spaces, H_0 (B). It follows that the path connected components of the boundary are in one-to-one correspondence with the path connected regions of configuration space (Hopcroft & Wilfong, 1984).

The above theorem can be generalized to cover a concept of motion more general than simple translation and rotation. In animation the parts of a moving object may change shape during a motion. Consider a person whose arm, leg, and body components are represented by abstract models. The components are filled out with a parameterized volume model where the precise surface shape is determined by the values of the parameters. As the person moves, the arm, leg, and body components must always stay in contact.

In this setting an object is a parameterized mapping from a canonical region of R^3 to R^3. The parameters determine the location, orientation and precise shape of a particular instantiation of the object. A motion for an object is any continuous path in parameter space. It can include not only translation and rotation but also certain continuous deformations of the object. Thus it could include surface appearance changes modeling the action of muscles under the skin.

Most of the interesting problems of motion planning remain to be solved. In particular, the problems of nonrigid objects have not been addressed at all. Reasonable heuristic or algorithmic solutions for even the simplest problems are urgently needed. The challenge of this area is obvious.

5.2.2. Fine Motion.   A key area of task reasoning is that involved in guaranteeing a certain behavior as moving objects come into contact. Situations where this occurs include gripping, mechanical assembly, and the use of tools. Techniques for solving these problems, that is, for obtaining a desired result, have been termed *fine motion strategies* (Mason, 1982).

For an example of a simple fine motion strategy (adapted from Mason, 1983), consider Fig. 1.11. It shows the very standard problem of inserting a peg in a hole. If the system uses the naive approach of simply placing the peg vertically into the hole, because of the tight tolerances, the positioning and motion of the

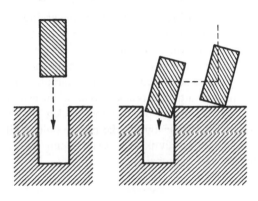

FIG. 1.11.   Two techniques for inserting a peg in a hole.

peg must be extremely accurate. If, on the other hand, the system has some degree of compliance, we can greatly improve the odds for a successful insertion. If the peg is set on the surface at an angle, slid over to the hole, and then raised to vertical (the right-hand illustration of Fig. 1.11), the insertion will succeed, even in the presence of fairly large positional errors.

While developing individual strategies of this sort is fun, and can even be valuable, the real problem is to automate techniques that can produce such strategies. A practical system for reasoning about fine motion strategies, and perhaps even automatically synthesizing them, would be the real prize.

The work being done in this area is an excellent example of the type of reasoning about physical objects that lies at the heart of stereophenomenology. In particular, the work of Lozano-Perez (1981), Mason (1983), and others on the automatic synthesis of fine motion strategies, correctness and completeness of a fine motion planner, and the use of configuration space in planning fine motion is very significant.

There are a number of closely related subjects that fall in the general class of fine motion. The topics of compliant motion and sensory feedback are two important examples. For instance, a person performing an assembly does not accomplish the task solely by position control. Instead the individual senses the forces as parts mate and uses this sensory information to achieve the desired goal. To test this hypothesis, observe that one can screw the bottom on a flashlight even if the parts are assembled in the dark.

We have already discussed, in Section 3.5, some of the problems of attempting to represent transformations and processes. The problem of interrelating sensory control information with descriptions of the task and representations for the objects and workspace is particularly difficult. At the moment there is no language for even describing these activities, much less reasoning about them. Much research, analysis and design remains to be done before we can hope to have practical systems for reasoning about or synthesizing fine motions.

5.3. Reasoning About Design

A unique area of reasoning is that concerning the function, shape and composition of objects. Rather than specifying the task and the tools available for performing it, we wish to consider the case where an abstract task or function is specified, and the system reasons about the nature of objects suitable for performing the task or function. For example, in assembly, instead of reasoning about how to attach a bracket, one might ask what the function of the bracket actually is. If the purpose of the bracket is to hold another part in place, one might consider redesigning the part to hold itself in place. In this environment, we can no longer think of a part as being determined by its size and shape but rather by its functionality.

FIG. 1.12.   Configuration space for
peg in hole problem.

As we have discussed throughout this chapter, the problems of representing concrete physical objects, to say nothing of tasks and processes, are far from solved. The problems of coming up with a representation for function that allows useful reasoning are harder still.

There has been some investigation of the ideas of automatic design from function within the artificial intelligence community. As early as 1971, Freeman and Newell (1971) proposed a model for functional reasoning in design, although in general rather than specific terms. In the area of perception, researchers have looked into the question of identifying objects or tools based on information about the function they are to perform (e.g., Winston et al., 1983). However, the problem of inferring desired shape from function is more difficult yet.

As an example of what might be possible, we describe one specific technique for trying to use a model of the task to determine object shape. Consider again the problem of trying to insert a peg in a hole. Suppose that the diameter of the peg is 99% of the diameter of the hole. Figure 1.12 illustrates the corresponding configuration space once the peg has been shrunk to a point. The hole into which we must move the point has a diameter of only 1% of the original diameter of the hole. If there were some way to widen the top of the configuration space hole, the insertion would be much simpler. Moving from configuration space back to real space, we see that widening the configuration space hole corresponds either to widening the top of the real hole or tapering the peg (Fig. 1.13). Thus, a model of the task serves to guide the design of the object.

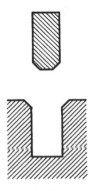

FIG. 1.13.   Tapering peg or hole to
simplify insertion.

This is a trivial example, but the underlying concept of inferring object design from the desired function is very important. A practical system for doing so would be of great benefit, although with our present understanding we can only speculate as to whether such a system could ever be constructed. Much more research in the area will be needed before this question can be answered.

6. CONCLUSIONS

We have outlined the basis for a new science, stereophenomenology, concerned with representing and reasoning about physical objects and processes. This science blends together elements from a variety of disciplines, including physics, mathematics, mechanical, electrical, and other branches of engineering, robotics, artificial intelligence, and many others. Clearly, computer science will play a decisive role in defining and solving the problems of this new area.

The science of object representation presents a vast array of exciting and challenging problems in abstract reasoning, data structures, algebraic geometry and topology, modeling, and many other areas. We have tried to give some feel for the richness and diversity of this new discipline. Its study will require important new types of interdisciplinary research and cooperation.

One of the most important points to make about this science is not its diversity, but rather its unity. We strongly believe that there is a unifying set of principles, tools, and methodologies, lying at the heart of stereophenomenology, that can be applied to a very broad range of important problems. Identifying these principles from the vast wealth of problems and results concerning physical systems lies at the heart of the challenge of robotics for computer science.

ACKNOWLEDGMENT

Parts of this paper are based on research supported by the National Science Foundation through its Computer Engineering program grant ECS 83-12096.

REFERENCES

Ballard, D. H., & Brown, C. M. (1982). *Computer vision*. Englewood Cliffs, NJ: Prentice-Hall.
Bates, J. L., & Constable, R. L. (1985). Proofs as programs. *ACM Transactions on Programming Languages and Systems, 7*(1), 113–136.
Blinn, J. F. (1982). A generalization of algebraic surface drawing. *ACM Transactions on Graphics, 1*(3), 235–256.
Borning, A. (1981). The programming language aspects of ThingLab: A constraint-oriented simulation laboratory. *ACM Transactions on Programming Languages and Systems, 3*(4), 353–387.

Brooks, R. A. (1981). Symbolic reasoning among 3-D models and 2-D imges. *Artificial Intelligence, 17*, 285–348.

Brooks, R. A., Greiner, R., & Binford, T. O. (1979). The ACRONYM model-based vision system. *Proc. Sixth Int'l. Joint Conf. Artificial Intelligence*, pp. 105–113.

Brooks, R. A., & Lozano-Perez, T. (1982). A subdivision algorithm in configuration space for Findpath with rotation. MIT AI Memo 684.

Brown, C. M. (1982). PADL-2: A technical summary. *IEEE Computer Graphics and Applications, 2*(2), pp. 69–84.

Brown, C. M., & Voelcker, H. G. (1979). The PADL-2 project. *Proc. Seventh NSF Conf. on Production Research and Technology*. Ithaca, NY: pp. F1–F6.

Burridge, R., Rajan, V. T., & Schwartz, J. T. (1984). *The peg-in-hole problem: Phenomena of sticking and jamming for nearly rigid bodies in frictional contact, and the transition from static to dynamic behaviour: I. Motions in two dimensions*. NYU-Courant Institute.

Clocksin, W. F., & Mellish, C. S. (1981). *Programming in Prolog*. Berlin: Springer Verlag.

Collins, G. (1975). Quantifier elimination for real closed fields by cylindrical algebraic decomposition. *Second GI Conference on Automata Theory and Formal Languages, 33*, 134–183, Springer Verlag.

Constable, R. L., & O'Donnell, M. J. (1978). *A programming logic*. Cambridge, MA: Winthrop.

Cook, R. L., & Torrance, K. E. (1982). A reflectance model for computer graphics. *ACM Transactions on Graphics, 1*(1), 7–24.

Dijkstra, E. W. (1976). *A discipline of programming*. Englewood Cliffs, NJ: Prentice-Hall.

Ferguson, J. (1964). Multivariable curve interpolation. *Journal of the ACM, 11*(2), 221–228.

Fitzgerald, W., Gracer, F., & Wolfe, R. (1981). GRIN: Interactive graphics for modeling solids, *IBM J. Research and Development, 25*(4), 281–294.

Floyd, R. W. (1967). Assigning meanings to programs. *Proc. Amer. Math. Soc. Symp. in Applied Math, 19*, 19–31.

Foley, J. D., & Van Dam, A. (1983). *Fundamentals of interactive computer graphics*, Reading, MA: Addison-Wesley.

Forbus, K. D. (1981). Qualitative reasoning about physical processes. *Proc. 7th International Joint Conf. on Artificial Intelligence*, pp. 326–330.

Freeman, P., & Newell, A. (1971). A model for functional reasoning in design. *Proc. Second International Joint Conf. on Artificial Intelligence*, pp. 621–633.

Hanrahan, P. (1983). Ray tracing algebraic surfaces. *Computer Graphics, 17*(3), 83–90.

Hoare, C. A. R. (1969). An axiomatic basis for computer programming. *CACM, 12*(10), 576–583.

Hopcroft, J., Joseph, D., & Whitesides, S. (1984). Movement problems for 2-dimensional linkages. *SIAM J. Computing, 13*(3), 610–629.

Hopcroft, J., Schwartz, J. T., & Sharir, M. (1984). On the complexity of motion planning for multiple independent objects: PSPACE hardness of the "Warehouseman's Problem". *International Journal of Robotics Research, 3*(4), 76–88.

Hopcroft, J., & Wilfong, G. (1984). *Reducing multiple object motion planning to graph searching*. (Cornell University Computer Science Technical Report # 84-616).

Leu, M. C., & Mahajan, R. (1984). Computer graphic simulation of robot kinematics and dynamics. *Proc. Robots 8 Conference, 1*, 4–80—4–101.

Levin, J. (1976). A parametric algorithm for drawing pictures of solid objects composed of quadric surfaces. *CACM, 19*(10), 555–563.

Levin, J. (1979). Mathematical models for determining the intersections of quadric surfaces. *Computer Graphics and Image Processing, 11*, 73–87.

Lozano-Perez, T. (1981). Automatic planning of manipulator transfer movements. *IEEE Transactions on Systems, Man, and Cybernetics, SMC-11*(10), 681–698.

Lozano-Perez, T. (1983). Spatial planning: A configuration space approach. *IEEE Transactions on Computers, C-32*(2), 108–120.

Lozano-Perez, T., & Wesley, M. A. (1979). An algorithm for planning collision-free paths among polyhedral obstacles, *CACM, 22*(10), 560–570.

Mason, M. T. (1982). *Manipulator grasping and pushing operations.* (MIT AI Technical Report #690).

Mason, M. T. (1983). *Automatic planning of fine motions: Correctness and completeness.* (CMU-RI Technical Report # 83-18).

Massey, W. S. (1967). *Algebraic topology: An introduction.* New York: Springer-Verlag.

McMahon, T. A. (1984). Mechanics of locomotion. *International Journal of Robotics Research, 3*(2), 4–28.

Meagher, D. J. (1982). Geometrical modelling using octree encoding. *Computer Graphics and Image Processing, 19*(2), 129–147.

Nelson, G. (1983). *How to use Juno.* (Xerox Palo Alto Research Center Report CGN-11).

Ocken, S., Schwartz, J., & Sharir, M. (1985). Precise implementation of CAD primitives using rational parameterization of standard surfaces. *Planning, geometry and complexity of robot motion.* Norwood, NJ: Ablex.

Popplestone, R. J., Ambler, A. P., & Bellos, I. M. (1982). An interpreter for a language for describing assemblies. In M. Brady et al. (Eds.), *Robot motion: Planning and control* (pp. 537–566). Cambridge, MA: MIT Press.

Raibert, M. H. (1984). Hopping in legged systems: Modeling and simulation for the 2D one legged case. *IEEE Trans. on Systems, Man and Cybernetics,* SMC-14(3), 451–463.

Requicha, A. A. G. (1983). Toward a theory of geometric tolerancing. *The International Journal of Robotics Research, 2*(4), 45–60.

Requicha, A. A. G., & Voelcker, H. B. (1982). Solid modelling: A historical summary and contemporary assessment. *IEEE Computer Graphics and Applications, 2*(2), 9–24.

Schwartz, J. T., & Sharir, M. (1983). On the 'piano movers' problem: II. General techniques for computing topological properties of real algebraic manifolds. *Advances in Applied Math, 4,* 298–351.

Sederberg, T. W., Anderson, D. C., & Goldman, R. N. (1985). Implicit representation of parametric curves and surfaces. *Computer Vision, Graphics, and Image Processing, 28,* 72–84.

Sommerville, D. M. Y. (1934). *Analytical geometry of three dimensions.* New York: Cambridge University Press.

VanWyk, C. J. (1981). A graphics typesetting language. *Proc. ACM SIGPLAN/SIGOA Symp. on Text Processing. SIGPLAN Notices, 16*(6), 99–107.

Vick, J. W. (1973). *Homology theory: An introduction to algebraic topology.* New York: Academic Press.

Voelcker, H. G., Requicha, A. A. G., Hartquist, E. E., Fisher, W. B., Shapiro, J. E., & Birrell, N. K. (1974). *An introduction to PADL: Characteristics, status, and rationale.* (Tech. Memo. No. 22). Production Automation Project, University of Rochester.

Walker, R. J. (1950). *Algebraic curves.* New York: Dover Publications.

Wesley, M. A., Lozano-Perez, T., Lieberman, L. I., Lavin, M. A., & Grossman, D. D. (1980). A geometrical modeling system for automated mechanical assembly. *IBM Journal of Research and Development, 24*(1), 64–74.

Winston, P. H., Binford, T. O., Katz, B., & Lowry, M. (1983). *Learning physical descriptions from functional definitions, examples, and precedents.* (MIT AI Memo 679).

2 Computational Geometry—A User's Guide

David P. Dobkin
Diane L. Souvaine
Princeton University

ABSTRACT

During the past decade, significant effort has been devoted to the design and analysis of efficient algorithms for computations which are geometric in nature. Most of this effort has been devoted to problems without direct applications in robotics. The methods used, however, should have a significant impact upon the development of algorithmic methods in robotics. A few of the more promising methods are described below along with some geometric applications.

1. INTRODUCTION

The past decade has brought significant progress in the development of algorithms for solving problems which are geometric in nature (see e.g., Lee & Preparata, 1984). Many of these algorithms depend on methods that were developed to solve problems in other branches of computer science. We believe that this trend will continue and that the methodology of computational geometry will ultimately find significant applications in robotics. Here we present and explain the methods used in developing three families of geometric algorithms. Two of these methods involve the decomposition of a problem into subproblems which can be more easily solved. The third involves transforming a problem in its entirety to a new format which may be more tractable.

The first of these methods is a hierarchical searching method. Here, a geometric problem is preprocessed to provide a coarse representation of the entire problem. Search queries upon the whole are then used to localize the region in which the problem is to be solved. Queries of this type alternate with computations which yield continually finer descriptions of these continually smaller

43

regions. Efficient algorithms result from balancing the two processes of localizing the search and of increasing the detail. Algorithmic efficiency is then balanced against preprocessing time and storage space requirements. A simple example of this process is the binary search of a sorted array. Here, the region is an interval of array indices and the finer description consists of increasing the density of information.

The second method presented is the divide-and-conquer method. Here, a problem is broken into subproblems which can be solved recursively. A method is then defined for combining solutions to subproblems in order to yield a solution to the entire problem. An extension of sort-merge techniques, this method has found significant application to all types of algorithm design. Attention here is focused on issues involved in applying it to problems which are geometric in nature. In some contexts, these first two methods often work together. For example, a divide and conquer method (i.e., sort-merge) might be used to order elements of an array to allow the use of a hierarchical search method (e.g., binary search). For instance, a Voronoi diagram (Shamos & Hoey, 1975), constructed via divide-and-conquer, yields a planar subdivision that we could search by the methods of section 2.

Finally, the duality method is one which has been used with success by mathematical programmers for the past 4 decades. We focus here specifically on problems involving 2- and 3-dimensional geometries, where duality is used as a transformation. Given two sets A and B and a query regarding their interrelationship, we apply the transform T and answer a different, hopefully easier, query about the relationship between T(A) and T(B). The form of T and its uses are described here.

The premise of this paper is that the algorithm for solving a particular problem is often less important than the methodology which led to the algorithm. We encourage the reader to look upon this paper as he/she would look upon a user's manual for a new programming system. We present methods both in their pure form and as a means to solving geometric problems. To continue the user's manual analogy, the reader might consider the former as an exposition of the macros which come with a particular system. The latter then provide examples of these macros in action. The applications we consider depend on geometrics principles: planar point location, convex hull construction and updating, and computation of polygon, disk, and half-space intersections. Combining these principles with relevant sets of facts for robotics should make both our methods and results useful to robotics workers.

2.   HIERARCHICAL SEARCH

2.1. Binary Search

Binary search is an effective technique for locating a particular object within a search domain, providing that the objects have some inherent ordering. A simple

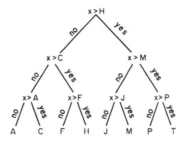

FIG. 2.1.   Sample binary search tree

example involves the search via random-access probes for an element in an ordered linear array. The procedure is modeled by the binary tree depicted in Fig. 2.1. The elements of the array are contained in order in the leaves and internal nodes of the tree. Figuratively, the search begins at the root of the tree comparing the desired element to the middle element of the array. Either answer will delete half of the array, and thus half of the tree, from further consideration. The search moves recursively to either the left subtree, comparing to the first quartile, or the right subtree, comparing against the third quartile. The search continues until the search element is found or the frontier of the tree is reached. We might say that the first question provides a coarse idea of the location of the number (vis à vis the median) and each succeeding question provides finer and finer detail; that is, in a binary sense, each question identifies one further bit of the array index of the number.

When we apply this technique to geometric problems, the process may become somewhat more complicated. Unlike the problem above, many geometric problems have no obvious one-dimensional ordering leading to an obvious progression of questions. Preprocessing is necessary to regularize the structure of the original problem. This yields a coarse description of the problem at which the search begins. Searching then involves answering questions based on this coarse description. The resultant answers point us toward further queries based on slightly more detail within a smaller instance of the problem, and so forth until an answer is reached. The preprocessing time must be balanced against the searching time as we describe below.

We consider here a problem to which this technique applies.

Input: A collection of N disjoint polygons in the plane.
Query: For a given point P, find all polygons to which it belongs.

The naive approach to this problem, merely testing the point P against each polygon in turn, quickly becomes impractical as both the number of polygons in search domain and the number of distinct query points increase. We develop a method of preprocessing the polygons to decrease individual query time for cases when the number of queries is large.

At first, we will restrict our attention to polygons which are axis-parallel rectangles. We represent these rectangles by their lower left and upper right vertices. Our development begins with a simple algorithm for this case and works toward a general procedure which applies to arbitrary polygons. All of the key ideas, the algorithmic "macro," can be seen in the simple rectangle case.

2.2. Rectangle Search I: Specific Example

We consider the simple example depicted in Fig. 2.2a where all vertices have integer coordinates. The search domain consists of four rectangles:

$$R_1 = \{(1, 5), (4, 7)\}$$
$$R_2 = \{(0, 0), (2, 4)\}$$
$$R_3 = \{(5, 3), (6, 6)\}$$
$$R_4 = \{(3, 1), (7, 2)\}$$

We begin by finding the smallest rectangle, Q, which encloses all of the search rectangles. In this case $Q = \{(0, 0), (7, 7)\}$. Together, the boundaries of Q and the four rectangles form a planar graph, S, with nineteen vertices and twenty-two edges which subdivides the plane into six regions: the unbounded region, R_5 which is exterior to Q; the four rectangular regions; and the region we call R_0 which is interior to Q but exterior to all four rectangles. Since all of the edges are finite line segments and thus only one region is unbounded, we can call this graph a "finite planar subdivision" (see Fig. 2.2b).

In order to determine which, if any, of the four rectangles contains the query point $P = (x, y)$, we first determine whether P lies in the unbounded region R_5 by making four comparisons to check whether either x or y is less than 0 or bigger than 7. If so, our search terminates. Otherwise, we need to determine the region interior to S which contains P. Ideally, a few quick comparisons would localize our search further, in a pattern similar to one-dimensional binary search. Unions of regions of the planar subdivision, however, produce awkward (i.e., nonrectangular) shapes against which P cannot easily be tested. We need to restructure the search domain. Since all current vertices have integer coordinates, we may refine the subdivision S, to a new planar graph S_0 which forms a complete grid of unit squares. As each of these squares is a subset of exactly one of the regions in S, the representation we choose for S_0 will include a pointer for each square face to its "parent" in S (see Fig. 2.2c).

At first, the task of locating P in one of the forty-nine regions of S_0 seems more complicated than the original problem. The regular shape of the regions, however, benefits us. If we delete an interior vertex, T, and all of the edges adjacent to it, four squares are merged to form one larger square. If we knew that P were included in the larger square and we knew the coordinates of the point T, we could quickly determine to which of the original four squares P belonged.

Therefore, we will create a new subdivision S_1 using the following procedure: sweep a vertical line left-to-right across the plane; stop every second time that it

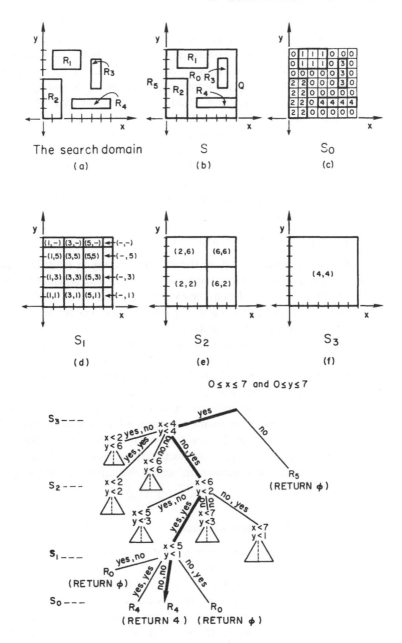

FIG. 2.2. Hierarchy for rectangle search I

intersects (contains) vertices and vertical edges of S_0; move bottom-to-top along the stopped line, deleting every second vertex and its adjacent edges; where necessary, add edges to preserve the boundary of Q. Here, both coordinates of a deleted vertex will be odd numbers. In most cases, four squares will be merged to form a larger square, but along the upper and right boundaries of S_0, two or even no squares may merge (see Fig. 2.2d). The representation of S_1 will associate a tag with each region which not only points to its parents, or originators, in S_0 but also identifies the vertex they all share.[1] This process cuts the size of the search domain from forty-nine internal regions to sixteen. If we repeat the process, we achieve a subdivision S_2 which has four internal regions (Fig. 2.2e) and then a subdivision S_3 with just one (Fig. 2.2f). The sets, S_0, S_1, S_2, S_3 constitute a four-level hierarchy. We measure the size of each subdivision, S_i, by the number of finite regions it contains:

$$|S_0| = 49; \ |S_1| = 16; \ |S_2| = 4; \ |S_3| = 1.$$

Our algorithm to determine which of the original rectangles contains P proceeds as follows: make at most four comparisons to determine whether P lies in the interior or exterior of S_3; if P lies in the exterior, return the null symbol $\emptyset$; if P lies in the interior of S_3, compare x and y against the two coordinates stored in S_3 to determine which region in S_2 contains P; then test P against the coordinate(s) stored there to determine which region in S_1 contains P; test P against the coordinate(s) stored in that region to determine which square in S_0 contains P; look up the original region, R, to which that square belongs; if R is one of the rectangles, R_1, ..., R_4, return the number of that rectangle; otherwise, return $\emptyset$. Fig. 2.2g demonstrates the path followed through the hierarchy for the case where $P = (11/2, 3/2)$. Four comparisons are executed at the root, followed by two more at each internal node in the tree for a total of ten comparisons.

As already mentioned, this algorithm successively considers finer details (rectangles of smaller area) as it narrows the search domain (by eliminating sets of rectangles).

2.3. Rectangle Search I: General Case

We now extend the algorithm to the more general case where the search domain consists of N axis-parallel rectangles, R_1, R_2, ..., R_N, whose vertices have real-valued coordinates. First, we project all of the vertices onto the x-axis, yielding at most $2N$ distinct values. Without loss of generality, we assume that the $2N$ values x_0, x_1, x_2, ..., x_{2N-1}, are distinct and represented in increasing order. Similarly, project all of the vertices onto the y-axis, achieving an ordered list y_0, ..., y_{2N-1}. Note that the values x_0 and y_0 need not originate with the same

---

[1] The term "parent" is used loosely, as a region in S_1 generally points to four distinct squares in S_0. In cases of multiple parenting, we shall henceforth use the term "originators."

vertex, and x_0 and x_1 need not define the x-extent of a single rectangle. The smallest rectangle which circumscribes R_1, ..., R_N is $Q = \{(x_0, y_0), (x_{2N-1}, y_{2N-1})\}$. S is the planar graph formed by the boundaries of Q and R_1, ..., R_N; R_{N+1} is the region exterior to Q; and R_0 is the region interior to Q but exterior to R_1, ..., R_N (see Fig. 2.3a).

To achieve the regular structure needed for S_0, proceed as follows:

1. for each x_i, $0 \le i \le 2N - 1$, draw the line segment from (x_i, y_0) to (x_i, y_{2N-1});
2. for each y_i, $0 \le i \le 2N - 1$, draw the line segment from (x_0, y_i) to (x_{2N-1}, y_i);
3. reinterpret the resulting picture as a planar graph, S_0.

Since Q was divided at each x-coordinate and y-coordinate which represented a vertex of the original rectangles, each one of the regions (faces) of S_0 is a subset of exactly one of the regions, R_0, ..., R_N. Thus, the representation of each face in S_0 will include a pointer to its parent in S.

Despite the fact that the regions in S_0 need not be squares, all interior vertices of S_0 will have degree four, as in the previous example. Thus we can use the procedure defined earlier to create subdivisions $S_1, S_2, S_3, \ldots$, each having fewer faces than the previous one. We continue until we reach an S_m which has a single finite region identical to the bounding box Q, (see Figs. 2.3b and 2.3c).

With the hierarchy, S_0, ..., S_m in place, we can determine which of the original rectangles contains P by the following procedure:

> Begin
> > 1. decide whether P lies in the bounded or unbounded region of S_m
> > 2. if P is in the unbounded region, return $\emptyset$ and stop
> > 3. $v_{m-1} \leftarrow$ the vertex associated with bounded region of S_m
> > 4. for $i \leftarrow m - 1$ to 1 step -1
> > > begin
> > > 5. test (x, y) against v_i to determine which of its originators in S_i contains P
> > > 6. $v_{i-1} \leftarrow$ the vertex stored with that originator
> > end;
> > 7. look up in S_0 the original region R_j to which v_0 points
> > 8. if $j = 0$, then return $\emptyset$, else return j.
> End.

The set-up phase of the algorithm, steps 1–3, and the interior of the loop, steps 5–6, each require constant time.

Thus, the time complexity of the algorithm is linear in m, the number of times the algorithm executes the inner loop. If we reconsider Fig. 2.3b, we can see that S_0 contains $(2N - 1)^2 = j^2$ regions ($|S_0| = j^2$). In the process of creating S_1, we eliminate the vertical divisions for $\lfloor j/2 \rfloor$ values of x. In like fashion, we delete the horizontal divisions for $\lfloor j/2 \rfloor$ values of y. Thus, S_1 must contain $(\lceil j/2 \rceil)^2 = k^2$

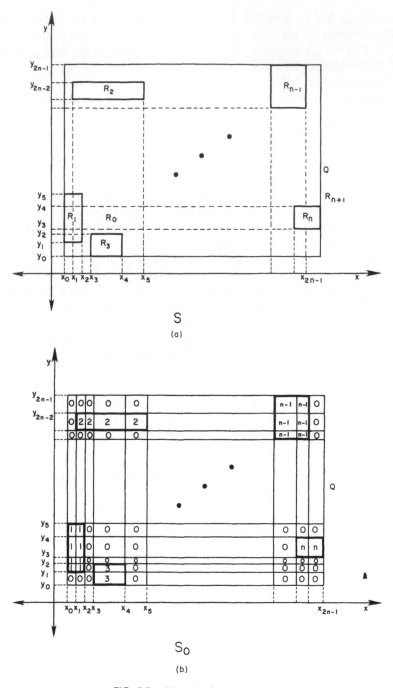

FIG. 2.3. Hierarchy for general case

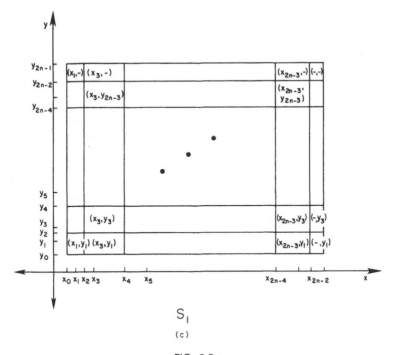

$$S_1$$

(c)

FIG. 2.3.c

regions. In the same manner, the process of creating S_2 removes $\lfloor k/2 \rfloor$ horizontal and vertical divisions, so ($|S_2| = (\lceil k/2 \rceil)^2$ finite regions. Since S_m contains but one finite region, $m = \lceil \log (2N-1) \rceil$.[2]

Thus, in determing the region containing P, this algorithm performs $4 + 2\lceil \log (2N - 1) \rceil$ comparisons, or $O(\log N)$ comparisons. The naive algorithm requires $4N$ comparisons, or $O(N)$ comparisons. Thus, even for small N (e.g. $N = 3$) this algorithm is preferable, provided the necessary hierarchy can be created easily. Given that S_0 contains $4N^2$ vertices and $(2N - 1)^2$ regions, however, just the representation of the hierarchy must take $\Omega(N^2)$ time. As a result, this algorithm has value only when a large number of points P need testing and when the size of N does not make N^2 prohibitively large.

2.4. Rectangle Search II

The hierarchical algorithm described above actually subdivides Q into more regions than necessary. Our second algorithm offers only a slight improvement in the worst-case space requirement, but is far more efficient on average and should make the later exposition clearer. In the above, all of the vertices of the N rectangles, $R_1, \ldots, R_N$, are projected onto both the x- and y-axes, creating $2N - 1$ finite intervals in each direction. Then, added horizontal and vertical

---

[2]Note that $\log x$ here always means $\log_2 x$.

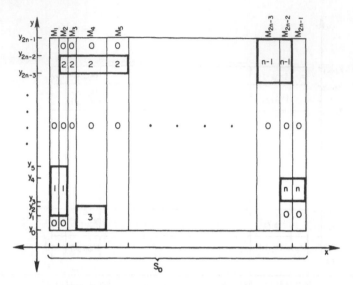

FIG. 2.4. Decomposition for rectangle search II

segments create a total of $(2N - 1)^2$ distinct regions (see Figs. 2.3a and 3b). Now we add only those vertical segments containing an edge of a rectangle. The new S_0 contains only the actual intervals along the x-axis, rather than the entire resulting graph. In other words, $S_0 = \{[x_0, x_1], [x_1, x_2], ..., [x_{2N-2}, x_{2N-1}]\}$. The sets, $M_1, ...M_{2N-1}$, consist of the distinct vertical intervals represented in the vertical slab defined by each respective horizontal interval. In the example of Fig. 2.4,

$$M_1 = \{[y_0, y_1], [y_1, y_5], [y_5, y_{2N-1}]\},$$
$$M_2 = \{[y_0, y_1], [y_1, y_5], [y_5, y_{2N-3}], [y_{2N-3}, y_{2N-2}], [y_{2N-2}, y_{2N-1}]\},$$
$$M_3 = \{[y_0, y_{2N-3}], [y_{2N-3}, y_{2N-2}] [y_{2N-2}, y_{2N-1}]\},$$
$$\vdots$$
$$M_{2N-1} = \{[y_0, y_3], [y_3, y_4], [y_4, y_{2N-1}]\}.$$

Each interval of M_i has a tag identifying the original region $R_0, ..., R_N$ which it represents. The size of M_i, $|M_i|$, is defined as the number of finite intervals it contains. $|S_0|$ is defined similarly.

Now we search the two dimensions sequentially. Instead of defining regions of positive area, the new S_0 contains only intervals. First, we need to define a hierarchy of sets $S_1, ..., S_m$, which will enable us to determine quickly which interval of S_0 contains the x-coordinate of P. Then we need to define a similar hierarchy for *each* of the $M_1, ..., M_{2N-1}$. Thus, when the correct x-interval has been chosen, we select the corresponding M_i. A search of that hierarchy will determine the y-interval containing P and thus the original region to which P belongs.

Since the processes are identical, we describe the merge and search methods for the S hierarchy only. Recall, that $S_0 = \{[x_0, x_1], [x_1, x_2], [x_2, x_3], \ldots, [x_{2n-2}, x_{2n-1}]\}$. Beginning with x_1, we select every other endpoint. In each case, we will merge the two intervals sharing that endpoint and insert the new interval into the set S_1. In the representation of S_1, however, each interval will have a pointer back to its originators in S_0 as well as a tag identifying the endpoint which the originators share. We will reiterate this process until we have a hierarchy of sets,

$$S_0 = \{[x_0, x_1], [x_1, x_2], \ldots, [x_{2N-2}, x_{2N-1}]\}$$
$$S_1 = \{[x_0, x_2], |x_2, x_4|, \ldots, |x_{2N-4}, x_{2N-2}|, |x_{2N-2}, x_{2N-1}|\}$$
$$\vdots$$
$$S_m = \{|x_0, x_{2N-1}|\}$$

As before, $m = \lceil \log (2N - 1) \rceil$.

The search process for a point $P = (x, y)$ begins by determining whether x lies in the interval $[x_0, x_{2N-1}]$ (two comparisons). If not, we are done. If so, then we look at the tag value stored in S_m and determine to which originator x belongs (one comparison). We repeat the process, until we know which basic interval in S_0 contains x. (So far we have completed $2 + \lceil \log (2N - 1) \rceil$ comparisons.) We select the appropriate M_i and perform a similar search in its hierarchy. Since M_i has at most $2N - 1$ intervals, this second phase search requires at most $2 + \lceil \log (2N-1) \rceil$ comparisons. When we have isolated the appropriate interval at the bottom layer, we can read off the original region which contains P.

This algorithm preserves the $O(\log N)$ query time of the previous example, but does not achieve the desired reduction in worst-case preprocessing time. The example depicted in Fig. 2.5 maximizes the total number of intervals defined over *all* of the M_i:

$$\sum_{i=1}^{2N-1} |M_i| = 2 \sum_{i=1}^{N-2} (2i + 1) + 2(2N - 2) + (2N - 1)$$

$$= 4 \sum_{i=1}^{N-2} i + 2(N - 2) + 2(2N - 2) + (2N - 1)$$

$$= 2(N - 2)(N - 1) + 8N - 9$$
$$= 2N^2 + 2N - 13$$

Since $|S| = 2N - 1$, the bottom level of this algorithm contains a total of $2N^2 + 4N - 14$ intervals. The previous algorithm has $(2N - 1)^2$ or $4N^2 - 4N + 1$ regions at the bottom level.

In the worst case, the second algorithm represents insignificant improvement; both algorithms may require $\theta(N^2)$ time to represent their bottom level objects and thus may require $\Omega(N^2)$ preprocessing. The example depicted in Fig. 2.5,

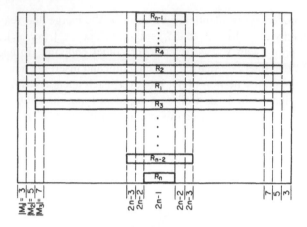

FIG. 2.5.   Instance of rectangle search II yielding $O(n^2)$ intervals

however, is quite unusual. On average, the second algorithm should substantially outperform the first. It also describes a ''slabbing technique'' of searching coordinates sequentially which is a valuable ''macro'' in computational geometry. In addition, it helps to clarify the criteria for effective hierarchies.

2.5. Rectangle Search III: Specific Example

In each of the cases above, the regularization process created too many new regions. The result was an algorithm using quadratic space. Our goal is to develop an algorithm which requires subquadratic preprocessing time and storage space while maintaining an $O(\log N)$ query time. What we need is a subdivision which creates as few new regions as possible while having uniform faces conducive to hierarchical abstraction. Euler's formula shows the feasibility of this approach: a planar graph on n vertices can have at most $2n - 4$ regions. Thus a regularization process which adds no vertices can add only a linear number of new regions.

Let us return to the four rectangle example of Fig. 2.2a. This time, we choose a large triangle Q which contains all of the rectangles in its interior. S is defined as the finite planar subdivision formed by the boundaries of Q and the four rectangles (see Fig. 2.6a). All vertices have alphabetic labels. Instead of adding vertical and horizontal segments which must create new vertices, we add a sequence of disjoint edges between preexisting vertices until every region is a triangle (see Fig. 2.6b). In the representation of the triangular subdivision, S_0, each face is associated with an integer which identifies its parent in S.

Limiting the number of new regions has sacrificed some of the uniformity. In the S_0 of algorithm I, each face was rectangular and every interior vertex had degree four. To create S_1, we chose a collection of independent (i.e., no two

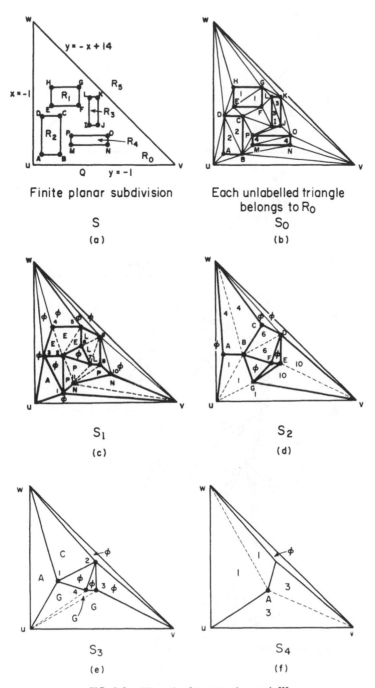

FIG. 2.6.   Hierarchy for rectangle search III

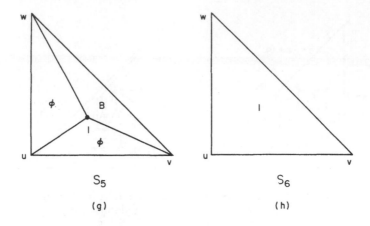

S_5

(g)

S_6

(h)

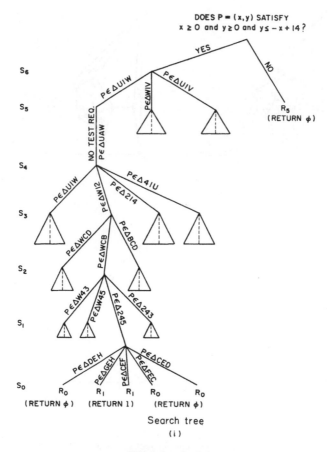

DOES P = (x,y) SATISFY
x ≥ 0 and y ≥ 0 and y ≤ − x + 14 ?

Search tree

(i)

FIG. 2.6.g, h, & i

vertices share an edge), interior vertices from S_0. The deletion of each of these vertices with its adjacent edges automatically produced a new rectangular subdivision. In Fig. 2.6b, consider the five independent vertices A, E, L, N and P. Vertex A has degree 3, while N has degree 4, E and L have degree 5, and P has degree 6. Thus, the deletion of these vertices and their incident edges will produce a new triangle, a quadrilateral, two pentagons and a hexagon (see Fig. 2.6c).

The success of the hierarchical approach depends on having the same type of subdivision at each level in the hierarchy. Consequently, we form a triangular subdivision S_1 by retriangulating the quadrilateral, hexagon and two pentagons using the dotted segments shown in Fig. 2.6c. Note that the interior vertices of S_1 have numeric labels, while the vertices of S_0 are labeled alphabetically. We shall continue to alternate alphabetic and numeric labels throughout the hierarchy to aid in distinguishing between adjacent levels.

The subdivision S_1 contains two kinds of triangles. The triangles in S_1 which are identical to triangles in S_0 are considered "old" triangles. Each "old" triangle will contain a "triangle pointer" to its duplicate, or "parent," representation in S_0. The subdivision S_1, however, has ten fewer regions than does S_0. The three triangles in S_0 sharing vertex A have been replaced by a single triangle in S_1; the four triangles sharing vertex N, by two triangles; the six triangles sharing vertex P, by four triangles; and the five triangles sharing E and L respectively, by three triangles. The representation of each of these "new" triangles contains a "vertex pointer" to the respective vertex in S_0.

The location of a point P in S_0 is easily derived from its position in S_1. Suppose P lies in the triangle of S_1 joining vertices 1, 2, and 3, denoted $\Delta 123$. Then P also lies in ΔBCD of S_0 and belongs to R_2. If P lies in $\Delta 234$ of S_1, then it also lies in the pentagon 23456 formed by the deletion of vertex E from S_0. We test P against the edges $\overline{EC}$, $\overline{ED}$, $\overline{EF}$, $\overline{EG}$ and $\overline{EH}$ to determine the angle at E containing P. Depending on P's location in $\Delta 234$, P lies in one of ΔCED, ΔDEH, ΔHEG, ΔGEF, or ΔFEC. Consequently, P might lie in either R_0 or R_1.

We repeat the process, deleting vertices 1, 4, 6, and 10 and retriangulating to form a subdivision S_2, with 8 fewer faces. The deletion of A, C, and G produces S_3; deleting 1 and 3 leads to S_4; deleting vertex B produces S_5; finally, deleting vertex 1 yields S_6 (see Figs. 2.6d–h).

Given $P = (3, 6)$, the algorithm follows the search tree depicted in Fig. 2.6i. First, we confirm that P lies in the interior of S_6. Next, P is found to lie in $\Delta U1W$ (by testing the ray $1P$ against the edges incident to vertex 1 in S_5). That triangle appears unchanged in S_4 except for the renaming of a vertex, so $P \in \Delta\ UAW$. $\Delta\ UAW$ in S_4 points to the vertex 1 in S_3. Consequently, we consider the four triangles in S_3 which share vertex 1 and determine that $\Delta 12W$ contains P. Moving to S_2, an examination of rays $\overrightarrow{CB}$, $\overrightarrow{CD}$, and $\overrightarrow{CW}$ proves that P belongs to ΔWCB. Subsequently, we determine that P lies in $\Delta 245$ of S_1 and in ΔEHG of S_0. Thus P belongs to R_1. This particular example required a total of 22 comparisons.

For the small example we have been considering, as we might expect, this new algorithm runs more slowly than either algorithm I or the naive algorithm. For a problem with only four original regions, the hierarchy contains seven levels and a cumulative total of 89 regions. Analysis of the algorithm in the general case, however, proves that it is asymptotically superior to the naive algorithm and the two other rectangular algorithms defined above. In addition, we note that nowhere did the algorithm depend on the rectangular shape of the original objects; thus the algorithm applies equally well for a search-domain consisting of general polygons. Furthermore, the searching done by the algorithm was hierarchical in nature, exactly as was the searching done by the first two algorithms.

2.6. General Polygon Search

In this case, our search domain consists of irregularly shaped polygons, rather than rectangles. Since a given polygon could have an unlimited number of sides, we will measure the search domain by the total number of vertices which it contains, rather than by the number of polygons. Similarly, at each level in the hierarchy, $|S_i|$ will represent the number of vertices S_i contains, rather than the number of regions.

The first task is to find a triangle, Q, which surrounds all of the polygons. We use the following approach: Scan all the vertices to determine x_0, the smallest x-coordinate; y_0, the smallest y-coordinate; and b_0, the largest value of $y + x$. The three sides of Q will lie on the lines $y = y_0 - 1$, $x = x_0 - 1$ and $y = -x + b_0 + 1$. Thus Q is an isosceles triangle with a right angle at $(x_0 - 1, y_0 - 1)$. The union of the bounding triangle Q and the given polygons forms a planar graph S of n vertices (three more than the number of original polygonal vertices). We triangulate S arbitrarily to form S_0 (see Figs. 2.6a and b). The triangulation of a planar subdivision on n vertices, S, to form a triangular subdivision, S_0, requires at most $O(n \log n)$ time.

The efficiency of this algorithm will depend heavily on both the number of vertices eliminated at each stage of the hierarchy (number of levels necessary) and the degree of each vertex eliminated (the number of originators a single triangle might have). We will need to make wise choices. Being fully triangulated, S_0 has $3n - 6$ edges and $2n - 4$ regions by Euler's relation. Since every edge is incident on two vertices, the average vertex degree must be $(6n - 12)/n$, or somewhat less than 6. To achieve this average value, at least half of the vertices must have degree smaller than 12. Let V include those vertices of S_0 with degree < 12. Thus, $|V| \geq n/2$. Let us choose an independent subset V' of V. A straightforward elimination procedure applies. Choose any vertex $v \in V$ and place it in V'. Then delete both v and any vertices adjacent to v from V. Since v has at most 11 neighbors, no more than 12 vertices are deleted from V. We can repeat this process at least $|V|/12$ times. Consequently, when V is empty, V' will contain more than $n/24$ independent vertices. Therefore, let each S_i be achieved by deleting independent vertices of degree ≤ 11 from S_{i-1}. The process stops at S_m which has 1 triangle.

With the hierarchy, $S_0, \ldots, S_m$ in place, which we will call an S-tree, we can determine which of the original polygons contains P by the following procedure:

Begin
    1.   decide whether P lies in the bounded or unbounded region of S_m
    2.   if P is in the unbounded region, return $\varnothing$ and stop
    3. $v_{m-1} \leftarrow$ vertex associated with bounded region of S_m
    4. $t_{m-1} \leftarrow \varnothing$
    5.   for $i \leftarrow m - 1$ to 1 step-1
          6. if $v_i \neq \varnothing$ then begin   ("new" triangle)
                  7) determine with triangle of S_i at v_i contains P
                  8) $v_{i-1} \leftarrow$ vertex stored at that triangle
                  9) $t_{i-1} \leftarrow$ pointer stored at that triangle
          end
        10. else begin   ("old" triangle)
            11. $v_{i-1} \leftarrow$ vertex stored at t_i
            12. $t_{i-1} \leftarrow$ pointer stored at t_i
          end
    13. look up in S_0 the original polygon R_j to which triangle t_0 belongs.
    14. if $j = 0$, then return $\varnothing$, else return j
End

The set-up phase of the algorithm, steps 1–4, can be completed in constant time. Since the degree of v_i is at most 11, each iteration of the loop, steps 6–12, can require at most 12 comparisons.

Thus, the time complexity of the algorithm is linear in m, the number of times the algorithm executes the inner loop. Recall, however, that S_m has 1 triangle and 3 vertices. Consequently, $|S_m| = 3 \leq (23/24)^m |S_0| = (23/24)^m n$. Therefore, $m \leq (\log n - \log 3)/(\log 24 - \log 23)$. Using the convenience of "O" notation, we may say that $m = O(\log n)$. Thus, the algorithm performs at most 12 comparisons at each of $O(\log n)$ levels. So the search process takes $O(\log n)$ time.

To evaluate the amount of preprocessing time the algorithm requires, first let us consider the amount of time and storage required just to represent the S_i. S_0 has n vertices, three more than the actual search domain. Thus the total number of vertices (which is linear in the number of edges and regions) represented throughout the hierarchy is approximately

$$|S_0| \sum_{i=0}^{m} (\frac{23}{24})^i < n \sum_{i=0}^{\infty} (\frac{23}{24})^i = 24\,n.$$

Thus, this portion of the problem requires only linear time and space.

The time required to prepare the representation of S_i from that of S_{i-1} or the representation of S_0 from S depends on the complexity of the triangulation process. As we stated above, the latter process, the triangulation of S to form S_0, requires at most $O(n \log n)$ time. In the former case, the vertices to be deleted can be determined in the manner described above in time linear in $|S_{i-1}|$. When a vertex is deleted, the resulting polygon of fewer than 12 sides can be retriangulated in constant time. Thus, given S_0, creating the remainder of the S-tree (S_1, ..., S_m) takes linear time.

Given a fixed collection of N disjoint polygons having a combined total of n vertices, the hierarchical algorithm we have described creates an S-tree in time $O(n \log n)$ and space $O(n)$ which enables any subsequent query to be answered in time $O(\log n)$. Throughout the development of this algorithm, no restriction has been placed on the number of vertices per polygon or on the shapes permissible. A naive approach to the problem would require only $O(n)$ time, so it remains preferable for cases where few queries are anticipated. As the number of queries grows, however, the hierarchical approach becomes vastly superior.

In the sections below, we shall assume that a constant bounds the number of vertices per polygon, making $n = \theta(N)$. In this case, the naive approach requires only time $O(N)$ to complete a query. In the general algorithm above, n may be replaced by N in each of the complexity results. The creation of an S-tree on N polygons, an S-tree of size N, requires preprocessing time $P_S(N) = O(N \log N)$ and space $S_S(N) = O(N)$. Although an S-tree is formed only after the triangulation of the entire search domain, we may say that each of the N polygons costs an average insertion time of $I_S(N) = P_S(N)/N = O(\log N)$. A query costs $Q_S(N) = O(\log N)$ time.

2.7. Dynamic Polygon Search

An interesting addition to the polygon search problem is to allow the search domain to change over time. The S-tree is unsuitable for this problem where polygons are added one at a time with queries interspersed. As the triangulation of S cannot be modified locally, the insertion of the Nth polygon would entail discarding the old S-tree on $N - 1$ polygons and building a new S-tree of size N. This is likely to be less efficient than the naive approach, since inserting N polygons sequentially into any initially empty structure would cost an excessive amount of time:

$$O\left(\sum_{i=1}^{N} i \log i \right) = O(N^2 \log N).$$

Polygonal search, however, does not require that all of the data be retained in a single structure. If some of the polygons were represented in one S-tree and the rest in another, the polygon containing P could be identified by forming the union of the results of searching the two S-trees independently. Thus instead of dismantling an entire S-tree in order to insert the Nth polygon, we may simply construct a new S-tree for that polygon. A subsequent query on the search domain will necessitate searching both structures in time $O(\log(N - 1) + \log 1)$ $= O(\log N)$.

A data structure for the insertion-dynamic polygon search problem may consist of a collection of S-trees, rather than a single such tree. Nonetheless, adding a new S-tree upon each insertion does not work. Although the insertion time would remain constant, queries after N insertions to an initially empty structure would require $O(N)$ time. What is needed is a method which balances the cost of creating large S-trees against the cost of searching a large number of trees.

An insertion strategy based on binary counting, which we will call a binary transform, produces a feasible data structure for insertion-dynamic polygon search, the DPS. The first polygon to join the search domain occupies its own S-tree. Upon the insertion of the second polygon, a single S-tree of size 2 is formed. The third polygon forms its own S-tree, but the insertion of the fourth polygon prompts the formation of a single S-tree of size 4. The histogram of Fig. 2.7 illustrates the continuing pattern. A DPS of size 7 consists of three S-trees of sizes 1, 2, and 4, respectively. All three structures are replaced by a single S-tree of size 8 upon the next insertion. In general, if N is represented in binary as $(b_n b_{n-1} \ldots b_0)$, a DPS of size N will contain an S-tree of size 2^i for those i with $b_i = 1$.

A DPS allows a balance between insertion time and query time. First of all, the DPS of size N contains no more than $\log (N + 1)$ S-trees. As no S-tree has size greater than N, a search of one tree requires at most $O(\log N)$ time. Hence, a search of the entire DPS can use at most $O(\log^2 N)$ time.

Analysis of the insertion time is more difficult. Adding the 15th polygon entails building an S-tree of size 1 at almost no cost, while the 16th polygon necessitates the construction of new S-tree of size 16. Instead of figuring the

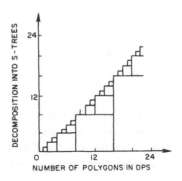

FIG. 2.7.   Binary transform

insertion cost of a polygon, $I_{DPS}(16)$, as the immediate expenditure of time associated with adding it to the DPS, however, we will assign to each polygon its share of the cost of building every S-tree to which it ever belongs. Of the polygons in a DPS of size 16, the first polygon to join the search domain has belonged to the largest number of distinct S-trees: one each of sizes 1, 2, 4, 8, and 16. Its cost must bound the average insertion time. $P_S(N)$, however, grows at least linearly in N, and thus $I_S(N) = P_S(N)/N$ is monotone nondecreasing. Therefore, we can bound $I_{DPS}(16)$ as follows:

$$I_{DPS}(16) \leq I_s(1) + I_s(2) + I_s(4) + I_s(8) + I_s(16)$$

$$= \sum_{i=0}^{4} I_s(2^i)$$

$$\leq \sum_{i=0}^{4} I_s(16)$$

$$\leq 5\, I_s(16).$$

In general, no element of a DPS of size N will have belonged to more than $(1 + \log N)$ different S-trees over the life of the algorithm. We may conclude:

$$I_{DPS}(N) \leq \sum_{i=0}^{\log N} I_s(2^i) \leq (1 + \log N)\, I_s(N) = O(\log^2 N);$$

$$P_{DPS}(N) = N\, I_{DPS}(N) \leq (1 + \log N)\, P_S(N) = O(N \log^2 N).$$

Thus the average insertion time is identical to the query time. As each polygon is represented in exactly one structure, the space requirements remain $O(N)$.

The creation of the DPS represents a macro applicable in other contexts. The next subsection presents a general development of this macro.

2.8. Decomposable Searching Problems

General polygon search is only one example of a class of searching problems suited to the dynamization process defined above. That process depended on the fact that the correct response to a query over a search domain D could be formed by merging the responses to separate queries over subsets of D using constant time. The problems having this attribute are called "decomposable searching problems."

Many common problems belong to this class. An element s belongs to a set $A \cup B$ if s belongs to A or s belongs to B. The nearest neighbor to a point x in a set $A \cup B$ is the closer of its nearest neighbor in A and its nearest neighbor in B. The smallest number in $A \cup B$ which is larger than a given number x is equal to the smaller of the two numbers satisfying this criterion separately in A and in B. The number of points in a set $A \cup B$ whose x-coordinate has a given value x_0 equals the sum of the number of such points in A with the number of such points in B.

Once we have a static data structure and a static algorithm for any one of these problems, we can apply the binary transform to achieve an insertion-dynamic structure and algorithm. At any point, the insertion-dynamic algorithm will query at most $\log(N + 1)$ of the static structures, implying that $Q_D(N) \leq Q_S(N)$ $\log(N + 1)$, as long as $Q_S(N)$ is monotone nondecreasing. When we have a insertion-dynamic structure of size N, each element of that structure has belonged to at most $(1 + \log N)$ structures during the course of the algorithm. Provided that $P_S(N)$ grows at least linearly with N, each element's cost is bounded above by $(1 + \log N) I_S(N)$, insuring that $P_D(N) \leq (1 + \log N) P_S(N)$. Each object is stored in only one static structure. If we assume that the static algorithm requires at least linear space, then we may conclude that the space requirements have remained unchanged: $S_D(N) \leq S_S(N)$. In summary, an insertion-dynamic algorithm achieved by applying the binary transform to a static algorithm achieves the following results:

$$Q_D(N) \leq Q_S(N) \log(N + 1)$$
$$P_D(N) \leq P_S(N) (1 + \log N)$$
$$S_D(N) \leq S_S(N)$$

Due to these performance results, we call the binary transform an admissible $(\log(N + 1), 1 + \log N)$ transform.

The chief asset of the binary transform is that it evenly assesses an $O(\log N)$ dynamization penalty to the query and average insertion times. Other transformations trade improved query time for slower overall processing, and vice versa. We will consider two examples: the triangular transform and the dual triangular transform.

The triangular transform allows the insertion-dynamic structure to contain at most two static structures at a time. Consequently, $Q_D(N) \leq 2 Q_S(N)$. To achieve this limit, it relies on the sequence of triangular numbers, $T_i = i(i + 1)/2$ for $i = 1, 2, \ldots$ If the search domain has size N, let T_j be the largest triangular number less than N. Then an insertion-dynamic data structure of size N contains one static structure of size T_j and one of size $N - T_j$. Suppose one more element is inserted. If $N + 1 = T_{j+1}$, then the previous two structures are discarded and a single structure of size T_{j+1} is created. If $N + 1 < T_{j+1}$, then just the structure of size $N - T_j$ is dismantled, and those objects together with the new one are formed into a new static structure of size $N + 1 - T_j$. [See Figure 2.8].

To assess the total processing time, first consider an example where $T_5 = 15$ elements belong to the data structure. Then, over the course of the algorithm, the first and second elements to join the search domain have belonged to five distinct structures: one each of sizes 1, 3, 6, 10, and 15. The third element has belonged only to structures of size 3, 6, 10, and 15. The fourth element, however, has

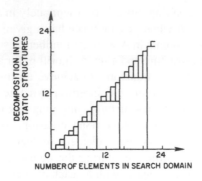

Triangular transform   FIG. 2.8.   Triangular transform

belonged to structures of size 1, 2, 6, 10, and 15. We can read off this history by tracing a horizontal line through the histogram of Fig. 2.8 at a given height. Examining the histogram shows that no element has belonged to more than five structures.

In general, after $T_j = N$ elements have been inserted, each element has belonged to at most j structures. By the definition of the triangular numbers, we may conclude that $j < \sqrt{2N}$, and thus, $P_D(N) \leq P_S(N) \sqrt{2N}$. Consequently, the triangular transform may be described as an admissible $(2, \sqrt{2N})$ transform. The query time has suffered only a constant dynamization factor, while processing time is slower by a factor of $O(\sqrt{N})$, making this transform a good choice for cases where we expect a large number of queries relative to insertions.

The reverse triangular transform, or the dual triangular transform, allows each element to belong to at most two static structures over the life of the algorithm. As each of these structures must have size less than or equal to N, $I_D(N) \leq 2 I_S(N)$ and $P_D(N) \leq 2 P_S(N)$. If the search domain contains $N = T_j$ objects for some j, then the data structure contains one structure of size i for all $1 \leq i \leq j$. If $N > T_j$, then the structure contains one structure of size i for all $1 \leq i \leq j$ in addition to $N - T_j (\leq j)$ structures of size 1 (see Fig. 2.9). Thus, at any time, the insertion-dynamic structure contains at most $2j < 2\sqrt{2N}$ structures, ensuring that $Q_D(N) \leq Q_S(N) 2\sqrt{2N}$. In summary, the dual triangular transform is an admissible $(2\sqrt{2N}, 2)$ transform.

The three transforms we have discussed have distinct advantages and disadvantages. The binary transform is best in cases where a comparable number of insertions and queries are expected. The triangular transform and the dual triangular transform are best, respectively, when a greater proportion of queries or a greater proportion of insertions are expected. There are an unlimited number of other transforms available. For example, in both triangular transforms, the number 2 was arbitrarily chosen as the limit for the number of active structures and as the limit on the number of structures per single element over time. The choice of the number 3 instead, or any other integer, would have led to different

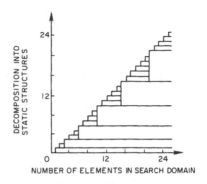

FIG. 2.9.   Dual triangular transform

transforms. In fact, we may have a transform isomorphic to any counting scheme.

These transforms may be applied to any decomposable searching problem, and the worst case complexity results we have derived will always hold. For some specific problems, however, the dynamization penalties will not in fact accrue. It may be possible, for example, to merge some of the static structures without completely dismantling them. Similarly, a tighter analysis of the query time may be possible. Nonetheless, the upper bounds readily available as a result of using a standard transform have great value. It is impossible, however, to create efficient transforms applicable to all decomposable searching problems which will convert static algorithms to dynamic algorithms allowing deletions.

2.9. Notes

Garey, Johnson, Preparata, and Tarjan (1979) produced an algorithm for triangulating monotone polygons in linear time. Chazelle and Incerpi (1984) have developed a new scheme for decomposing a simple polygon into a collection of monotone polygons. A modification of that approach leads to an $O(n \log n)$ algorithm for decomposing all of the faces of a planar subdivision on n vertices into monotone polygons (Chazelle, 1984b).

The problem solved here by hierarchical search was first stated by Knuth (1973). A first solution was proposed in Dobkin and Lipton (1976). This paper introduced the notion of x-axis and y-axis projection which have proved useful in other contexts (Bentley & Ottmann, 1979; Brown, 1981; Shamos & Hoey, 1976). The topic of planar subdivision search was extended by Preparata, 1981; Lipton & Tarjan, 1979; Lipton & Tarjan, 1980; Kirkpatrick, 1983. Our development is most similar to that of Kirkpatrick (1983). A similar technique was used in (Dobkin & Kirkpatrick, 1985).

The dynamization techniques and the classification of decomposable searching problems are due to Bentley and Saxe (Bentley, 1979; Bentley & Saxe,

1980). Some of the methodology introduced here is also seen in the order-decomposable methods mentioned in the next section.

3.   HIERARCHICAL COMPUTATION

3.1. Divide-and-Conquer Computation

The most powerful technique in algorithm design is the divide-and-conquer technique, as it can be applied to problems of vastly different structures. The sorting of natural numbers is a classic example of the divide-and-conquer procedure. The goal is to sort a set of n integers. To do so, we divide the original set into two equal or nearly equal subsets. Each subset is sorted (by applying the method recursively) with the results (two sorted sets) merged to form a sorted sequence describing the entire set.

A simple example involves producing an ordered list of grades from an alphabetized list of students and their averages. A divide-and-conquer tree begins as a binary search tree, with the original objects ordered in the leaves and with each non-leaf node containing a pointer to the largest element in its left-subtree (see Fig. 2.10). In binary search, however, we begin at the root and descend the tree, considering one node at each level and deciding to progress to either the right or the left subtree. The leaf reached determines the solution. A divide-and-conquer algorithm involves both descending and ascending the tree, visiting each node in each direction. The division is done as we pass down the tree, distinguishing different levels in the hierarchy. For any nonleaf node, the algorithm sorts its left subtree and then sorts its right subtree. The conquering part of the computation is done as we pass up the tree. The algorithm merges the sorted sequences stored at the children and stores the result of the computation at the given node. At the conclusion, the complete sorted sequence is stored at the root.

We may choose how much space we wish to utilize during the course of this algorithm. When we merge two sequences, we create new storage for the result. We may either retain the original sequences or destroy them. In this first case,

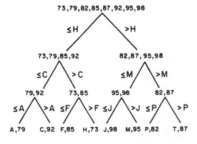

FIG. 2.10.   Sample divide-and-conquer tree

the algorithm will need $O(n \log n)$ space, whereas the second case uses only linear space. Investing the extra space allows easy retrieval of previous states of the computation. In some applications, the ability to retrace previous steps easily has enough significance to outweigh the cost in space. In general, however, the linear-space approach is preferable, as it is here.

The efficiency of the algorithm depends on the amount of work required at the leaves of the tree, the complexity of the merge phase, and the overall size of the tree. Typically, this yields a recurrence of the form $T(n) = 2\ T(n/2) + M(n)$ with $T(1) = C$. In this case, no work is necessary at the leaves ($T(1) = 0$), and the merge algorithm is linear in the size of the subtree: $M(n) = n - 1$. Thus $T(n) = 2\ T(n/2) + n - 1 = O(n \log n)$.

Having briefly described the divide-and-conquer method, we now consider its specialization to the hierarchical methods best suited to geometric problems. Not only do they apply to static geometric problems, but they also serve as a basis for dynamization techniques which apply to a large subclass of geometric problems.

3.2. Two-dimensional Convex Hulls

Over the years, numerous algorithms have been developed for the computation of the convex hull of a set of N points in the plane (see Chapter 5). The algorithm we describe here demonstrates the hierarchical strategy of computing. If the points have been sorted according to the value of their x-coordinate, the set can be divided in two at the median value. The division step now involves computing the hulls of the left and right collections of points. Conquering involves merging the left and right hulls.

Merging two disjoint convex hulls requires determining two distinct segments which are tangent to both hulls and then deleting all vertices and edges within the quadrilateral defined by those tangent segments (see Fig. 2.11a). To simplify our exposition, we consider only the problem of finding the lower tangent and maintaining the "bottom convex hull" or "bc-hull." The bc-hull contains the bottom part of the convex hull, the portion extending from the leftmost vertex to the rightmost vertex in the counter-clockwise direction. The bc-hull, however, augments the bottom part of the hull with vertical rays in the positive direction at

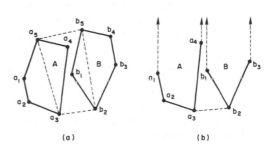

FIG. 2.11. Merging two convex
hulls versus merging two bc-hulls   (a)   (b)

its two endpoints.[3] (See Fig. 2.11b.) Once two bc-hulls have been merged and the corresponding two top convex hulls (tc-hulls) have been merged, the complete convex hull can be formed in constant time.

Given two bc-hulls A and B, each consisting of N vertices with A lying to the left of B, we divide the merge process into three parts. First, we seek a "bridge," i.e., a segment tangent to both A and B. Any vertex on either hull could be an endpoint of the bridge, or a "bridge point." A variant of binary search is used to locate the actual bridge points. Once the bridge points a and b have been identified, we split A at a and B at b. Finally, we form a new hull by concatenating the left portion of A, including a, and the right portion of B, including b. In the chain of vertices representing the new hull, a and b are adjacent to each other, indicating the bridge segment.

The vertices of each bc-hull, C, are represented in the leaves of a concatenable queue. The term "concatenable queue" denotes a binary search tree which has been implemented in such a way as to enable efficient searching, splitting and concatenating. The concatenable queue, or Q-structure, retains all of the features of a binary search tree; each nonleaf node α still contains a pointer, $V[a]$, to that vertex in its left subtree having the largest x-coordinate. Thus the middle vertex of the portion of a bc-hull represented by a single subtree can be identified in constant time. We associate with Q_c, the Q-structure representing the bc-hull C, two additional values: M_c stores the maximum x-coordinate of any vertex of C; m_c contains the minimum x-coordinate.

The search process begins by assigning to a and b the vertices represented at the roots of Q_A and Q_B, the middle vertices of the respective hulls. Define $\overrightarrow{a_r}$ and $\overrightarrow{b_r}$ as the rays extending from a and b rightward toward the next vertex on the hull. Similarly, $\overrightarrow{a_l}$ and $\overrightarrow{b_l}$ represent the rays extending from a and b leftward toward the previous vertex on the hull. The angles formed at a and at b guide our search. Until a bridge is found, we repeatedly remove a fraction of the vertices of one or both hulls from consideration as a bridge endpoint. The size of the four angles, $\angle(\overrightarrow{ab}, \overrightarrow{a_l})$, $\angle(\overrightarrow{ab}, \overrightarrow{a_r})$, $\angle(\overrightarrow{b_r}, \overrightarrow{ba})$, $\angle(\overrightarrow{b_l}, \overrightarrow{ba})$, determine whether $\overline{ab}$ is itself the bridge and, if not, which vertices may be removed.[4]

We consider four cases determined by the presence or absence of reflex angles. If there are no reflex angles (Fig. 2.12a), then the segment $\overline{ab}$ is itself the bridge and the search is finished. The second category includes all cases where one or both of $\angle(\overrightarrow{ab}, \overrightarrow{a_l})$ and $\angle(\overrightarrow{b_r}, \overrightarrow{ba})$ is reflex. If $\angle(\overrightarrow{ab}, \overrightarrow{a_l})$ is reflex, then neither a nor any vertex to the right of a can be a bridge point. Thus the bridge point must lie in the left subtree of Q_A (see Fig. 2.12b). If $\angle(\overrightarrow{b_r}, b_*)$ is reflex, then both b and all vertices of B left of b can be removed from future consideration and the bridge point must lie in the right subtree of Q_B (see Fig. 2.12c).

---

[3]Alternately, we can say that the bc-hull of a set P is defined as the convex hull of the union of P with the point $(0, \infty)$. The tc-hull is the convex hull of $P \cup \{(0, -\infty)\}$.

[4]All angles are measured in the counter-clockwise direction from the first ray to the second.

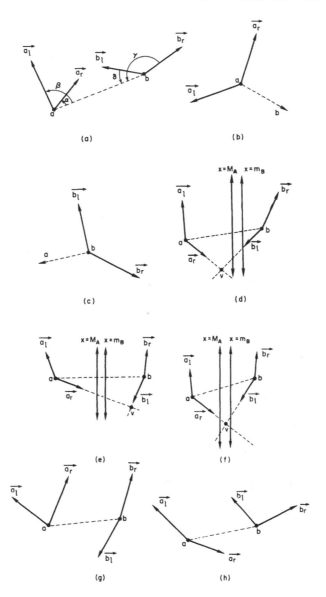

FIG. 2.12. Case analysis for merging bc-hull *A* with bc-hull *B*

The third category involves both $\angle(\overrightarrow{ab}, \overrightarrow{a_r})$ and $\angle(\overrightarrow{b_l}, \overrightarrow{ba})$ being reflex. In this case, we determine the intersection point, $v = (x, y)$, of a_r and b_l. We let M_A represent the maximum x-coordinate of A, and let m_B represent the minimum x-coordinate of B. If $x < m_B$, then no vertex of B could lie below $\overrightarrow{a_r}$. Consequently, neither a nor any point left of a could be a bridge point and they can be removed from future consideration (see Fig. 2.12d). Similarly, if $x > M_A$, then no vertex of A may lie below $\vec{b_l}$ and so b and all vertices to the right of b can be rejected (see Fig. 2.12e). If $M_A < x < m_B$, then we may delete both sets (see Fig. 2.12f).

In the final category, only one of $\angle(\vec{ab}, \vec{a_r})$ and $\angle(\vec{b_l}, \vec{ba})$ is reflex. If $\angle(\vec{b_l}, \vec{ba})$ is reflex, $\vec{ab}$ forms an angle smaller than $180°$ with both $\vec{a_l}$ and $\vec{a_r}$ and thus a remains a potential bridge point. In other words, no point of A lies below $\overleftrightarrow{ab}$. But since $\vec{ba}$ is interior to the angle from $\vec{b_r}$ to $\vec{b_l}$, neither b nor any point right of b can lie on the bridge (see Fig. 2.12g). If $\angle(\vec{ab}, \vec{a_r})$ is reflex, then neither a nor any point left of a can be a bridge point (see Fig. 2.12h).

In each of the last three categories, we cut in half the number of potential bridge points on at least one of the bc-hulls. We may repeatedly set a and b to the midpoints of the remaining vertices of A and B, respectively, by moving to the leftson or rightson of the current node in Q_A or Q_B. We then rerun the algorithm. Since the height of both Q_A and Q_B is $O(\log N)$, at most $O(\log N)$ iterations are required.

We may formalize the algorithm as follows:

MERGE (A,B,C)
Begin
   1. $\alpha \leftarrow$ the root of Q_A
   2. $\beta \leftarrow$ the root of Q_B
   3. *done* $\leftarrow 0$
  repeat
    begin
    4. $a \leftarrow V|\alpha|$
    5. $b \leftarrow V|\beta|$
    6. determine which of $\angle(\vec{ab},\vec{a_l}), \angle(\vec{ab},\vec{a_r}), \angle(\vec{b_r},\vec{ba}), \angle(\vec{b_l},\vec{ba})$, is reflex
    7. if none are, then *done* $\leftarrow 1$
    8. else if $\angle(\vec{ab},a_l)$ or $\angle(\vec{b_r},\vec{ba})$ is reflex, then
      begin
      9. if $\angle(\vec{ab},\vec{a_l})$ is reflex, then $\alpha \leftarrow$ *leftson*(α)
      10. if $(\vec{b_r},\vec{ba})$ is reflex, then $\beta \leftarrow$ *rightson*(β)
      end
    11. else if $\angle(\vec{ab},\vec{a_r})$ and $\angle(b\vec{ba})$ are reflex, then
      begin
      12. $(x, y) \leftarrow$ the intersection of $\vec{a_r}$ and $\vec{b_l}$
      13. if $x < m_B$, then $\alpha \leftarrow$ *leftson*(α)
      14. if $x > M_A$, then $\beta \leftarrow$ *rightson*(β)
      end
    15. else if either $\angle(\vec{ab},\vec{a_r})$ or $\angle(\vec{b_l},\vec{ba})$ is reflex, then
      begin
      16. if $(\vec{ab},\vec{a_r})$ is reflex, then $\alpha \leftarrow$ *rightson*(α)
      17. if $\angle(\vec{b_l},\vec{ba})$ is relfex, then $\beta \leftarrow$ *leftson*(β)
    end
  until *done*

18. $Q_C \leftarrow$ *concatenate*(*leftsplit*(Q_A,a), *rightsplit*(Q_B,b))
19. $M_C \leftarrow M_B$
20. $m_C \leftarrow m_A$
End

The set-up phase of the algorithm, steps 1–3, the interior of the loop, steps 4–17, and the final two steps each require constant time. As the loop is iterated $O(\log N)$ times, the algorithm through step 17 requires only $O(\log N)$time. Step 18 requires a choice similar to the one we mentioned in section 3.1. We may leave Q_A and Q_B intact and merely copy the left portion of Q_A and the right portion of Q_B before concatenating the two sections. As the copying itself requires $O(N)$ time, the merge algorithm as a whole requires linear time. Alternatively, we may split the actual representations of Q_A and Q_B and then concatenate the appropriate sections. In this case, we lose the ability to recapture immediately the previous states of the computation, but step 18 now requires only $O(\log N)$ time, preserving an $O(\log n)$ worst-case complexity for the complete merge algorithm.

Having fully detailed the merge process, we return to the hierarchical algorithm as a whole. The entire process can be modeled by a divide-and-conquer tree, a T-tree. Suppose we wish to determine the bc-hull of the set of points, P, depicted in Fig. 2.13a. First, we sort the points in ascending order of x-coordi-

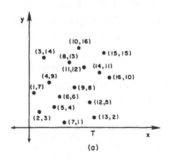

(a)

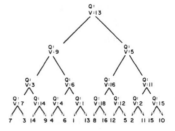

Beginning of divide-and-conquer tree

(b)

FIG. 2.13. An instance of the convex hull problem

nate. This example includes one point for every x-coordinate from 1 to 16 inclusive, and henceforth each point will be identified by its y-coordinate. The points are incorporated into the leaves of a balanced binary search tree (e.g., a BB[a]-tree). Mirroring the binary search tree, each nonleaf node a includes a pointer $V[a]$ to the rightmost vertex of its left subtree. The T-tree for P, T_P, is formed by augmenting each nonleaf node, a, with a Q-structure, Q_a, which represents the bc-hull of all points in the subtree at a (see Fig. 2.13b).

The Q-structures are formed gradually as we move up the tree. At the bottom level, the tree induces a natural pairing of points. The bc-hull of a pair of points consists of a segment between them with a positive vertical ray added at each endpoint (see Fig. 2.14a). The Q-structures on the first level of the tree describe the bc-hulls of these pairs. At each succeeding level of the tree, sibling bc-hulls are merged according to the process described above to form the Q-structure at the parent node (see Figs. 2.14b-e). Figure 2.14f represents T_P at the termination of the algorithm. The root contains the bc-hull of the entire set P. We can then repeat the same process to form the tc-hull. The merger of these hulls gives the convex hull of the original set of points.

For a point set of size n, this algorithm requires $\theta(n \log n)$ preprocessing time to sort the points. The recurrence relation, $T(n) = 2 T(n/2) + M(n)$ describes the time complexity of the remainder of the algorithm. The bc-hull of a set of $n/2$ points must have no more than $n/2$ vertices. Thus, when the partial bc-hulls remain intact, each level of the T-tree requires $O(n)$ storage and the merge technique uses linear time. Consequently, the overall algorithm will run in $\theta(n \log n)$ time, using $O(n \log n)$ space. If we choose not to preserve the partial bc-hulls, the final T-tree contains only the complete bc-hull. At intermediate stages in the algorithm, the T-tree contains only the bc-hulls of disjoint sets. Consequently, the algorithm uses only linear space. Each merge requires only logarithmic time. Consequently,

$$T(n) = 2 T\left(\frac{n}{2}\right) + M(n) = \sum_{i=1}^{\log n} \frac{n}{2^i} \log 2^i = O(n)$$

In order words, after the initial sorting this algorithm will require only linear time. Since the problem of finding convex hulls corresponds to sorting and thus any algorithm must require $\Omega(n \log n)$ time, either algorithm can be considered optimal.

3.3. Dynamic Convex Hulls

In the previous example, we determined the bc-hull of a fixed set of points. In many applications, however, we may wish to maintain the bc-hull of a set which allows insertions and deletions. In other situations, balanced binary search tree accommodates insertions and deletions easily. The appropriate location is found

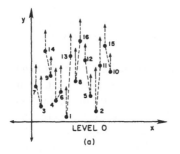

LEVEL 0

(a)

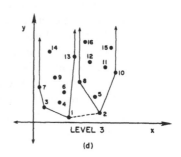

LEVEL 3

(d)

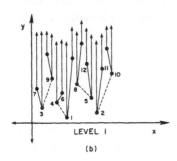

LEVEL 1

(b)

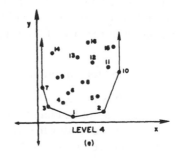

LEVEL 4

(e)

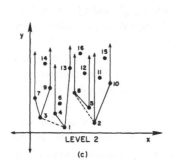

LEVEL 2

(c)

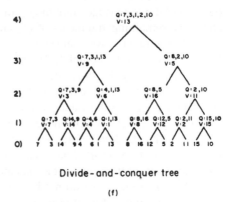

Divide-and-conquer tree

(f)

FIG. 2.14.   Building the bc-hull of
the instance in Fig. 2.13

by beginning at the root and descending the tree until the relevant location is found. After inserting or deleting the point, the path to the root is retraced, rebalancing the tree and updating any pointers as necessary. The entire process requires $O(\log n)$ time. This case is more complicated.

Suppose that we have the complete T-tree for the original set of points and that all partial hulls have been retained. After locating the appropriate leaf position in the T-tree for the insertion or deletion, we can rebuild the be-hull on level 1 in constant time.. Its sibling hull remains unchanged, so we can merge the two hulls to form the parent Q-structure. At each node as we ascend the tree, the sibling hull is already available. Provided no rebalancing is necessary, a single merger will move us one level higher in the tree. Suppose we arrive at a node and determine that rebalancing is required. A BB(α)-tree can be rebalanced using no more than two local rotations per node as we ascend the tree. To update the Q-structures following these rotations, we will have to reform at most two of the bc-hulls by merging some of the partial bc-hulls in a new order. Nonetheless, only a constant number of merges will be required at each node. Each merge requires time linear in the size of the subtree.

Therefore, the total insertion/deletion time can be $ID(n) = O\left(\sum_{i=1}^{\log n} \frac{n}{2^i} \right) = O(n)$, which is too large to be considered efficient.

Suppose that in building the original T-tree we had chosen to conserve space and had thus destroyed the partial hulls. After each insertion or deletion, we would then have to process the entire tree from level 0 to the root, again using $O(n)$ time.

We must try a new approach. In a new T^* structure, the new Q-structure at a node α will be formed from the actual portions of the children Q-structures, Q_γ and Q_δ, using logarithmic time. At α however, we retain a pointer, $B[\alpha]$, to the position of the left bridge point in Q_γ. In addition, we associate with the nodes γ and δ the remaining fragments of Q_γ and Q_δ, denoted Q_γ^* and Q_δ^* (see Fig. 2.15). Consequently, we could split Q_α after the left bridge point, as indicated by $B[\alpha]$, and reform the complete Q_γ and Q_δ in logarithmic time.

We can modify the insert/delete algorithm as follows: begin at the root and split the current bc-hull and form the bc-hulls of its two children; compare the desired point against the values of V at the root and move to the left or right son accordingly. At each interior node, again split the Q-structure and form the Q-structure of its children before moving to one child or the other. The descent will require at most $O(\log n)$ time at each node, or $O(\log^2 n)$ time overall. In the ascent, we will perform at most a constant number of merges at each node, using only $O(\log n)$ time as we do not have to preserve entire partial hulls. Thus the ascent also requires at most $O(\log^2 n)$ time. Consequently, we can maintain the bc-hull of a set of n points at a cost of $O(\log^2 n)$ per insertion/deletion.

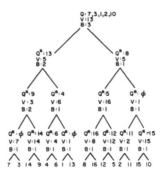

FIG. 2.15. Dynamic T-tree

3.4. Order Decomposable Problems

The two dimensional bc-hull problem is a single example of a class of set problems suited to the dynamization process defined above. That process entailed ordering the set and then forming the solution over the entire set by merging the solutions of the first i and the last $n - i$ elements, for any $1 \le i \le n$, in time $M(n)$. This suggests the name "$M(n)$-order decomposable" for any set problem having an $M(n)$ merge algorithm dependent on some ordering scheme, ORD. A static divide-and-conquer algorithm takes time $O(ORD(n) + T(n))$ where $T(n) = 2\ T(n/2) + M(n)$.

A number of common problems belong to this class. Suppose we wish to compute the maximal elements of the planar set $A \cup B$ where all points a_i in A are left of all points b_j of B. Recall that p is a maximal element of a set C iff for all $q \in C$, either $x_p > x_q$ or $y_p > y_q$. $MAX(C)$ represents the sequence of maximal elements of C in decreasing order of y-coordinate and in ascending order of x-coordinate. Figure 2.16a highlights the maximal elements within the sample sets A and B by linking them together with a polygonal chain of horizontal and vertical segments. Let y_B represent the maximum y-coordinate in B. Then $MAX(A \cup B) = concatenate(\{p \epsilon MAX(A) | y_p > y_B\}, MAX(B))$. Splitting $MAX(A)$ at the appropriate spot and concatenating both require $O(\log n)$ time. Consequently, the two-dimensional maximal element problem is $O(\log n)$-order decomposable.

Consider the problem of determining the intersection of a collection of n lower half-planes: for $1 \le i \le n$, a line $l_i : y = a_i x + b_i$ bounds the half-plane $H_i = \{(x, y) | y < a_i x + b_i\}$ from above. Sort the half-planes according to the slopes of the bounding lines. Suppose we have the border A of the intersection region defined by the first i half-planes and the border B of the region defined by the last $n - i$ (see Fig. 2.16b). Find the intersection point of the two contours in time $O(\log n)$. Split them there, and concatenate the front piece of the first with the

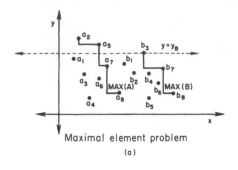

Maximal element problem

(a)

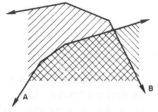

Half - plane intersection problem

(b)

FIG. 2.16. Other examples of order-decomposable problems

tail piece of the second, also using $O(\log n)$ time. Consequently, the lower half-plane intersection problem is $O(\log n)$-order decomposable.

In either case, a dynamic algorithm operates on a balanced T-tree. The leaves contain the ordered elements of the set, and each nonleaf node α contains a pointer $V[\alpha]$ to the largest element in its left subtree. At each nonleaf node, we compute the solution over its subtree. At a node α with children γ and δ, we form Q_α from Q_γ and Q_δ using $O(M(n))$ time. As we merge the two children Q-structures, we will store a string of information describing the steps we take as $B[\alpha]$, a tag at node α. As the merge process itself took only $O(M(n))$ time, writing a string which describes that process cannot require more than $O(M(n))$ time or $O(M(n))$ space. The unused pieces of Q_γ and Q_δ are stored at γ and δ as Q_γ^* and Q_δ^*, respectively. Only the root will retain a complete Q-structure.

To insert or delete an object, we locate the appropriate leaf using binary search. At every node α on the path from the root to the leaf, we use the information stored in $B[\alpha]$ to split the Q-structure and recreate the complete Q-structures at the children, using only as much time as the original merge. Thus the process of descending the tree requires time

$$DOWN(n) = O(M(n) + M(n/2) + M(n/4) + \ldots) = O\left(\sum_{i=0}^{\lfloor \log n \rfloor} M\left(\frac{n}{2^i}\right)\right).$$

If $M(n) = \Omega(n^\epsilon)$ for $\epsilon > 0$, then $DOWN(n) = O(M(n))$. Otherwise, $DOWN(n) = O(M(n) \log n)$. After updating level 0 of the tree, we ascend the tree reforming Q-structures on the path from the leaf to the root. Even with rebalancing, only a constant number of merges will be performed at each level. Consequently, $UP(n)$ = $O(DOWN(n))$.

We may conclude that any $M(n)$-order decomposable problem has a dynamic algorithm which accommodates both insertions and deletions in time $ID(n)$ = $O(M(n))$, if $M(n)$, = $\Omega(n^\epsilon)$ for $\epsilon > 0$, and in time $ID(n)$ = $O(M(n) \log n)$ otherwise. In other words, whenever there is a "cheap" merging algorithm for a set problem, the problem can always be dynamized efficiently.

3.5. Notes

Divide-and-conquer techniques are as prevalent in computational geometry as they are elsewhere in the design of algorithms. A common data structure which derives from this approach is the Voronoi diagram which is used for closest point problems (see Chapter 3).

The first convex hull algorithm was proposed by Graham (1972). His method is simple, but does not easily allow dynamization. We follow the development of Preparata and Hong (1977) here. This algorithm also extends to three dimensions. The problem of computing hulls has been widely studied (see Chapter 5).

Overmars and van Leeuvwen (Overmars & van Leeuwen, 1981; van Leeuwen & Overmars, 1981) were the first to consider methods of computing dynamic convex hulls. Their work also led to the generalization to order-decomposable problems (Overmars, 1981, 1983). Our development here follows theirs. Similar techniques are used in (Dobkin & Munro, 1985). For details on the use of the BB(α) data structure, see (Reingold, Nievergelt, & Deo, 1977).

4.   GEOMETRIC TRANSFORMATIONS

4.1. Introduction

In this section, we discuss methods of transformation. Typically, transformations change geometric objects into other geometric objects (e.g., take points into lines) while preserving relations which held between the original objects (e.g., order or whether they intersected). A number of geometric problems are best solved through the use of transformations. The standard scheme is to transform the objects under consideration, solve a simpler problem on the transformed objects, and then use that solution to solve the original problem. No single transformation applies in all cases; a number of different transformations have been used effectively. Here, we describe three commonly used transformations in 2-space and in 3-space and demonstrate their applications.

4.2. Mathematical Background

The domain of all of our two-dimensional transformations will be the projective plane. The projective plane is an enhanced version of the Euclidean plane in which each pair of lines intersects. The projective plane contains all points of the Euclidean plane (the "proper points"). We introduce a set of "improper points" with one point associated with every direction in the plane. Two parallel lines, then, intersect at that "improper point" indicated by the direction of the parallel lines (this can be thought of as a point at infinity). All "improper points" are considered to lie on the same line: the "improper line" or the "line at infinity." Thus, any two lines in the projective plane intersect at exactly one point: two nonparallel proper lines intersect at a proper point (i.e., one of the Euclidean plane); two parallel proper lines intersect at the improper point bearing the same direction; and a proper line intersects the improper line at the improper point defining the direction of the proper line. Likewise, between every two points passes exactly one line: There is a proper line passing through every pair of proper points; a proper line passes through a given improper point and a given proper point; and the improper line passes through any two improper points.

In three dimensions, projective space will be obtained similarly. All points in E^3 are called "proper points." Define one "improper point" for every direction in space. Collectively, the "improper points" form the "improper plane." Therefore, each proper plane will intersect the improper plane in an improper line; a pair of parallel proper planes intersect in an improper line; and a pair of nonparallel proper planes intersect in a proper line. Given any two lines, either they are skew or else they intersect in a single point. Any line and any point not on the line define a plane.

In general, the actual algorithms used to solve problems rely solely on Euclidean geometry. Therefore, although all the transformations will map the projective plane (space) onto itself, we will wish to choose a transformation which maps the objects under consideration to "proper" objects. Thus, the parameters of the original problem will dictate which transformations are appropriate. Throughout the section, we will use the following notation: The images of a point V, a line h, and a triangle XYZ under a transformation B will be denoted by V_B, h_B, and XYZ_B, respectively.

4.3. Point/Line Duality I

The appropriate transformation is selected based on relationships among the original objects and the given problem to be solved. Our choice is guided by the properties to be preserved under the transformation.

We first return to the two-dimensional problem of determining the intersection of a collection of N lower half-planes where for $1 \leq i \leq N$, a line $l_i : y = a_i x + b_i$ bounds the half-plane $H_i = \{(x, y) \mid y < a_i x + b_i\}$ from above. A redundant half-plane is defined as one which contains the entire intersection region in its interior. One approach to the problem consists of identifying and eliminating the redundant half-planes. For each remaining half-plane H_i, a portion of l_i lies on the boundary of the intersection region. To determine the boundary, we then sort the l_i in descending order of slope in $O(n \log n)$ time; determine the intersection points between pairs of adjacent lines in linear time; now the boundary consists of the left half of the first line up to its intersection with the second line, the segments from one intersection point to the next, and the half of the final line extending to the right from its intersection with the second to last line.

To complete the algorithm, we must define a process for determining redundant half-planes. A half-plane H_i is redundant iff there are half-planes H_j and H_k such that the slopes satisfy the inequality $a_j < a_i < a_k$ and such that l_i lies above the intersection point of l_j and l_k. We say that the pair l_j, l_k forms a certificate of redundancy for l_i. Given a collection of N half-planes, the naive method of deciding the redundancy of a given half-plane entails comparing its bounding line against all other pairs of lines in time $\theta(N^2)$. Determining all redundancies requires $\theta(N^3)$.

Clearly, a transformation which allows for efficient identification of redundant half-planes must in some way preserve both the above/below relationship and the concept of slope. We show that the transformation T which maps the point (a, b) to the line $y = ax + b$ satisfies these criteria. To do so, we must first determine the image of a line $l: y = cx + d$ under the transformation T. As defined above, T maps the points $(n, cn + d)$ and $(m, cm + d)$, both of which lie on l, to the lines $y = nx + cn + d$ and $y = mx + cm + d$, respectively. Both of these lines pass through the point $P = (-c, d)$ (see Fig. 2.17). In other words, T maps the set of all points lying on l to the set of all lines passing through P, or the "pencil of lines" through P. More simply, we just say that T maps $y = cx + d$ to the point $(-c, d)$.

That T maps $y = cx + d$ to the point $(-c, d)$ rather than to the point (c, d) destroys the symmetry of the transformation. We prefer transformations which are self-inverting. Nonetheless, T remains valuable because it preserves both the above/below relationship and the concept of slope. First of all, the negative of the slope of a line survives as the x-coordinate of the image point. Consequently, if the slope of l exceeds the slope of k, then the point l_T lies to the left of k_T. Secondly, the vertical distance from $P = (a, b)$ to the line $l: y = cx + d$ is defined as the distance from P to the point $(a, ac + d)$ which lies on l, $b - (ac + d)$. The

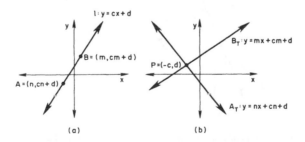

FIG. 2.17.   A line (a) and its dual (b) under the transformation T

vertical distance from the line $P_T : y = ax + b$ to the point $l_T = (-c, d)$ is defined as the distance from the point $(-c, b - ac)$ lying on P_T to the point $l_{T'} (-ac + b)$ $- d$. The two quantities are identical in both sign and magnitude.

Before applying T to the half-plane intersection problem, we must determine the image of a vertical line under T. Each point (m, a) on the line $x = m$ is mapped to a line $y = mx + a$, a line with slope m. All lines with slope m, however, intersect at the improper point P_m associated with m. Thus, T maps the vertical line $x = m$ to the pencil of lines through P_m, or just to P_m. Conversely, each line $y = mx + a$ which passes through P_m is mapped to a point $(-m, a)$ which lies on the line $x = -m$. Consequently, the image of P_m under T corresponds to $x = -m$. In general, we prefer to ignore improper points. Whenever a given problem involves vertical lines, we either choose to use a transformation other than T or we rotate the object space slightly until no vertical lines remain.

The transformation T preserves the notion of slope and vertical distance and thus can help to identify the redundant half-planes within a collection where none is bounded by a vertical line. Consider the collection of lower half-planes bounded above by the lines h, j, k, l, m, and n and their images or "duals," under T: h_T, j_T, k_T, l_T, m_T, and n_T (see Figs. 2.18a and b). Portions of m, k, h, and l together form the boundary of the intersection region R. Thus, these four lines cannot correspond to redundant half-planes. The line j does correspond to a redundant half-plane: the slope of j has a value between the slopes of k and h, and j lies above the intersection point of k and h. Consequently, j_T must lie between the horizontal values of k_T and h_T. in addition, j_T must lie above the line $\overleftrightarrow{k_T H}_T$ the dual of the intersection point of k and h. Similarly, since h, l, forms a certificate of redundancy for n, n_T must lie between the horizontal values of h_T and l_T and above the line $\overleftrightarrow{h_T l_T}$.

We may conclude that a line p in the object plane has certificate of redundancy q, r if, and only if p_T lies directly above the line segment $\overline{q_T r_T}$ in the image plane. A line p which has no such certificate corresponds to a significant half-plane. The only points p_T which lie above no segment $\overline{q_T r_T}$ are those points on

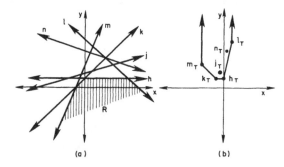

(a)   (b)

FIG. 2.18. Computing the intersection of a set of lower half-planes (a) by finding the bc-hull of the dual set of points (b)

the bottom convex hull, or bc-hull, of the set of all image points. In the example of Fig. 2.18a, we observe that m_T, k_T, h_T and l_T form the bc-hull, while j_T and n_T lie the interior.

The transformation T enables us to distinguish the significant and redundant half-planes in a collection of size N in time $O(N \log N)$: finding the dual of each bounding line takes time $O(N)$; determining the bc-hull of the daul points takes time $O(N \log N)$. Those half-planes corresponding to points on the bc-hull are significant. The half-planes corresponding to points in the interior are redundant.

We note that our algorithm for using a geometric transformation to solve the half-plane intersection problem consisted of three parts. We first identified the geometric techniques we might use (in this case redundancy). Next, we identified the invariants required by a transformation (notion of above/below and slope). Finally, we found an appropriate transformation and solved the problem. This is a classic example of how geometric transformations are used.

Had we wished to solve the lower half-space intersection problem, we could have used a three dimensional version of T:

$$(a,b,c) \xrightarrow{T} z = ax + by + c$$

$$z = ax + by + c \xrightarrow{T} (-a, -b, c)$$

Here, T still preserves vertical distance, distance with respect to the x-coordinate, as well as the directional cosincs, which represent the rates at which z changes with respect to x and y. The points on the bottom part of the convex hull of the transformed problem will correspond to the nonredundant half-spaces within the original problem. The details are left to the reader.

A number of other intersection problems depend on the above/below relationship and/or the concept of slope. A line segment intersects a line if, and only if one endpoint lies above the line and the other lies below the line. A ray $\vec{r}$ intersects a line l if, and only if one of the following conditions holds:

1. $\vec{r}$ extends to the right with greater slope than l and the vertex lies below l;
2. $\vec{r}$ extends to the right with smaller slope than l and the vertex lies above l;
3. $\vec{r}$ extends to the left more steeply than l and the vertex lies above l;
4. $\vec{r}$ extends to the left with smaller slope than l and the vertex lies below l.

The intersection of two segments, two rays, or a ray and a segment can be determined by a sequence of tests involving slopes and/or vertical separation.

To show how T can be applied to the problems above, we must first determine the image under T of both rays and segments. Recall that the image of a line is the set of all lines through a point P. It is reasonable to assume that the image of a line segment should be a set of *some* of the lines through a point P. The set may

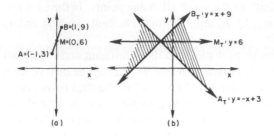

(a)   (b)

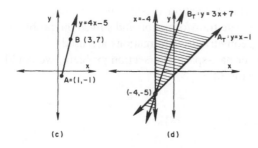

(c)   (d)

FIG. 2.19. A line segment (a) and its dual (b); a ray (c) and its dual (d)

not contain a vertical line since a line segment $\overline{AB}$ does not pass through a point at infinity. Thus, we may conclude that the image of $\overline{AB}$ consists of the set of all lines lying between A_T and B_T in the counter-clockwise direction (see Figs. 2.19a and b). This region is aptly described as a double wedge. A ray $\overrightarrow{AB}$ may be considered a line segment extending from A through the point B and ending at the improper point having direction $\overrightarrow{AB}$. The image of $\overrightarrow{AB}$, then, is a double wedge formed by A_T and the vertical line to which the improper point is mapped. As demonstrated in Figs. 2.19c and d, testing a point on $\overrightarrow{AB}$ determines which of the two double wedges corresponds to $\overrightarrow{AB}$.

Under the transformation T, we may determine the intersection of segments, rays and lines as follows. A line segment $\overline{AB}$ intersects a line l iff one of A_T and B_T lies above l_T and the other lies below. More simply, $\overline{AB}$ intersects l if l_T lies within the double wedge $\overline{AB}_T$ (see Figs. 2.20a and b). A ray $\overrightarrow{AB}$ extending rightward with slope m intersects a line l iff l_T lies to the right of $x = m$ and above A_T, or l_T lies to the left of $x = m$ and below A_T (see Figs. 2.20c and d). A ray $\overrightarrow{AB}$ extending leftward with slope m intersects l iff l_T lies to the right of $x = m$ and below A_T or to the left of $x = m$ and above A_T. Either case simplifies to $\overrightarrow{AB}$ intersects l iff l_T lies in the double wedge $\overrightarrow{AB}_T$. The segments, $\overline{AB}$ and $\overline{CD}$ intersect iff the image of the line $\overleftrightarrow{AB}$ lies in the interior of the double wedge $\overline{CD}_T$, and vice versa. This is equivalent to saying that $\overline{AB}$ and $\overline{CD}$ intersect iff $\overleftrightarrow{AB}$ intersects $\overline{CD}$ and $\overleftrightarrow{CD}$ intersects $\overline{AB}$.

This concludes our survey of the transformation T. As we noted earlier, there is a general scheme for applying both this and other transformations to geometric problems. While this one is particularly useful for some low-level intersection

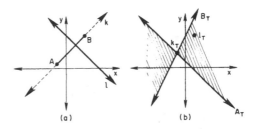

(a)   (b)

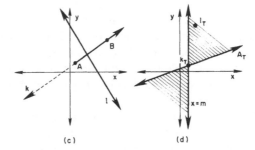

FIG. 2.20. Using duality to check
for the intersection of: a line segment
and a line (a & b); a ray and a line (c
& d)

(c)   (d)

problems, the other transformations given in this section apply equally well in other contexts.

4.4. Point/Line Duality II

The second transformation we consider is well suited for polygon inclusion/intersection problems. We note that T is not particularly useful in these contexts. Applying T to the problem of whether a line segment s intersected $\triangle ABC$ would, in fact, complicate the solution process. T maps $\triangle ABC$ and its interior to the region R which is infinite but contains no vertical ray (see Figs. 2.21a, b). A segment s intersects $\triangle ABC$ if and only if the following conditions are satisfied: The vertex of the double wedge s_T lies in R and the intersection of s_T and R contains at least one infinite line (see Fig. 2.21c). As neither R nor its complement form convex regions, these conditions are difficult to test.

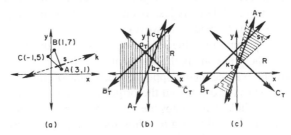

(a)   (b)   (c)

FIG. 2.21.   The effect of the transformation T on a triangle and an intersecting line segment

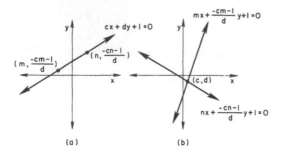

FIG. 2.22. A line (a) and its dual (b) under the transformation S

Our second transformation, S, maps the point (a, b) to the line $ax + by + 1 = 0$. We observe in Fig. 2.22 that S maps each point on the line $cx + dy + 1 = 0$ to a line passing through the point (c, d). Thus S maps the line $cx + dy + 1 = 0$ to the pencil of lines through (c, d), or the point (c, d). Like T, S maps a line segment to a double wedge, this time to one which does not contain the origin. The image of a ray is a double wedge formed by one line passing through the origin and one not.

The transformation S preserves the concept of the distance of an object from the origin. The line l: $ax + by + 1 = 0$ is perpendicular to the line m: $bx - ay = 0$ at the point $(-a/(a^2 + b^2), -b/(a^2 + b^2))$. The line l is precisely $1/\sqrt{a^2 + b^2}$ units away from the origin along the line m. The point (a, b), however, which is dual to l, is precisely $\sqrt{a^2 + b^2}$ units away from the origin along the line m in the opposite direction. Thus, points further from the origin are mapped to lines closer to the origin. That S is also its own inverse (i.e., $(a, b) \xrightarrow{S} ax + by + 1 = 0 \xrightarrow{S} (a, b)$) contributes to its usefulness.

We note that each improper point is mapped to the line through the origin having the same direction and vice versa. Similarly, the origin and the improper line are duals. Consequently, S should not be applied to lines or to segments of lines which pass through the origin. Nonetheless, the mere fact that the domain of a problem contains a line through the origin should not make us abandon S. By translating the axes in one direction or another, we may be able to insure that S will map every object in the domain of our problem to another proper object.

Since the transformation S preserves the concept of distance from the origin, it is most useful for problems where the origin serves a focal role. The image under S of an arbitrary triangle, $\triangle ABC$, is just as unwieldy as the image produced by T (see Figs. 2.23a and b). Consider, however, a convex polygon which contains the origin. Each segment from the origin to a point A on the boundary lies completely within the polygon. If the segment has slope m, then S maps each point on the segment to a line on the opposite side of the origin with slope $-1/m$. The line A_S is closest to the origin and the origin itself corresponds to the line at infinity. We may conclude that S maps the segment to the half-plane bounded by A_S which does *not* include the origin. Thus the interior of a convex polygon containing the origin is mapped to the union of a set of half-planes *not* containing the origin, or the exterior of a convex polygon.

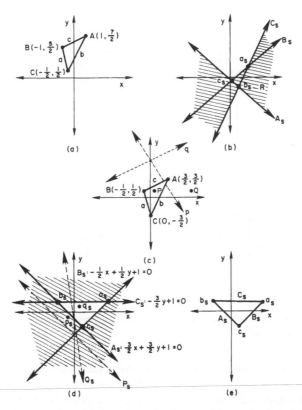

FIG. 2.23. The effect of the transformation S on an arbitrary triangle (a & b) and on a triangle enclosing the origin (c, d, & e)

We can demonstrate this phenomenon on the triangle of Fig. 2.23a, by first translating the axes 1/2 to the left and 2 units up, as shown in Fig. 23c. Now the union of the double wedges a_S, b_S and c_S consists of the entire plane except for the triangle formed by the vertices of a_S, b_S and c_S, which we call $\triangle ABC_S$ (see Fig. 2.23d). $\triangle ABC$ contains a point P if and only if P_S is exterior to $\triangle ABC_S$. A line p intersects triangle $\triangle ABC$ if and only if p_S lies outside $\triangle ABC_S$. A point Q is exterior to ABC if and only if Q_S lies in ABC_S. And a line q is disjoint from $\triangle ABC$ if and only if $\triangle ABC_S$ contains q_S. This tidy symmetry allows us to simplify our definition of the transformation much as we did when we said that the image of a line is a point rather than a pencil of lines. When S is applied to a triangle containing the origin, we say that the image of each segment is a point, rather than a wedge, and the image of each vertex is a segment, rather than a line. In other words, S maps one triangle containing the origin to another triangle containing the origin (see Fig. 2.23e). The significance of the duality, however, is that the interior of one triangle is mapped to the exterior of the other.

This particular duality transformation can be useful in cases where we wish to

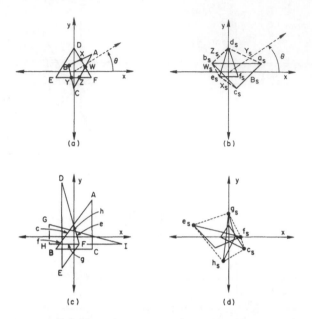

FIG. 2.24.   Computing the intersection of triangles (a) by finding the convex hull of the dual triangles (b)

determine the intersection of triangles. In particular, suppose that, having already identified a common point, we wish to compute the intersection of two triangles. We translate the axes so that the common point lies at the origin. Let R represent the set of all points lying on the boundary of either triangle. Then the border of the intersection region can be defined as follows: $I = \{p \in R|$ at some angle $\theta \in [0, 2\pi)$, p is the closest point in R to the origin}. The region of intersection forms a polygon which can be identified by listing a sequence of vertices. In Fig. 2.24a, the sequence of vertices describing the intersection region is $WXBYZ$. At θ, the segment XW is closest to the origin. In the dual problem, we wish to identify the polygon containing those points p which are the farthest from the origin for some θ. The process is simple: Finding the convex hull of ABC_S and DEF_S will yield a sequence of vertices e_S, c_S, a_S, d_S, b_S. The segments joining these vertices, X_S, B_S, Y_S, Z_S, W_S, correspond to the vertices of the intersection region of the original problem (see Fig. 2.24b).

Thus, the transformation S enables us to convert an intersection problem to a convex hull problem. The technique succeeds for any number of convex polygons as long as each contains the origin (see Fig. 2.24c and d). It works in a similar fashion in three dimensions. The three-dimensional version of S maps a point $P = (a, b, c)$ to the plane $ax + by + cz + 1 = 0$ which is perpendicular to line passing through both P and the origin $1/\sqrt{a^2 + b^2 + c^2}$ units on the opposite side of the origin from P. Suppose we apply S to a pair of intersecting planes.

Each plane is mapped to a point. The line formed by the intersecting planes maps to the line passing through the two image points. Consequently, the image of a convex polyhedron containing the origin will be another convex polyhedron containing the origin. The interior of the first, however, will have been mapped to the exterior of the second. Thus to compute the intersection of two or more convex polyhedra, we first find a common point, translate the axes so that this point becomes the origin; apply S; find the convex hull of all of the dual polyhedra; reapply S to obtain the vertices and faces of the intersection polyhedron.

4.5. Inversion (Point/Point Duality)

We turn now to the problem of determining the intersection of a collection of disks. Each of the transformations defined so far has been geared towards problems of straight line geometry, so we need a different type of transformation here. We note, however, that problems in straight line geometry are generally easier to solve than those involving curves. Consequently, a transformation which maps circles to lines could simplify the disk intersection problem.

We define a new transformation G which is similar to S in that it functions as its own inverse and preserves the concept of distance of an object from the origin. In this case, however, instead of mapping points to lines on opposite sides of the origin, G maps the point $P = (a, b)$ to a point which is $1/\sqrt{a^2 + b^2}$ units away from the origin in the same direction as P. Using polar coordinates, G maps (r, θ) to $(1/r, \theta)$. In fact, for G to be a one-to-one mapping of the plane onto itself, we must use polar coordinates and consider the origin as an infinite number of points of the form $(0, \theta)$, $\theta \epsilon (0, 2\pi)$, all of which lie on a circle of radius zero centered at the origin. In this way, the point at the origin denoted by $(0, \theta)$ is mapped to (∞, θ), the improper point having direction θ. The entire null circle is then mapped to the improper line. A circle centered at the origin with finite radius r is dual to a concentric circle with radius $1/r$.

The transformation G pairs each circle passing through the origin with a straight line which does not pass through the origin. As an example, we consider the line $l: y = -3x + 1$ (see Fig. 2.25). We determine the images of the points A, B, C, D, E, F by drawing a ray from the origin through each one of them and picking the point at the appropriate distance. In addition, we note that the rays $\theta = \theta_6$ and $\theta = \theta_0$, which are parallel to the line l, will intersect l at the point at infinity and only at the point at infinity. Thus the origin, or th point $(0, \theta_6) = (0, \theta_0)$, lies on the dual of l.

As demonstrated above, a line whose closest point is R units from the origin in direction θ is transformed to a circle passing through both the origin and $(1/R, \theta)$. A line $\theta = \theta_0$ passing through the origin, however, is mapped to itself: the segment from $(-1, \theta_0)$ through the origin to the point $(1, \theta_0)$ is mapped to the segment from $(-1, \theta_0)$ through the point at infinity to the point $(1, \theta_0)$, and vice

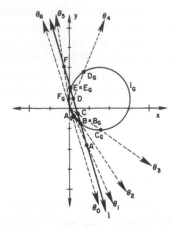

FIG. 2.25.   A line and its dual under the transformation G

versa. A circle containing the origin is mapped to another circle containing the origin (see Fig. 2.26a), and a circle not containing the origin is mapped to another circle not containing the origin (see Fig. 2.26b). In all cases of circle/circle duality, the interior of one circle is mapped to the exterior of the other: the interior of a circle through the origin is mapped to a half-plane *not* containing the origin; a half-plane bounded by a line through the origin is mapped to itself.

   G enables us to determine easily the intersection of a collection of disks whose boundaries all contain the origin. In the example shown in Fig. 2.27, three arcs bound the intersection of disks D, E, and F: the arc along circle D running in the clockwise direction from x to y; the arc along circle F from y to z; and the arc along circle E from z to x. G can help to determine this boundary. G transforms each of the disks to a half-plane bounded below by a straight line not through the origin. The intersection of these half-planes is bounded by the vertices x_G, y_G, z_G, where x_G is the point of infinity first along D_G and secondly along E_G, and by the segments of D_G, F_G, and E_G which join the vertices. These correspond

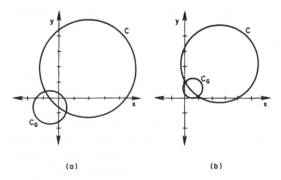

(a)   (b)

FIG. 2.26.   Two circles and their duals

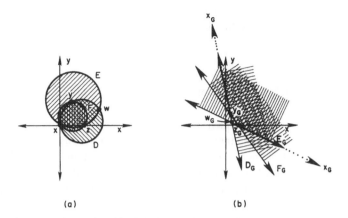

(a)   (b)

FIG. 2.27. Computing the intersection of a set of circles (a) by finding the
intersection of the dual half-planes (b)

exactly to the vertices and arcs of the intersection region in the original. Similarly, the union of the half-planes would correspond to the union of the disks.

A problem involving only disks which all pass through a single point would be quite rare. We may, however, extend G to three dimensions: (r, θ, ϕ). Given a collection of disks in a plane, we can choose an arbitrary point P not in the plane. Then for each disk d, we determine the unique ball which passes through P and which contain d as a cross-section. Next we translate the axes so that the origin coincides with P. G will now map each ball to a half-space not containing the origin. The intersection of the half-spaces will correspond to the intersection of the balls which we can then intersect with the original plane to determine the

TABLE 2.1
Summary of the Transformations Considered in this Section.

	T	S	G
(a,b)	$y=ax+b$	$ax+by+1=0$	$\left(\dfrac{a}{a^2+b^2}, \dfrac{b}{a^2+b^2} \right)$
$y=ax+b$	$(-a,b)$	$\left(\dfrac{a}{b}, \dfrac{-1}{b} \right)$	$\left(x+\dfrac{a}{2b} \right)^2 + \left(y-\dfrac{1}{2b} \right)^2 = \dfrac{a^2+1}{4b^2}$
$ax+by+1=0$	$\left(\dfrac{a}{b}, \dfrac{-1}{b} \right)$	(a,b)	$\left(x+\dfrac{a}{2} \right)^2 + \left(y+\dfrac{b}{2} \right)^2 = \dfrac{a^2+b^2}{4}$
$(x-h)^2+(y-k)^2=h^2+k^2$	—	—	$hr+ky=\dfrac{1}{2}$
preserves	above/below, slope	distance	distance
makes improper	vertical lines	the origin	the origin
useful for	half-plane,ray $\cap$	polygon $\cap$	disk $\cap$

intersection of the disks. The interested reader may find that these or other transformations will simplify his/her favorite problem.

4.6. Notes

The mathematical background section reflects *The VNR Concise Encyclopedia of Mathematics.*

Geometric transformations such as those mentioned here have their roots in the mathematics of the early nineteenth century (Boyer, 1968). Transformations appear in the work of Steiner, and projective geometry originated with Poncelet. Their application to problems of computing dates back to the concept of primal and dual problems in the study of linear programming (see e.g., Papadimitriou & Steiglitz, 1982).

Brown (1979) gives a systematic treatment of transformations and their applications to problems of computational geometry. Since his dissertation, these methods have found vast application (Chazelle, 1983; Chazelle, Guibas, & Lee, 1983; Dobkin & Edelsbrunner, 1984; Edelsbrunner, Kirkpatrick, & Maurer, 1982; Edelsbrunner, O'Rourke, & Seidel, 1983).

The examples given here are derived from Brown's work (1979) as well as from Muller and Preparata (1978). The latter reference gives a complex application of duality to the polyhedron intersection problem, extending their work on halfspace intersection (Preparata & Muller, 1979). The problem is divided there into the subproblems of finding an intersection point and then using that as a basis for computing the entire intersection. The problem of finding the first point was then further studied in Chazelle and Dobkin (1980) and Dobkin and Kirkpatrick (1983).

5.   CONCLUSION

We have presented here three different techniques which have proved useful in deriving algorithms for geometric problems: hierarchical search, hierarchical computation, and geometric transformations. Although in each case we supply specific examples, each technique has a wide variety of applications.

Our ability to produce an efficient algorithm using one or more of the given techniques derives from some understanding of the underlying geometry. Each of the first two depends on finding an appropriate data structure or basic unit. For the general polygon search algorithm, we chose a triangular subdivision. For the convex hull algorithm, concatenable queues held the partial hulls. To choose an appropriate geometric transform, the problem must be carefully analysed to determine the properties or characteristics which must be preserved. To solve the half-plane intersection problem, we select a transformation which preserves vertical distance and slope.

Although the three techniques appear to be distinct, they are mutually dependent in many applications. For example, the dynamic convex hull algorithm above depends on hierarchical search for insertions and deletions. In general polygon search, the initial triangulation of S to form S_0 depends on a hierarchical computation of an ordered list of vertices. Geometric transforms can convert a whole problem or part of a problem to a format more accessible to hierarchical decomposition. Together or separately, these three techniques span a wide field of application.

Although all applications here have been geometric in nature, we believe that these methods will have profitable applications in the field of robotics. We invite feedback.

ACKNOWLEDGMENTS

This research was supported in part by the National Science Foundation under Grant MCS83-03926. The second author was supported by an Exxon Fellowship.

We thank D. Barbara, B. Chazelle, S. Friedman, R. Horn, J. Kent, and K. Supowit for reading earlier drafts of this paper and contributing to a more readable document. We are especially indebted to T. Agans for his superb drafting of the numerous figures.

REFERENCES

Many of the references listed here are not cited in the text. We include them here to broaden our survey.

Aho, A. V., Hopcroft, J. E., & Ullman, J. D. (1974). *The design and analysis of computer algorithms.* Reading, MA: Addison-Wesley.
Bentley, J. L. (1979). Decomposable searching problems. *Inf. Proc. Lett., 8,* 244–251.
Bentley, J. L., & Ottmann, T. A. (1979). Algorithms for reporting and counting geometric intersections, *IEEE Trans. Comp., C-28,* 643–647.
Bentley, J. L., & Saxe, J. B. (1980). Decomposable searching problems I: Static-to-dynamic transformation. *Journal of Algs., 1,* 301–358. A preliminary version appeared in *Proc. 20th IEEE FOCS,* 1983, 148–168.
Bentley, J. L., & Wood, D. (1980). An optimal worst-case algorithm for reporting intersections of rectangles. *IEEE Trans. Comp., C-29,* 572–577.
Boyer, C. B. (1968). *A history of mathematics.* New York: Wiley.
Brown, K. Q. (1979). *Geometric transforms for fast geometric algorithms.* Carnegie-Mellon. (Tech. Report CMU-CS-80-101.)
Brown, K. Q. (1981). Comments on "Algorithms for reporting and counting geometric intersections," *IEEE Trans. Comp., C-30,* 147–148.
Chazelle, B. (1983). The polygon containment problem. *Advances in computing research* (Vol. 1, pp. 1–34). Greenwich, CT: JAI press.
Chazelle, B. (1984a). Intersecting is easier than sorting. *Proc. 16th ACM STOC,* pp. 125–134.
Chazelle, B. (1984b, October). Personal communication with D. Souvaine.

Chazelle, B., Cole, R., Preparata, F., & Yap, C. (1984). *New upper bounds for neighbor searching,* Brown University. (Tech. Report CS-84-11.)

Chazelle, B., & Dobkin, D. P. (1980). Detection is easier than computation. *Proc. 12th ACM STOC,* pp. 146–153.

Chazelle, B., & Dobkin, D. P. (1985). Optimal convex decompositions. In G. T. Toussaint (Ed.), *Machine Intelligence and Pattern Recognition 2: Computational Geometry.* Amsterdam, The Netherlands: Elsevier Science Publishers.

Chazelle, B., Drysdale, R. L., & Lee, D. T. (1986). Computing the largest empty rectangle. *SIAM J. Comp.,* 15, 300–315.

Chazelle, B., Guibas, L. J., & Lee, D. T. (1983). The power of geometric duality. *Proc. 24th IEEE FOCS,* pp. 217–225.

Chazelle, B., & Incerpi, J. (1984). *Triangulation and shape-complexity. ACM Trans. Graphics, 3,* 135–152.

Cole, R., & Yap, C. K. (1983). Geometric retrieval problems. *Proc. 24th IEEE FOCS,* pp. 112–121.

Dobkin, D. P., Drysdale, R. L., & Guibas, L. J. (1983). Finding smallest polygons. *Advances in computing research* (Vol. 1, pp. 181–214). Greenwich, CT: JAI Press.

Dobkin, D. P., & Edelsbrunner (1984). Space searching for intersecting objects. *Proc. 25th IEEE FOCS,* pp. 387–391.

Dobkin, D. P., & Kirkpatrick, D. G. (1983). Fast detection of polyhedral intersections. *Theo. Comp. Sci., 27,* 241–253.

Dobkin, D. P., & Kirkpatrick, D. G. (1985). A linear algorithm for determining the separation of convex polyhedra. *Journal of Algs., 6,* 381–392.

Dobkin, D. P., & Lipton, R. J. (1976). Multidimensional searching problems. *SIAM J. Comp., 5,* 181–186.

Dobkin, D. P., & Lipton, R. J. (1979). On the complexity of computation under varying sets of primitives. *J. Comp. Syst. Sci., 18,* 86–91.

Dobkin, D. P., & Munro, J. I. (1985). Efficient uses of the past. *Journal of Algs., 6,* 455–465.

Dobkin, D. P., & Reiss, S. (1980). The complexity of linear programming. *Theo. Comp. Sci., 11,* 1–18.

Dobkin, D. P., & Snyder, L. (1979). On a general method for maximizing and minimizing among certain geometric problems. *Proc. 20th IEEE FOCS,* 9–17.

Dyer, M. E. (1983). A geometric approach to two-constraint linear programs with generalized upper bounds. *Advances in computing research* (Vol. I, pp. 79–90). Greenwich, CT: JAI Press.

Edelsbrunner, H., Kirkpatrick, D. G., & Mauer, H. A. (1982). Polygonal intersection searching. *Inf. Proc. Lett., 14,* 74–79.

Edelsbrunner, H., O'Rourke, J., & Seidel, R. (1983). Constructing arrangements of lines and hyperplanes with applications. *Proc. 24th IEEE FOCS,* pp. 83–91.

Edelsbrunner, H., Overmars, M. H., & Wood, D. (1983). Graphics in flatland: A case study. *Advances in computing research* (Vol. I, pp. 35–60). Greenwich, CT: JAI Press.

Garey, M. R., Johnson, D. S., Preparata, F. P., & Targan, R. E. (1979). Triangulating a simple polygon. *Inf. Proc. Lett., 7,* 175–179.

Graham, R. L. (1972). An efficient algorithm for determining the convex hull of a planar set. *Inf. Proc. Lett., 1,* 132–133.

Greene, D. H. (1983). The decomposition of polygons into convex parts. *Advances in computing research* (Vol. I, pp. 235–260). Greenwich, CT: JAI Press.

Grünbaum, B. (1967). *Convex polytopes.* New York: Wiley Interscience.

Guibas, L. T., & Yao, F. F. (1983). On translating a set of rectangles. *Advances in computing research* (Vol. I, pp. 61–78). Greenwich, CT: JAI Press.

Karmarker, N. (1984). A new polynomial-time algorithm for linear programming. *Proc. 16th ACM STOC,* pp. 302–311.

Kirkpatrick, D. G. (1980). A note on Delaunay and optimal triangulations. *Inf. Proc. Lett., 10,* 127–128.

Kirkpatrick, D. G. (1983). Optimal search in planar subdivisions. *SIAM J. Comp., 12,* 28–35.

Knuth, D. E. (1973). *Sorting and searching.* Reading MA: Addison-Wesley.

Lee, D. T., & Preparata, F. P. (1977). Location of a point in a planar subdivision and its applications. *SIAM J. Comp., 6,* 594–606.

Lee, D. T., & Preparata, F. P. (1979). An optimal algorithm for finding the kernal of a polygon. *Journal of the ACM, 26,* 415–421.

Lee, D. T., & Preparata, F. P. (1984). Computational geometry—a survey. *IEEE Trans. Comp., C-33,* 1072–1101.

Lipton, R. J., & Tarjan, R. E. (1979). A separator theorem for planar graphs. *SIAM J. App. Math., 36,* 177–189.

Lipton, R. J., & Tarjan, R. E. (1980). Applications of a planar separator theorem. *SIAM J. Comp., 9,* 615–627.

Lloyd, E. L. (1977). On triangulation of a set of points in the plane. *Proc. 18th IEEE FOCS,* pp. 228–240.

Muller, D. E., & Preparata, F. P. (1978). Finding the intersection of two convex polyhedra. *Theo. Comp. Sci., 7,* 217–236.

Overmars, M. H. (1981). Dynamization of order decomposable set problems. *Journal of Algs., 2,* 245–260.

Overmars, M. H. (1983). *The design of dynamic data structures.* Doctoral thesis presented at the University of Utrecht, Druk: Sneldruk Boulevard Enschede.

Overmars, M. H., & van Leeuwen, J. (1981). Maintenance of configurations in the plane. *Journ. Comput. Sys. Sci., 23,* 166–204. A preliminary version appeared in Proc. 12th ACM STOC, pp. 135–145.

Papadimitriou, C. H., & Steiglitz, K. (1982). *Combinatorial optimization: Algorithms and complexity.* Englewood Cliffs, NJ: Prentice Hall.

Preparata, F. P. (1979). An optimal real-time algorithm for planar convex hulls. *C. ACM, 22,* 402–405.

Preparata, F. P. (1981). A new approach to planar point location. *SIAM J. Comp., 10,* 473–482.

Preparata, F. P., & Hong, S. J. (1977). Convex hulls of finite sets in two and three dimensions. *Comm. ACM, 20,* 87–93.

Preparata, F. P., & Muller, D. E. (1979). Finding the intersection of n half-spaces in time $O(n \log n)$. *Theo. Comp. Sci., 8,* 45–55.

Reingold, E. M., Nievergelt, J., & Deo, N. (1977). *Combinatorial algorithms.* Englewood Cliffs, NJ: Prentice Hall.

Seidel, R. (1984). *A method of proving lower bounds for certain geometric problems.* Cornell University. (Tech. Report 84-592.)

Shamos, M. I. (1975). Geometric complexity. *Proc. 7th ACM STOC,* pp. 224–233.

Shamos, M. I., & Hoey, D. (1975). Closest-point problems. *Proc. 16th IEEE FOCS,* pp. 151–161.

Shamos, M. I., & Hoey, D. (1976). Geometric intersection problems. *Proc. 17th IEEE FOCS,* pp. 208–215.

Supowit, K. J. (1983). Grid heuristics for some geometric covering problems. *Advances in computing research* (Vol. I, pp. 215–234). Greenwich, CT: JAI Press.

THE VNR Concise Encyclopedia of Mathematics. (1977). W. Gellert, H. Kustner, M. Hellwich, & H. Kastner (Eds.). New York: Van Nostrand Reinhold Company, pp. 146–203, 547–561.

van Leeuwen, J., & Overmars, M. H. (1981). The art of dynamizing. *Mathematical Foundations of Computer Science* (pp. 121–131). Berlin; New York: Springer Verlag.

3 Algorithmic Motion Planning

Chee-Keng Yap
Courant Institute of Mathematical Sciences
New York University

Recently, there have been significant theoretical advances in the algorithmic study of motion planning in robotics. We present a unifying perspective of the area. In particular, we give an exposition of two techniques (which we call the *decomposition* and *retraction* approaches) that have been used to solve such problems. We will show that both techniques are in fact "universal" in a precise sense. Open questions and new directions are posed.

Motion planning, i.e., the problem of computing a path for a robot, is widely recognized as a fundamental problem in robotics. A typical formulation of this problem is as follows: given a robot and a description of its environment, to plan a path of the robot between two specified positions, which avoids collision with obstacles in the environment. This is a "global" planning problem; to use its results, one would also have to solve low level control theory problems or kinematic and dynamics equations. These latter issues are not considered here: We refer to (Brady et al., 1983) for a discussion of these matters.

In view of its centrality, it is not surprising that the motion planning problem has been addressed in several disciplines and is known by other names in the literature: findpath problem, obstacle-avoidance, collision-avoidance, or movers' problem. Recently, several papers addressing this problem represent a marked departure from the earlier approaches. The characteristic marks in this line of work are those of computational geometry:

- Motion planning is formulated purely in geometric terms. This allows a deeper study of the inherent mathematical structure of the problems.

- Various asymptotically efficient techniques drawn from the study of algorithms and data-structures are employed. Complexity theory also sheds considerable light on the inherent complexity of the motion planning problem.
- The algorithms are combinatorial (nonnumeric) in nature and exact (non-heuristic).

These characteristics also imply certain limitations: posing the problem in purely geometric terms allows us to specify precise solution criteria but it also rules out nongeometric considerations which may be important. Next, the usual caution about asymptotic analysis applies here: although the asymptotic behaviour of an algorithm may be favorable, the algorithm may not perform well for small-sized problems that arise in practice. Finally, although we emphasize the combinatorial behavior of algorithms, there may be subtle interactions between the combinatorial and the numerical properties of algorithms about which very little is known.

Thus, our main purpose is to describe the recent work on algorithmic motion planning which adopts the view-point of complexity theory and computational geometry. We shall try to give a unifying picture of the subject. Most results will be stated without proof: instead, by giving the underlying motivations, and through examples, we hope to facilitate a reading of the original papers. Just in case our treatment creates the impression that motion planning in this setting has only theoretical interests, we remark that efforts have started with the aim of implementing these theoretical algorithms. The present chapter is organized as follows:

Section 1: contains a short history of the motion planning problem
Section 2: introduces the basic terminology
Section 3: gives a taxonomy of problems considered in this paper
Section 4: discusses some side issues
Section 5: shows how to move a disc, which serves to illustrate the retraction approach
Section 6: shows how to move a ladder, illustrating the decomposition approach
Section 7: contains a general account of the two approaches
Section 8: outlines the result of Schwartz and Sharir stating that the general (algebraic) motion planning problem is solvable using the decomposition approach. This shows that the decomposition approach is "universal"
Section 9: observes that the retraction approach is also universal. As an application, we consider a problem of Hopcroft and Wilfong concerning the coordination of many moving objects
Section 10: describes some simple lower bound results in motion planning
Section 11: concludes with some open problems and new directions

1. HISTORY

The following short account is not exhaustive, but it will hopefully put some perspective on the current interest in robotics. There are two archetypes of a robot in practice: one is a multijointed robot arm fixed to some base, and the other is an autonomous mobile automaton. (The anthropomorphic robot in the popular imagination is a combination of these two). Although the engineering issues for the arm and the mobile automaton are different, their motion planning problems fall naturally under a common mathematical framework. We see three lines in the historical development: The first discussion of motion planning arose in connection with work on mobile automata. The work on manipulator arms came somewhat later.[1] Finally, we have the current interest from the complexity point of view.

One of the earliest works on mobile automata was Shannon's in the late 1940s (Sutherland, 1969). Shannon constructed an electronic mouse that moves in a maze, namely a checkerboard in which there may be a wall segment separating any pair of adjacent squares. The maze-searching algorithm of Shannon is extremely simple: Located at each square are two bits of memory indicating one of the four compass directions. After a square has been visited, these bits indicate the direction in which the mouse last left the cell. Shannon's mouse always attempts to leave a square in the direction 90° to the left of the recorded direction (turning by additional 90° each time a wall segment prevents it from leaving). It is easy to show that the mouse will surely find its goal (cheese) if it is at all possible. E. F. Moore (1959) reformulated Shannon's maze problem in terms of graphs, and showed a simple algorithm for finding the shortest path in the graph from a source to a goal node. Further refinements of the approach were made by C. Lee (1961) and S. Akers (1967); in particular Lee described his algorithm using the physical analogy of diffraction patterns. Several later results along this line characterize when a finite automaton (perhaps equiped with some auxiliary memory such as "pebbles" that can be placed or picked up) can search certain graphical mazes; for instance, see Blum and Kozen (1978) for a recent paper. In the 1960s, I. Sutherland (1969) considered the problem of searching an environment that does not easily fall under a graph theoretic formulation, i.e., the geometry of the environment does not naturally divide the space into passageways. He describes an algorithm for tracking spurs: A spur is simply a point of visual discontinuity from a current position. Similar ideas were given in Nilsson (1969), who introduced the idea of the visibility graph. For his Ph.D thesis at Stanford, Moravec (1981) modified an existing roving robot, endowing it with a motion planning ability based on an algorithm for moving a disc amidst a set of obstacles modeled as discs. An influential paper of Lozano-Pérez and Wesley (1979) studied path planning for a polyhedral object amidst polyhedral

---

[1]This is not to say that the construction of manipulator arms comes after mobile automata, but only that the *algorithmic study* of arm motion seems to have come later.

obstables. In the planar case, when the body has a fixed orientation, the planning problem is easily reduced to the motion of a point by using the "growing obstacle" idea (see below). They present some heuristics for generalizing this technique to more complex cases.

According to Shimano (1978) manipulator arms were first developed to handle radioactive materials, or more generally any dangerous material. These are tele-operated, actively controlled by a human. Argonne National Laboratory started one such arm probject in 1947. We shall not be concerned with tele-operated arms since motion planning is not at issue there. For the same reason, we are also not concerned with artificial limbs in prosthetic applications. The current literature on more "intelligent" manipulator arms seems to have begun in the 1960s. Initially, the emphasis was on lower level control questions in kinematics, dynamics, trajectory and servo control (Ernst, 1961; Pieper, 1968; Paul, 1972). Among the early papers on global issues in motion-planning are Widdoes (1974) and Udupa (1977). Udupa studied in detail a planar 2-link arm consisting of a large "boom" that rotates as well as slides back and forth at a fixed "base" point, and a smaller "forearm" that is attached by a hinge-joint to the free end of the boom. The idea of growing obstacles is usually attributed to Udupa: Intuitively, path planning for a robot amidst obstacles can be reduced to moving the robot, shrunken by some amount, amidst the obstacles grown by some corresponding amount. Despite the attraction of this analogy, the concept of "grown obstacles" is essentially limited to the simplest cases of circular or spherical robot bodies.

The current algorithmic interest in motion planning is usually traced to the previously noted work of T. Lozano-Perez and M. Wesley. More recent and continuing work along such lines may be seen in Lozano-Pérez, (1981, 1983). The first paper from the complexity point of view is Reif (1979), which outlined a polynomial time algorithm for moving a polyhedral body, with the error tolerance as a parameter. Reif showed a *PSPACE*-hard lower bound on the inherent complexity of a certain three dimensional motion planning problem. The papers by Schwartz and Sharir (1983a, 1983b) may be said to have put the growing obstacle idea on firm and general mathematical foundations. Intuitively, the "grown obstacle" corresponding to an actual obstacle C may be identified with the positions in the space of robot positions where the robot intersects C. Schwartz and Sharir proceeded to show that the boundary of these grown obstacles are semialgebraic (provided the original problem is semialgebraically defined) and reduce the motion planning problem to a decision problem for the language of Tarski for real closed fields. In subsequent papers (Schwartz & Sharir, 1983c; Sharir & Ariel-Sheffi, 1983; Schwartz & Sharir, 1984), they extended their analysis to the case of several disjoint discs, "k-spiders" (k line segments placed in a star-like configuration), and a rod in three dimensions. The work by Hopcroft, Joseph, & Whitesides, 1982) contains *PSPACE*-hard and *NP*-hard lower bounds for various motion planning problems in the plane (extending the lower bound of Reif) and studied an interesting problem of moving an m-link arm in a circular region. The efficiency of the latter problem has been improved in Kantabutra and Kosaraju (1984).

The use of Voronoi diagrams for motion planning first appeared in the thesis of P. Rowat (1979), who used them as a heuristic for motion planning in a digitized space. Ó'Dúnlaing and Yap (1982), Ó'Dúnlaing, Sharir, and Yap (1984a, 1984b) introduced the idea of retraction as a general method for motion planning and rediscovered the use of generalized Voronoi diagrams as retracts. Heuristic ideas similar to retraction were proposed by R. A. Brooks (1982, 1983). Brooks has implemented his algorithms which are reported to be extremely fast in practice. Yap (1984a) presented retraction algorithms for coordinating the motion of two or three discs that improves on the decomposition algorithms in Schwartz and Sharir (1983c); Yap also points out that the retraction approach is not restricted to the use of Voronoi diagrams. Thus, the work of Hopcroft and his coworkers can also be construed as falling within the retraction framework. See Hopcroft and Wilfong (1984a, 1984b) for studies on the coordinated motion of many rectangles. In particular, their result showing the existence of certain forms of motion may be seen as a theorem about the retraction method (see section 9 of this chapter and also the chapter by Hopcroft and Krafft). Lower bounds on such coordinated motion planning can be found in Hopcroft, Schwartz, and Sharir (1984), Spirakis and Yap (1984).

There are several papers that deal with the problem of finding the shortest path. For instance, Lozano-Pérez and Wesley (1979), Sharir and Schorr (1984), Baker (1985). These results concern "shortest" in terms of distance. In general, the concept of shortest is meaningful only in the special cases where the robot is a point, disc or sphere. A possible generalization is to consider "minimum time" paths subject to velocity or other constraints; but this encounters the difficult issues of kinematics and dynamics.

2. BASIC CONCEPTS

We establish some basic terminology used in this chapter. Elementary concepts of point set topology will be assumed.

Physical Space S.   It is assumed that our robot system B lives in a physical space S. The term "physical" need not be taken literally. Mathematically S corresponds to the Euclidean space of dimension two or three. In general, let $\mathscr{E}^k$ denote the k-dimensional Euclidean space. S is assumed to be $\mathscr{E}^2$ or $\mathscr{E}^3$ here since these are the most interesting cases. We are interested in the set $\mathfrak{R}$ of rigid transformations of S, namely transformations that preserve distance and orientation. $\mathfrak{R}$ is also known as the Euclidean Group. It is well known that a rigid transformation can be written as a translation followed by a rotation.

Example 1. In $\mathscr{E}^2$, the rigid transformations form a 3 dimensional space where each transformation Z can be specified by a triple (x, y, θ) where $(x, y) \in \mathscr{E}^2$ and $0 \le \theta < 2\pi$ is an angle. Thus (x, y) specifies a translation and θ a rotation. A useful way of representing rotation by angle θ is the 2×2 matrix

$$rot(\theta) = \begin{pmatrix} \cos\theta & -\sin\theta \\ \sin\theta & \cos\theta \end{pmatrix}.$$

Then a point $p = (a, b) \in \mathscr{E}^2$ is transformed by (x, y, θ) to the point

$$(x, y) + rot(\theta) \cdot p = (x + a\cos\theta - b\sin\theta, y + a\sin\theta + b\cos\theta).$$

In general, a rotation in $\in^k$ can be represented by an orthogonal[2] $k \times k$ matrix R with real entries and having determinant 1. Then a rigid transformation Z can be represented by a pair (t, R) where $t \in \mathscr{E}^k$ represents a translation and R is a rotation matrix. A point $p \in \mathscr{E}^k$ is transformed by $Z = (t, R)$ to $t + Rp$. In $\mathscr{E}^3$, the set of pure rotations (resp. translations) has three degrees of freedom so that $\mathfrak{R}$ is 6 dimensional. It is often convenient to represent each point $(x, y, z) \in \mathscr{E}^3$ by its homogenous coordinates $p = (wx, wy, wz, w)$, $w \ne 0$, and the rigid transformation (t, R) by a certain 4×4 matrix M such that the transformation of p by M is given by Mp (e.g., see Paul (1981). •

Rigid Bodies and Their Placements. A *rigid body* B in S is a compact connected subset of S. Let Z be a rigid transformation. For any point $p \in S$, write $p[Z]$ for the Z-transformation of p. Then for any body (or any subset of S) B we write $B[Z]$ for the set

$$\{p[Z]; p \in B\}.$$

Example 2. In all our examples, including the present one, S is assumed to be $\mathscr{E}^2$ unless otherwise noted. Suppose B is a disc centered at the origin. For any point $p \in \mathscr{E}^2$ and placement $Z = (x, y, \theta)$ we have $p[Z] = (x, y) + rot(\theta) \cdot p$ (see example 1). It is plain that $B[Z]$ is a disc centered at (x, y). Since $B[Z]$ is independent of θ we may as well regard $\mathfrak{R}$ in this case as consisting of pure translations. •

We will use the somewhat unorthodox terminology of *placement* to refer to rigid transformation Z: The rationale is that a transformation Z is usually seen as a way of locating a body B in physical space. Thus we read "$B[Z]$" as "the placement of B at Z."

Robot System and Admissible Placements. We define a *robot system* to be a pair $(B, \mathscr{A})$ where $B = (B_1, B_2, \ldots, B_m)$ is an m-tuple of $m \ge 1$ rigid bodies, and $\mathscr{A}$ is a subset of $\mathfrak{R}^m$. Usually $\mathscr{A}$ is implicit, and we refer to B as the robot

---

[2]A square matrix $R = (a_{ij})$ is orthogonal if its inverse is equal to its transpose $R^T = (a_{ji})$.

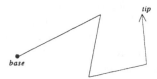

FIG. 3.1. A 4-link arm.

system itself. A *placement* of B is an m-tuple $Z = (Z_1, \ldots, Z_m) \in \Re^m$. Each $Z \in \mathscr{A}$ is an *admissible placement* of B; and $\mathscr{A}$ is the *space of admissible placements* of B. If $Z = (Z_1, \ldots, Z_m)$, we write $B_i[Z]$ ($i = 1, \ldots, m$) for $B_i[Z_i]$ and $B[Z]$

for the union $\bigcup_{i=1}^{m} \cup B_i[Z]$. Each B_i is a *component* of B and if B has only one

component then B is a *rigid* robot.

Example 3. An *arm* is a nonrigid robot system consisting of a sequence $(B_1, \ldots, B_m)$ of line segments called *links*. One end of B_i is attached to B_{i-1} (with the other end to B_{i+1}) by a *joint* which is of the hinge (or revolute) type. Another common joint is the sliding (or prismatic) type. The unattached ends of the first and last links are called the *base* and *tip* (or *hand*) of the arm. It is assumed that the base is fixed at the origin. The space of admissible placements $\mathscr{A} \subseteq \Re^m$ consists of those placements that obey the joint constraints and also the property that the links do not intersect each other except at the joints. This is more carefully described next. (See Fig. 3.1.) •

It is important to note that $\mathscr{A}$ is typically specified by a conjunction of constraints or *couplings* between pairs of components B_i and B_j. Constraints that simultaneously involve more than two components are rare. The following are some possible couplings.

(i) "A point p in B_i is attached to a point q in B_j." This simply says that $Z = (Z_1, \ldots, Z_m) \in \mathscr{A}$ must satisfy the constraint $p[Z_i] = q[Z_j]$. Observe that in $\mathscr{C}^2$, this coupling specifies a hinge joint. In $\mathscr{C}^3$, this is a universal joint; furthermore, it is easy to see how we can use two such couplings to specify a hinge joint (pick any two points p in the axis of the joint).

(ii) "An axis L in B_i coincides with an axis L' in B_j." (Here, an axis is simply a line.) This says that $L[Z_i] = L'[Z_j]$ and in $\mathscr{C}^3$ this amounts to a sliding-cum-rotating joint. Two parallel lines would specify a purely sliding joint.

(iii) "A point p in B is immobilized." This says $p[Z] = c$ for some constant point c. Thus this constraint can be used to fix the base of an arm at the origin.

(iv) "B_i and B_j must not intersect." This says $B_i[Z_i] \cap B_j[Z_j] = \emptyset$.

(v) "B is confined to a region $R \subseteq S$." This says $B[Z] \subseteq R$.

We might add that the couplings (i–iii) correspond to what are called holonomic constraints in mechanics, but (iv–v) are nonholonomic. The holonomic constraints effectively reduce the degrees of freedom of the set of admissible placements. Our formulation of admissible placements $\mathcal{A}$ as a subset of $\mathfrak{R}^m$ may not always be the most convenient. For example, in case of a m-link arm connected by hinge joints, it is more convenient to parametrize the admissible placements by m angles. Thus the $3m$ parameters of $\mathfrak{R}^m$ are replaced by m parameters. The use of angles here is an example of 'generalized coordinates' (again, borrowing a term from mechanics). In general, instead of using the parameters of $\mathfrak{R}^m$, we may parametrize the space $\mathcal{A}$ using generalized coordinates. The advantage is that, if the generalized coordinates are appropriately chosen then all or some of the holonomic couplings can be automatically taken care of. In the ideal case, where all the couplings are automatically satisfied in the generalized coordinate system, the set of placements in the generalized coordinates is equal to the set of admissible placements. We will not try to formalize this alternative formulation of $\mathcal{A}$ in terms of generalized coordinates but it will be clear how this is done in each particular case.

Environment and Space FP of Free Placements.   By an *environment* we mean a closed subset E of S. Each connected component of E is call an *obstacle*. The definition of "environment" is different from that for a "body" in that we do not require E to be a connected or bounded set. Indeed, we often insist that the complement of E be bounded. The most commonly analyzed environments in $\mathscr{C}^2$ are polygonal, i.e., those whose boundary ∂E can be decomposed into straight line segments.

Relative to E and a robot system B, we define an admissible placement Z of B to be *free* if $B[Z] \cap E = \varnothing$. Let $FP = FP(B, E)$ denote the set of free placements. By definition, $FP \subseteq \mathcal{A}$. Since S is a metric space, we can endow FP with a metric induced by the Hausdorff metric[3] on closed subsets of S. More precisely, the distance $d(Z, Z')$ between $Z, Z' \in FP$ is defined to be the Hausdorff distance between $B[Z]$ and $B[Z]$.

Sometimes it is convenient to consider the set of *semifree* placements, defined as those Z such that the interior of E has empty intersection with $B[Z]$. Let $SFP = SFP(B, E)$ be the set of semifree placements. It is clear that the closure of FP is contained in SFP. The reverse containment does not generally hold: in illustration, if B is a disc and Z is a semifree placement such that $B[Z] \cap E$ consists of three points forming an equilateral triangle then Z is not in the closure of FP.

It should be noted that the present formulation assumes the environment to be fixed (immobilized), but the general theory can easily allow mobile obstacles

---

[3]If X is any metric space with distance function d, then we can define the Hausdorff distance $d(A, B)$ between two non-empty closed subsets A, B of X. Note that we have used the symbol d for both the Hausdorff metric as well as the original metric because the original metric may be regarded as the special case of the Hausdorff metric restricted to singleton sets. For any set $S \subseteq X$ and $\epsilon > 0$, let S_ϵ denote $\{p : (\exists q \in S)\ d(p, q) < \epsilon\}$ where $d(p, q)$ is the metric on X. Then the Hausdorff distance is $d(A, B) = inf\{\epsilon > 0 : B \subseteq A_\epsilon \text{ and } A \subseteq B_\epsilon\}$.

(such as chairs). An even more general view is to allow the environment and robot to undergo continuous deformations (Hopcroft & Wilfong, 1984a).

Motion and Clearance.   A *path* or *motion* of B between two placements Z and Z' is defined to be a continuous function on the unit interval, $\pi : [0, 1] \rightarrow FP$, such that $\pi(0) = Z$ and $\pi(1) = Z'$. Occasionally it is convenient to treat semifree motion, i.e., $\pi : [0, 1] \rightarrow SFP$. It is important to realize that in the present work we do not necessarily view the domain of π as a "time" parametrization (but see section 4). The *clearance* of a placement Z of B with respect to an environment E is defined as $inf \{d(p, q): p \in B[Z] \text{ and } q \in E\}$, and denoted by $Clearance_{B,E}(Z)$ or simply $Clearance(Z)$. The *clearance* of a motion π is defined as

$$Clearance (\pi) = inf \{Clearance(\pi(t)) : t \in [0, 1]\}.$$

Complexity.   We make the usual assumptions of the field of computational geometry with regards to the use of infinite-precision real numbers in algorithms, and counting as unit cost each arithmetic or other simple operations. References to complexity theory are mainly isolated to section 10, so they will not disrupt the understanding of the rest of the article. In any case, we refer the reader to the introduction to this volume where some further background to the subject is given. In particular, we will mention the tractable-intractable dichotomy several times in this chapter.

Reading Guide.   Sections 5, 6, and 7 do not depend on the next two sections; the reader may skip directly to section 5 and following, if she so desires, in order to read some real results (rather than more definitions and discussions). Section 8 depends on section 3 only to the extend that it describes a solution to the general motion planning problem (P1) as defined in section 3.

3.   A TAXONOMY OF MOTION PLANNING PROBLEMS

The *general motion planning problem* can be given the following schema:

(P1)   Given a robot system B, an environment E, and a motion objective O, find a motion π for B amidst E that achieves O subject to some optimality criterion C (or report that no motion exists.)

The criterion C is fixed for the problem and hence the input to the algorithm is a triple (B, E, O), suitably restricted to a class of possible inputs. Indeed, the criterion C is often not imposed at all. The motion objective O can be regarded as a set (perhaps implicitly defined) of motions, and the task of the algorithm is to choose some motion from O satisfying C. To make (P1) into a precise algorithmic problem, we must describe the class of possible inputs (B, E, O) and also C.

Example 4. Suppose B is an m-link arm specified by an m-tuple of numbers denoting the length of each link in the arm. A placement of B will be specified by a sequence of m orientations. We consider polygonal environments E where the boundary of E is a disjoint collection of closed simple (i.e., not self-intersecting) polygons; each polygon is specified as a sequence of vertices. The motion objective O is specified by a pair (Z_1, Z_2) of initial and final placements of B, indicating that we want to find a motion from Z_1 to Z_2. The criterion C specifies that the total distance traversed by the tip of the arm is to be minimized. •

There are many ways to make (P1) explicit and they lead to different families of problems. The following is an attempt to classify the main distinctions.

(1) *Dimension of the physical space S.* Perhaps the most fundamental dichotomy is between problems whose underlying physical space S is $\mathscr{E}^2$ versus those whose space is $\mathscr{E}^3$. Despite the fact that real robots live in three dimensional space, two dimensional problems are often sufficiently realistic. For instance, in planning the motion of roving robots, it is usually sufficient to represent the environment by a two dimensional "floor-plan."

(2) *Geometry of the space of admissible placements $\mathscr{A}$.* The dimension of $\mathscr{A}$ is clearly a critical characteristic. However, the geometry of $\mathscr{A}$ is profoundly affected by the physical nature of the robot B. By this we refer to the geometry of the components of the robot system and how they are coupled. We note three main types:

   (a) *Rigid robot:* this is the simplest case where B has only one component. The commonly investigated bodies (in 2-dimensions) are where B is a disc, a line segment (called a "ladder" or "rod"), a rectangle or a convex polygon.
   (b) *Serially coupled robot:* here the components can be ordered as $B_1, \ldots, B_m$ where B_i is coupled to B_{i-1} and B_{i+1} only. The main example is an arm, already seen in previous examples.
   (c) *Loosely coupled robot:* the robot components are essentially independent of each other in that the only mutual constraint is non-intersection among themselves. Of course, this can be generalized to groups of components. Motion planning of this sort will be called *coordination* problems, since we imagine several independent robots trying to cooperate on a common task.

(3) *Boundaries of bodies and environment.* To represent the bodies and environment we shall assume that their boundaries are composed of finitely many segments of algebraic curves (in $\mathscr{E}^2$) or patches of algebraic surfaces (in $\mathscr{E}^3$). This is of sufficient generality to cover all cases that have been studied. Indeed, in the majority of cases, the bodies and obstacles are polygonal or polyhedral.

(4) *Motion objective.* The simplest is where one is given an initial placement Z_1 and a final placement Z_2 of B, and it is required to find a motion from Z_1 to Z_2. A more general version is where Z_2 is replaced by a set G of placements. Here it is required to find a motion from Z_1 to *any* placement in G (the goal set). For example, the typical goal of motion planning for a robot arm is the positioning of its tip (or hand). The goal set G in this case can be specified by a point $p \in \mathscr{E}^2$ denoting the set of all placements of B in which the hand of B is at p.

(5) *Optimality criteria.* We mention three natural criteria for the solution motion π:

  (a) The trivial case is where no criterion is imposed, so any π satisfying the motion objective is acceptable.
  (b) The semifree motion π is 'shortest' among those satisfying the motion objective. As noted in section 1, this criterion is mainly useful for robots that are discs or spheres. Also note that we resort to 'semifree motion' in this case because under typical assumptions, the space of free motions are not closed.
  (c) The motion maximizes the minimum clearance, among those motions satisfying the motion objective.

More complicated criteria may be posed. E.g., the motion is the shortest among those achieving the maximum possible clearance.

(6) *The basic planning problem and local planning.* In (P1), B ranges over some class of robot systems. We shall see that if the robot systems in the class have arbitrarily many degrees of freedom, there is strong evidence from complexity theory that such problems are computationally intractable. These negative results need not be interpreted as insurmountable difficulties. The simple reason is that there is little practical need to do motion planning for general classes of robot systems. In view of this, the restriction of (P1) where the robot system B is fixed is called the *basic planning problem.* An important result in this subject (see section 8) is that, under rather general conditions, the basic problem is computationally tractable. Unfortunately, the degree of the polynomial bounding the running time increases linearly with the degrees of freedom of B. To cope with this in practical systems, we may use the following observations:

  (a) For most environments, motion planning for a complex robot B can usually be reduced to that of a simpler system B'. For example, if B is a rigid body moving in a relatively uncluttered environment (such as the middle of a large room) we can usually enclose B by a disc B' and plan for B' instead. We call this the "simplification heuristic."
  (b) When the heuristic of (a) fails to find a motion when in fact one exists, it can be attributed to the failure of finding a motion for B' through a low clearance

locality. Such localities often correspond to stereotype situations such as at a door or a turn in a corridor. Thus we can complement the simplification heuristic with further heuristics that attempt to recognize certain stereotype situations and invoking more sophisticated algorithms to deal with them. We may call this the *local expert heuristic* (to borrow a term of high currency in artificial intelligence).

This justifies consideration of what we call the *local planning problem:* instead of fixing *B,* we fix the environment *E.* The resulting algorithm is thus an expert for moving in the environment *E.* (Note the duality between this and basic planning problems.) Thus we can reduce the planning problem for a complex system *B* to the planning problem for a simpler *B'* together with several 'local planning experts'. The important point is that the combined complexity of the simplified planner for *B'* and the local planners is *additive* rather than multiplicative, which should be considerably less than a full-scale algorithm for *B.* An example of a local planning problem is (Yap, 1984b) where we consider the problem of moving a chair (simple polygon) through a door (a gap in an infinite wall). Of course, this raises the interesting new algorithmic issue of identifying the occurrences of the stereotype localities in a given *E.*

(7) *Decision problems.* The schema (P1) has been formulated as a path finding problem. We can also pose the problem as asking for a yes-no response: instead of asking for a path, we ask if there exists such a path. These are called *decision problems* and complexity theory is normally set up to treat such problems only. Hence when we discuss lower bounds in chapter 10 (using the tools of complexity theory), we treat decision problems. Clearly the decision problem is no harder than the path finding version; on the other hand, the converse has not been shown. For instance, if the complexity of the decision problem is $t(n)$, does it follow that the path finding problem is $O(t(n))$?

4. ISSUES RELATED TO ALGORITHMIC MOTION PLANNING

All the preceding motion planning problems are *off-line* in the sense that planning takes place with complete *a priori* knowledge of the environment. In the *on-line* (or *real-time*) version of the problem, the environment might be partially unknown, and the algorithm would have to discover or interact with the robot's environment by periodically issuing movement and sensing commands, while subsequent actions would be based on the feedback from these commands. Vision and other sensory capabilities would serve the purpose of discovering the environment. For a precise algorithmic problem of this kind, see Cole and Yap

(1983). In any case, there are a variety of situations where our off-line formulation is the appropriate one. For instance, in planning the layout of a factory environment the (human) planner has complete control of the placements of obstacles and robots; with an efficient off-line algorithm, the planner can easily test robot motion for alternative layout plans.

A second remark is that our algorithms find "physical" paths only; the question of timing or velocity is not considered. To realize motion along the given path, one must be able to translate this into actuator commands. Typically (see Brady et al., 1983), this is done by sending to a *feedback controller* a stream of (position, velocity)-values at a servorate of 20 to 200 Hz. One way to organize this is to postulate a *trajectory planning algorithm* that transforms the physical path given by our motion planning algorithm into the controller input stream:

In this framework, we face another algorithmic problem that we may call "trajectory planning":

(P2)   Given a physical path in S, plan a trajectory that approximates the path while achieving some optimality criteria.

A *trajectory* (like a motion) is defined as a map $\pi : [0, 1] \rightarrow FP$ except that the domain is now interpreted as time and the first time-derivative $\dot{\pi}$ exists, is continuous and piecewise differentiable. The solution of kinematic equations, etc., becomes essential in the trajectory planning. To make the trajectory realistic, we may restrict the trajectory to some reasonable class of functions (say, cubic splines). See, for instance, Rajan (1985). Since we consider motion planning to be a global issue, in some sense, we may rule out (P2) as our main concern; but see Brady et al. (1983) for recent progress in solving (P2). However, these considerations raise the following global question which may be called the "kinematic motion-planning problem":

(P3)   Given a robot system B and an environment E, plan a trajectory (or kinematic path) between two specified initial and final kinematic states of B, subject to some kinematic criteria C.

By a "kinematic state" of B we refer to a (placement, velocity) pair; if the trajectory is parametrized as $(x_1(t), \ldots , x_m(t))$ then the kinematic state at time t is $(x_1(t), \ldots , x_m(t), \dot{x}_1(t), \ldots , \dot{x}_m(t))$. The criteria C may specify that the kinematic path has a certain form, has bounded (min-max) velocity and acceleration, etc. We can view (P3) as a combination of the first two boxes ("motion planning" and "trajectory planning") in the above diagram. Undoubtedly this is an important question although little is known about its computational complexity.

Finally, recall that we assume an idealized world where numerical computations are error-free, i.e., infinite precision numbers are used. Under this model, we prove that, for example, our algorithms are guaranteed to find a solution if one exists. In practice there are several sources of imprecision: (i) imprecision of the input data, (ii) actual numerical computations are carried out to a bounded precision, (iii) the robot control and sensors have limited accuracy. These arguments do not dismantle the validity of our approach: ours is a fiction justified in the name of theoretical understanding, much as physics justifies the use of point masses. However they advise caution when applying our results. The inaccuracies of these algorithms are a legitimate topic of study in their own right but beyond our present scope. Even if an account of the inaccuracies is not readily available, one can obtain empirical data on the performance of these algorithms.

5. MOVING A DISC

In this section, we consider a simple nontrivial motion planning problem: moving a circular disc B amidst an environment E of polygonal obstacles. One of our aims here is to illustrate the retraction approach to motion planning. The algorithm for a disc also has independent interest. In particular, the relative efficiency of the disc algorithm means that the algorithm can be used as an important simplification heuristic (see section 3) for moving more complex bodies in the plane. We impose no optimality criteria, so any path will do. Our exposition follows that in Ó'Dúnlaing and Yap (1982).

We assume that B is a fixed disc of radius $r > 0$. A placement p of B may be uniquely identified by the location of the center of the disc.[4] The path objective is specified by a pair p_0 and p_1 denoting the initial and final placements of B. The environment E is polygonal and can be represented by the set of the polygonal paths forming its boundary. We may take the size of the input to be n if there are n edges in the boundary of E. (See Fig. 3.2.)

Let $\Omega = \mathscr{E}^2 \sim E$ denote[5] the complement of E. For simplicity we assume Ω to be bounded although it is easy to remove this assumption if desired. The boundary $\partial\Omega$ of Ω is partitioned into a set S of pairwise disjoint sets consisting of open line segments (called *walls*) and singleton sets (called *corners*).[6] Hence each wall is incident to two corners. We also assume E is "full-bodied" (i.e., E is the closure of its interior); this implies that each corner is incident to at least two walls. An *object* refers to a *wall* or a *corner*. A placement p is *free* if $B[p] \subseteq \Omega$: It

---

[4]In general, recall that a placement is a rigid translation; but here, the translation that moves a disc from its 'standard' position to location p is identified with p. This is because, as noted in example 2, the rotational component is irrelevant for a disc.

[5]We write $X \sim Y$ for the set theoretic difference of two sets.

[6]By abuse of language, we often treat a corner $\{p\}$ as a point p.

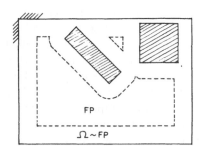

FIG. 3.2. Walls and corners and
the boundary of *FP*.

easily follows that the set *FP* of free placements may may regarded as a subset of
Ω. In fact, the boundary ∂FP of *FP* is seen to be equal to the set of points p such
that p is at distance r from E.

To motivate the retraction approach, let us examine a very simple local situation:

1. Suppose s_1, s_2 are two (sufficiently long) walls in S. If the disc were to "navigate between s_1 and s_2" then it seems natural that the center of the disc should
move along the bisector of the angle formed by the two lines containing s_1 and s_2.
(See Fig. 3.3.)

Again:

2. To navigate between two corners s_1 and s_2, it is natural to expect the center of
the disc to trace out part of the bisector of the segment joining s_1 and s_2.

To make these observations into a general remark, let us introduce the concept of
the *clearance circle* at a point p: this is the circle centered at p with radius equal
to the (Hausdorff) distance $d(p, E)$. If the clearance circle at p intersects an object
s then we say that p is *near* to s. We may characterize the locus traced by the
center of the disc in (1) and (2) as the set of points p that is near to both s_1 and s_2.
Using this characterization, it is not hard to extend observations (1) and (2) to:

3. If the disc were to navigate between a corner s_1 and a wall s_2, then the path
would be part of the parabola defined by s_1 as the focus and the line containing s_2 as
the directrix.

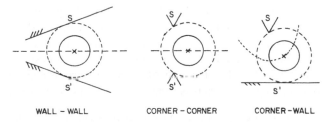

WALL – WALL   CORNER – CORNER   CORNER –WALL

FIG. 3.3. Navigating between two objects S and S'

In general, consider the points $p \in \Omega$ such that the clearance circle at p intersects at least two objects: the set of such p is called the *Voronoi diagram* of Ω and denoted $Vor(\Omega)$.[7] This diagram has been studied in many other contexts and its importance in computational geometry was first pointed out in Shamos and Hoey (1975). We refer the reader to the chapter by Leven and Sharir in this volume for the basic properties of Voronoi diagrams and other applications. From the preceding discussion, it is not surprising to know that the Voronoi diagram can be decomposed into a set of straight or parabolic curve segments called *Voronoi edges*. Each Voronoi edge is a component of the set of those points p such that, for some objects s_1, $s_2 \in S$, $p \in Vor(\Omega)$ is near to s_1 and s_2. The endpoints of a Voronoi edge are called *Voronoi vertices*. For computational purposes, the diagram is represented as a planar graph $G = (V, A)$ where V is the set of Voronoi vertices and A the set of Voronoi edges. With each edge of A we associate the two objects that are near to that edge; this also gives us the equation of the line or parabola containing the edge. The importance of $Vor(\Omega)$, both here and in its other numerous applications, comes from the following two facts:

(i)   It is small: there are only $O(n)$ edges in $Vor(\Omega)$. See Kirkpatrick (1979), Shamos and Hoey (1975).

(ii) It can be efficiently computed: Recently, Yap (1985) describes a relatively simple $O(n\log n)$ algorithm. In Kirkpatrick (1979), a more complicated $O(n\log n)$ algorithm is outlined. An earlier $O(n\log^2 n)$ algorithm is also available (see Lee & Drysdale, 1981).

The local observations (1)–(3) above are based on the intuition that each point on the Voronoi diagram has clearance that is "locally maximal." A point p *not* on the Voronoi diagram is, by definition, near to some unique object s; thus we could move p directly away from s thereby increasing the clearance. More pertinently, if π is a path in *FP* then we can locally perturb the path in this way until π (except for the initial and final sections of the path) is contained in the Voronoi diagram. This suggests the idea that finding a general path between two placements can be reduced to finding a path in $Vor(\Omega)$. We proceed to formalize this idea.

Let p be a point not on the Voronoi diagram and let $s \in S$ *be the unique object near to* p. Further, let $q \in s$ be the point such that $d(p, q) = Clearance\ (p)$. Consider the half-line from q through p. If we move B along this half-line away from q then the clearance is initially increasing: in fact this is the direction where the clearance increases at maximum positive rate. Since we assumed that Ω is bounded, the clearance cannot increase indefinitely. So there must be a first point denoted $Im(p)$ at which either the rate of increase becomes negative or the direction of maximum positive increase changes. It is not hard to see that $Im(p)$ is

---

[7]This is sometimes called a "generalized" Voronoi diagram since the usual definition of the diagram is when E (complement of Ω) is a finite set of points.

near to some other object $s' \neq s$. This implies that $Im(p)$ is on the Voronoi diagram. We call $Im(p)$ the *image* of p. We may extend this definition to all of *FP* by defining $Im(p)$ to be p if $p \in Vor(\Omega)$. We have therefore introduced a function

$$Im: P \rightarrow Vor(\Omega)$$

which is the identity on $Vor(\Omega)$. A fundamental property of this map is:

Lemma 1. The map *Im* is continuous.

Thus the image map is a *retraction* of *FP* and the Voronoi diagram is a *retract*.[8] We can now apply the image map to any motion $\pi : [0, 1] \rightarrow FP$ to obtain another motion $Im \circ \pi : [0, 1] \rightarrow Vor(\Omega)$. Thus if π is a motion from p to q then $Im \circ \pi$ is a motion from $Im(p)$ to $Im(q)$. It is not hard to see that if p is free, then there is a straight line motion in *FP* from p to $Im(p)$. Thus we have shown:

Theorem 2. Let p, q be free placements for a disc B in a polygonal environment E. Then there is a motion for B from p to q if and only if there is a motion within the Voronoi diagram $Vor(\Omega)$ from $Im(p)$ to $Im(q)$.

This justifies the following ''retraction'' algorithm which takes as input the quadruple (r, E, p, q) where r specifies the radius of B.

(1) (Preprocessing) Compute the Voronoi diagram $Vor(\Omega)$ where $\Omega = \mathscr{E}^2 \sim E$.

(2) (Check freeness) First verify that p and q are free, otherwise there is no motion. The naive method to check whether p is free is to compute the distance of p to each wall and take the minimum: p is free if this minimum is greater than the radius of the disc.

(3) (Retract to the Voronoi diagram) Compute the points $Im(p)$ and $Im(q)$. If the previous step was done using the suggested naive method then we also know the objects near p (say). Suppose p is not on the Voronoi diagram. Hence it is nearest to a unique point q on some object s. Then $Im(p)$ is the first intersection of the ray R from q through p with $Vor(\Omega)$. The straightforward method to compute this is to intersect each Voronoi edge with R, taking the intersection closest to p.

(4) (Locate in Voronoi diagram) Determine the edges e and e' in $Vor(\Omega)$ that respectively contain $Im(p)$ and $Im(q)$. [If $e = e'$ then either there is a direct motion in e connecting $Im(p)$ and $Im(q)$ or else we should split the edge e into two parts containing respectively $Im(p)$ and $Im(q)$. We leave this detail to the reader.]

---

[8]In topology, a function $f : X \rightarrow Y$ from a topological space X to a subspace $Y \subseteq X$ is called a *retraction* if f is continuous and is the identity on Y. Y is a *retract* of X if there exists a retraction from X to Y. As we shall see, the continuity requirement is not critical for motion planning.

(5) (Graph search) Do a depth-first or breadth-first search of the edges of $Vor(\Omega)$ to find a motion from $Im(p)$ to $Im(q)$. During the search, an edge e'' of $Vor(\Omega)$ can only be traversed if the minimum clearance along e'' is greater than the radius of B. For each e'', this is easily checked in constant time.

Except for the preprocessing step, all the others clearly take $O(n)$ time; $O(n\log n)$ time is sufficient for the preprocessing step. An important feature recommending this algorithm is that the preprocessing step depends on the environment alone. Hence, if the environment is fixed, subsequent motion planning for different values of r, p, q takes only linear time. Furthermore, if the environment changes only slowly or incrementally (say the change involves the insertion or deletion of $O(1)$ objects at a time) then there are known methods of updating the Voronoi diagram in linear time.

Other results: We can generalize the environment E such that the boundary of E is composed of circular arcs as well as line segments. The Voronoi diagram is now made up of conic segments and can still be computed in $O(n \log n)$ time (Yap, 1985). If E is defined by a set of n arbitrarily intersecting circles (as in Moravec, 1981) then the Voronoi diagram can be computed in $O(n\log^2 n)$ (Sharir, 1983). In both these extensions, the search time remains $O(n)$.

Shortest paths. It is interesting to note an even simpler problem than that of moving a disc: moving a point p amidst an environment E of polygonal obstacles. To find *any* obstacle-avoiding path between two specified points p_0, p_1 can easily be done in $O(n\log n)$ time as follows: using a standard technique in computational geometry, sweep a *scan-line* across the plane to determine for each corner x on the boundary of E, the nearest walls $a(x)$ and $b(x)$ of E that are (respectively) immediately above and below x. The scan-line sweep process takes $O(n\log n)$ time. It is easy to construct a path between p_0 and p_1 using this information in linear time. To make the problem for moving a point more interesting, we usually ask for the *shortest* path between p_0 and p_1. Unfortunately this seems to make the complexity quadratic in general. In fact, if we consider the shortest path problem in three-dimensions, there is no known polynomial-time algorithm (Sharir & Schorr, 1984). We have already noted that the notion of "shortest" path is a limited one in motion-planning since it is in general not meaningful. The appropriate generalization is to consider the minimization of physical quantities such as time or energy subject to other physical constraints; unfortunately this brings us into direct contact with difficult problems in kinematics and dynamics.

6. MOVING A LADDER

The next problem we consider is that of moving a closed line segment B between two specified placements amidst polygonal obstacles. The aim is to illustrate the decomposition approach to motion planning which appeared in the previously

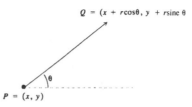

FIG. 3.4. A ladder at placement
(x, y, θ).

mentioned series of papers by Schwartz and Sharir.[9] Assume that the ladder B has length $r > 0$. One end of the ladder will be called the P-end and the other the Q-end. A placement of the ladder is denoted by (ix, y, θ) or (p, θ) where $p = (x, y)$ gives the P-end of B at placement (x, y, θ) (see Fig. 3.4). Again we have the polygonal environment E, the complement Ω of E, the set S of objects (= walls and corners) derived from E, and the space of free placements $FP = FP(B, E)$. The basic idea of the decomposition approach is to divide FP into relatively simple *cells*. We only require these cells to be (pathwise) connected. Let us proceed to see one way of defining these cells.

We first define the *plane projection* of a placement $Z = (x, y, \theta)$ to be the point $P(Z) = (x, y)$. If p is any point in Ω, the 'fibre' at p is the set of free placements whose plane projection is p. An orientation θ is *admissible* at p if (p, θ) is in this fibre. Let $\Theta(p)$ denote the set of admissible orientations at p. Clearly $\Theta(p)$ is partitioned into a finite number of open angular intervals (θ_1, θ_2). It is intuitively clear that in general, $\Theta(p)$ varies continuously as p varies in a small local neighborhood. To analyze this carefully, we consider the set $\Theta^*(p)$ of *semi-admissible* orientations at p where θ is semi-admissible at p if (p, θ) is semi-free. Clearly $\Theta^*(p)$ is partitioned into a finite number of closed angular intervals $[\theta_1, \theta_2]$. An orientation in the finite set $\Theta^*(p) \sim \Theta(p)$ is a *limit orientation* at p. Clearly, the ladder at a limit orientation θ touches some object s.

Definition.

(i)   Let θ be a limit orientation at p. If the ladder at (p, θ) touches s then we call s a *clockwise* (resp. *counterclockwise*) *stop* of p (at θ) if for all sufficiently small clockwise (resp. counterclockwise) rotation of the ladder about its P-end, the ladder is no longer semifree.

(ii) A limit orientation θ is *exceptional* at p if the ladder at (p, θ) does not touch a unique stop.

(iii) A point p is *regular* if p has no exceptional orientations.

Thus θ can be exceptional in one of two ways: either (a) by touching more than one stop, or (b) by touching no stops. Observe that the ladder touches no stops if

[9]The presentation and terminology here departs somewhat from those in the original paper Schwartz and Sharir (1983a). For example, one possible point of confusion for readers familiar with the original paper is this: "stops" are objects here while in the original paper they denote orientations.

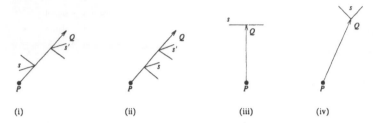

FIG. 3.5.   Exceptional orientations.

the ladder at (p, θ) is perpendicular to an object with the Q-end touching the object (note: the ladder is perpendicular to a corner s if the line perpendicularly through the Q-end of the ladder intersects E only at s for sufficiently small neighborhoods of s.) Fig. 3.5 illustrates this.

It is not hard to prove:

Lemma 3. If p is regular then there is an open neighborhood U about p such that

(i) U consists of regular points,

(ii)   for each p' in U, there is a 1-1 onto correspondence between the angular intervals of $\Theta(p')$ with those of $\Theta(p)$ satisfying the following: if $(\theta_1, \theta_2) \subseteq \Theta(p)$ corresponds to $(\theta'_1, \theta'_2) \subseteq \Theta(p')$ then the clockwise and counterclockwise stops of both intervals are identical.

(iii) $\Theta(p)$ varies continuously (in the Hausdorff metric of sets) as p ranges over U.

In particular, this shows that the set of regular points is open. The two ways for an orientation to be exceptional ((a) and (b) above) lead to (A) and (B) in the next definition.

Definition.

(A)   Let $s \in S$ be an object (wall or corner). Define a point $p \in \Omega$ to be s-*critical* if there is a semifree placement Z whose plane projection is p where $B[Z]$ intersects s at the Q-end; furthermore, if s is a wall then $B[Z]$ is perpendicular to s.

(B)   Let $s, s' \in S$ be a pair of distinct objects. Define $p \in \Omega$ to be (s, s')-*critical* if there is a semifree placement Z whose plane projection is p where $B[Z]$ intersects both s and s'.

It is easy to see that, for each $s, s' \in S$, the set of $s-$ *or* (s, s')-critical points is a disjoint union of curve segments.[10] There are two types of class (A) curve segments:

---

[10]Class (A) (resp. (B)) curve segments correspond to Types (I) and (II) (resp. Types (III) and (IV)) in Schwartz and Sharir (1983a).

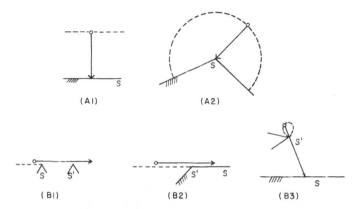

FIG. 3.6.   Class (A) and (B) critical curves.

(A1)   If s is a wall with endpoints s_1, s_2, then an s-critical curve is a segment of a line at distance r from s.

(A2)   If s is a corner, then an s-critical curve is a circular arc of the circle centered at s with radius r.

Similarly, class (B) curves divide into three types:

(B1, B2)   For two corners s, s' or a wall s and one of its endpoint s', the curve is easily seen to be a straight line.

(B3)   For a wall s and a corner s' not incident on s, the curve segment is part of a more esoteric fourth degree curve called the "conchoid of Nicomedes."

Figure 3.6 illustrates these.

It easily follows from the foregoing development that Ω is partitioned by the critical curves into open *regular regions* consisting of regular points. The key definition is the following:

Definition. Let R be a regular region and s, s' be two objects in S. The *cell* denoted $Cell(R, s, s')$ is the set of all free placements Z whose plane projection is in R such that if we place B at Z then:

(i)   if we rotate B clockwise about its P-end point until the first semifree placement Z', then B will touch s at Z'.

(ii)   if we rotate B counterclockwise instead, B will touch s' first.

Note that s and s' may be equal in this definition. The cell $Cell(R, s, s')$ is empty unless s and s' are clockwise and counterclockwise stops for some p in R. To accommodate those placements $Z = (p, \theta)$ where all orientations are admissible, ie., $\Theta(p) = [0, 2\pi)$, we say such a Z belongs to the cell $Cell(R, \lambda, \lambda)$ where λ is a distinguished symbol. A free placement is critical or regular according to

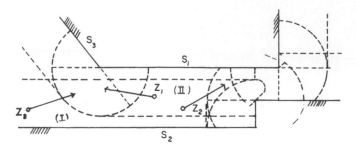

FIG. 3.7.   Critical curves in a simple environment.

whether its plane projection is critical or regular. It is now clear that each regular placement belongs to a unique cell. As we would expect:

Lemma 4. Each cell is a connected open set.

Example 5. Consider the environment shown in Fig. 3.7 (the reader should try to identify all the types of curves). (Note that curves of all classes occur in this example.) Consider the two regular regions labeled (I) and (II): The placements Z_1 and Z_2 have plane projections in region (II) while Z_3 projects into region (I). Consider the cells:

$$C_1 = Cell(\text{II}, s_1, s_2), \quad C_2 = Cell(\text{II}, s_2, s_1), \quad C_3 = Cell(\text{I}, s_2, s_2).$$

Clearly $Z_1 \in C_1$, $Z_2 \in C_2$, and $Z_3 \in C_3$. It is not hard to see that there is no motion from Z_1 to Z_2 without the P-end of the ladder leaving region (II).[11] As we shall see next, this means that C_1 and C_2 are not "adjacent." On the other hand, Z_1 can move to Z_3 then to Z_2 such that the P-end crosses directly from region (II) to region (I) and back again. This means C_3 is adjacent to both C_1 and C_2.

The plane projection of a motion $\pi : [0, 1] \to FP$ is a curve $\tau : [0, 1] \to \Omega$ that we will call the *trace* of π. Suppose there is some regular region R such that the trace of π remains continuously in R. Then it follows from the definitions that π remains continuously in the cell $Cell(R, s, s')$, for some s, s' (possibly $s = s' = \lambda$). Next consider what happens when the trace crosses from one regular region to another. To rule out pathologies, one can appeal to general principles (such as "Sard's Lemma"—see Schwartz and Sharir (1983a) in differential geometry to show the following:

Lemma 5. If there is a motion π' between two regular placements then there is another motion π between the same placements that is *normal* in the sense of satisfying:

(i)   The trace $\tau(t)$ of $\pi(t)$ has a well-defined tangent everywhere along its length.

(ii)   A critical point is called a *vertex* if it is adjacent to more than two regular regions. The trace $\tau(t)$ avoids all vertices and is critical only finitely often.

---

[11]Note that if we allow semifree motion, then it is sufficient for the P-end of the ladder to move to one of the horizontal boundaries of region (II) and back again.

(iii)   From (ii), if $\tau(t_0)$ is a critical point then there is a neighborhood $U \subseteq \Omega$ of
         $\tau(t_0)$ such that the set of critical points in U form a smooth curve β that is
         adjacent to exactly two regular regions. We then require that the trace inter-
         sects β transversally and the orientation of $\pi(t)$ is constant for a sufficiently
         small neighborhood (t_1, t_2) of t_0.

Intuitively, two cells are "adjacent" to each other if there is a motion that
crosses from one to the other. But if we restrict outselves to normal motions, we
can get a more constructive criterion for adjacency as shown in the next definition. But first we need a notation. Let $C = Cell(R, s, s')$ for some R, s, s'. For
each point $p \in R$, let $\phi_C(p)$ and $\psi_C(p)$ denote the limit orientations such that

$$\theta \in (\phi_C(p), \psi_C(p)) \text{ iff } (p, \theta) \in C.$$

In other words, $\phi_C(p)$ and $\psi_C(p)$ define the clockwise and counterclockwise
limits of placements in C whose plane projection is p. We can uniquely extend
the domain of ϕ_c and ψ_c to the closure $\bar{R}$ of R. It is not hard to construct an
example such that there exists a point p on the common boundary of two regular
regions R and R', and there exists two cells C and C' that project onto R and R'
(respectively), such that $\phi_C(p)$ and $\phi_{C'}(p)$ are not equal.
    Definition. Let C_i ($i = 1, 2$) be a cell whose plane projection is R_i. The two
cells are *normal adjacent* if

(i) R_1 and R_2 share an open portion β of a critical curve,
(ii)   for all points $p \in \beta$, the open intervals $(\phi_{C1}(p), \psi_{C1}(p))$ and $(\phi_{C2}(p), \psi_{C2}(p))$
       overlap.

It can be shown that two cells are normal adjacent if there is a normal motion that
crosses from one cell to the other. *Note:* although the condition (ii) is required for
all points in β, it can be shown to be equivalent to requiring it for *some* such
point.
    Example 6. To give a simple illustration of normal adjacency, consider two
regular regions R, R' which are separated by an s-critical curve β. Assume s is a
wall and R' is the region that is next to the wall (see Fig. 3.8). Consider a normal
motion whose trace $\tau(t)$ crosses from R to R'. Let θ_0 be the orientation of the

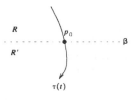

FIG. 3.8. Illustrating a crossing
rule.

motion at some instant t_0 when the trace is at β. There must be two walls s_1, s_2 such that as the trace approaches $p_0 = \tau(t_0)$ the motion lies in $C = Cell(R, s_1, s_2)$. Then as t approaches t_0, the interval $[\phi_C(\tau(t)), \psi_C(\tau(t))]$ is well-defined and tends to $[\theta_1, \theta_2]$. When the trace crosses into R', this interval 'splits' into one of two intervals $[\phi_{C'}(\tau(t)), \psi_{C'}(\tau(t))]$ or $[\phi_{C''}(\tau(t)), \psi_{C''}(\tau(t))]$ where $C' = Cell(R', s_1, s)$ and $C'' = Cell(R', s, s_2)$. Note that

$$[\phi_C(p_0), \phi_C\psi(p_0)] = [\phi_C](p_0), \psi_{C'}(p_0)]\cup[\phi_{C''}(p_0), \psi_{C''}(p_0)].$$

Therefore, the cell C is normal adjacent to both C' and C''.

We have just described what Schwartz and Sharir, 1983a) call a "crossing rule." To compute the normal adjacency relationship among cells we must categorize all possible such rules as a trace crosses a critical curve; we refer to the original paper for the other cases. •

We now define the (normal) *adjacency graph* $G = G(B, S)$ whose nodes correspond to the cells (labelled by triples (R, s, s')) and whose edges connect pairs of normal adjacent cells. The main theorem is now stated:

Theorem 6. Let Z_1, Z_2 be regular free placements. There is a motion in FP between Z_1 and Z_2 iff there is a path between nodes C_1 and C_2 in G where C_i is the cell containing $Z_i(= 1, 2)$.

Proof. Suppose there is a normal motion π from Z_1 to Z_2. Then π traverses a sequence of cells. It is clear that two consecutive cells in this sequence are adjacent, hence there is a path in the adjacency graph G from C_1 to C_2. The converse follows from the fact that each cell is connected and two normal adjacent cells belong to the same connected component of FP.

We will now sketch an algorithm for motion planning: since B is fixed, the input is the set S of objects and the initial and final placements Z_1, Z_2. ∎

(1) (Preprocessing) Compute the adjacency graph G:

   (a) We first compute the set of critical curves and their intersections. It is also necessary to sort the intersections along each curve that they occur in—this can be done by some suitable parametrization of the critical curves. This defines the set of regular regions and their adjacency relationship where the information can be represented by an embedded planar graph; the complexity is $O(n^4)$ since there are n^2 critical curves (all algebraic of degree at most four) and hence $O(n^4)$ critical regions and intersection points.

   (b) For each regular region R, we compute the set of pairs (s, s') of clockwise-counterclockwise stops associated with R: one way to do this is to pick an arbitrary point $p \in R$ and then rotate a half-line emanating from p. By a variant of well-known scan-line algorithms (e.g., Shamos & Hoey, 1976) in computational geometry, we can keep a balanced binary tree which is updated during the rotation yielding the desired information. This costs $O(n\log n)$ per region or $O(n^5\log n)$ overall. Now we have the set of cells $Cell(R, s, s')$.

(c) To compute adjacency graph G, we consider in turn each portion of a critical curve β that separates two incident regular regions R, R'. For each pair (s, s') of objects, if both $Cell(R, s, s')$ and $Cell(R', s, s')$ are defined then the two cells are adjacent. Other types of adjacency depend on more detailed considerations and the reader is referred to the original paper which enumerates them under various "crossing rules." This completes the construction of G.

(2) (Locate initial and final nodes) We first locate the regular regions R_i ($i = 1$, 2) containing the points $P(Z_i)$ (the P-end of the ladder in placement Z_i). A straightforward method is to check if $P(Z_i)$ is in each of the regions: if the region has k critical segments forming its boundary, this takes $O(k)$ time. Since each segment is examined twice and there are $O(n^4)$ segments, this would take $O(n^4)$ time. Next we determine the angular interval (θ_1, θ_2) containing $\theta(Z_i)$, and the stops s_1, s_2 at θ_1, θ_2. Thus we have determined the cell C_i containing Z_i. Note that in the process of this computation we could detect the case where Z_i is not free.

(3) (Path search) Finally, we search for a path in the graph G from C_1 to C_2 using standard (depth-first or breadth-first) graph searching techniques. If such a path exists then it is easy to transform it into a motion by concatenating simple steps that moves from one crossing of a cell to the next.

The above complexity is $O(n^5 \log n)$ but it can be reduced to $O(n^5)$. The $O(n^4)$ estimate for the number of regular regions is probably excessive; an improvement of this estimate would give a corresponding improvement in the time bound for this algorithm. Recently, Leven and Sharir (1984) has obtained an alternative decomposition algorithm which runs in time $O(n^2 \log n)$: roughly speaking, instead of projecting $Z = (x, y, \theta)$ to (x, y), they project it to θ, and study the corresponding decomposition of FP. This algorithm has the merit of being relatively simple compared to another algorithm (Ó'Dúnlaing, Sharir, & Yap, 1983), which has approximately the same time-bound but which is based on the retraction method. Finally, we note that FP for a ladder can have $\Omega(n^2)$ many connected components so that the last two algorithms is close to optimal among algorithms that must compute some representation of FP. It remains open whether subquadratic time bounds can be achieved.

7. TWO APPROACHES TO MOTION PLANNING

It is instructive to give a more general account of the methods underlying the two previous sections. Abstractly, the motion planning problem amounts to computing a path between two points Z_1, Z_2 in a given topological space FP. The corresponding decision problem amounts to checking if the points belong to the same connected component of FP. Here and throughout this paper, 'connected component' always means pathwise-connected component or simply, path-component. This can be viewed as an instance of the general 'extension problem' in topology (see the introduction of Cooke and Finney, 1967). As pointed out in Cooke and Finney (1967), many famous problems of topology are also instances of this problem; furthermore the topological questions can often be reduced to purely algebraic ones. Hence it is not surprising that the two approaches to motion planning described below use the topological concepts of cell complexes and retraction, and that the underlying techniques in both cases are algebraic.

In the decomposition approach, we partition FP into a finite collection of connected subsets called cells, and then construct the graph G representing the adjacency relation between these cells. Motion planning is then reduced to searching in G. We begin with a very general notion of decomposing a space:

Definition. A *cell decomposition* of a topological space S is a finite partition K of S such that each set $c \in K$ is a connected subset of S. We call c a *cell*. Two cells b, c are *adjacent* if $(\bar{b} \cap c) \cup (b \cap \bar{c})$ is nonempty where $\bar{c}$ is the closure of c. The *adjacency graph* of K is the graph $G(K)$ with vertices corresponding to the cells and with edges connecting two vertices whenever the corresponding cells are adjacent.

Note that in the above definition, we do not require a cell to be homeomorphic to a ball (of some dimension); in general, this freedom can lead to considerably fewer cells. The following is the basis of the decomposition approach:

Lemma 7. (*decomposition method*) Let K be a cell decomposition of FP. Then there is a motion between two placements Z_1, Z_2 in FP iff there is a path in $G(K)$ between c_1, c_2 where c_i is the cell containing Z_i.

For this theorem to be effective, we need (a) a decomposition algorithm A to compute K and $G(K)$ from FP, (b) a cell location algorithm A' to compute a cell $c \in K$ containing a given $Z \in FP$, and (c) a path transformation algorithm A'' to convert paths in $G(K)$ into a continuous motion of B. If we are interested in the decision version of motion planning, then we can dispense with (c). The next section will show that the algorithms A and A' exist under rather general conditions. The decomposition approach is sometimes called the 'projection approach' because the decomposition of FP is typically carried out by projection onto lower dimensional spaces.

In the retraction approach, we define a lower dimensional subspace $FP' \subseteq FP$ and show that there is a "retraction-like" function Im from FP to FP'. We say retraction-like because in general, unlike the case in section 5, we can no longer hope for a continuous function. Using the retraction-like function, we show that motion planning in FP can be reduced to motion planning in FP'. By repeated retractions of FP' to $FP'' \subseteq FP'$, etc., we eventually obtain a one-dimensional network $N \subseteq FP$ such that any motion in FP can be retracted to a motion in N. By representing N as a graph, we have again reduced motion planning to graph

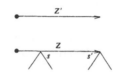

FIG. 3.9.   A discontinuity of *Im*.

searching.[12] An alternative view of the network N is to think of it as a system of highways, and the retraction map as a canonical method for getting on to the highway. The next example shows a retraction-like map that is discontinuous.

Example 7. In Ó'Dúnlaing, Sharir and Yap (1983), we consider the problem of section 6, viz., moving a ladder B in a polygonal environment E. Recall (section 2) that the clearance of $Z \in FP$ is $Clearance(Z) = inf\{d(p, q) : p \in B[Z]$, $q \in E\}$. Define $Racetrack(Z)$ to be the planar set

$$\{p \in \mathscr{E}^2 : d(p, B[Z]) = Clearance(Z)\}.$$

Its peculiar name is derived from the shape of the set. Note that $Racetrack(Z)$ intersects E in a disjoint union of isolated points and closed line segments. The following subspace $FP' \subseteq FP$ is considered:

$$FP' = \{Z \in FP : Racetrack(Z) \cap E \text{ is disconnected}\}$$

We consider the function $Im : FP \rightarrow FP'$: if $Z \in FP'$ then $Im(Z) = Z$; otherwise choose any $p \in B[Z]$, $q \in E$ such that $d(p, q) = Clearance(Z)$. Then we move B directly away from q along the ray through p, preserving the orientation of B, until the first $Z' \in FP'$. $Im(Z)$ is defined to be equal to Z'. (If FP is unbounded then Z' may be undefined but we may assume that FP is bounded for our purposes.) The clearance of $Im(Z)$ is at least that of Z. Notice that Im is generally continuous except at placements such as illustrated in Fig. 3.9. To see that the depicted horizontal placement Z is a discontinuity of Im, note $Racetrack(Z)$ intersects both corners s and s'. Let Z' be the horizontal placement vertically above Z such that $Racetrack(Z')$ intersects s and s', as well as some other object. Any horizontal placement Z'' slightly to the left of Z would not be in FP' and its image $Im(Z'')$ would be some distance vertically above. As Z'' approaches Z, $Im(Z'')$ approaches Z'. However, at Z, the image of Z is itself. Since $Im(Z) \neq Im(Z'')$, we get a discontinuity.

Despite these discontinuities, we can reduce motion planning in FP to motion planning in FP'. The important thing to observe is that these discontinuities are rare (measure zero) and they are well-behaved in the following sense: suppose $\{Z_n : n \geq 1\}$ (resp. $\{Im(Z_n) : n \geq 1\}$) is a sequence that approaches Z_0 (resp. Z'),

---

[12]It is interesting to note that we can combine retraction with decomposition: for instance, instead of retracting FP' further, we can decompose it. However, we know of no actual algorithm that is a hybrid of this form.

written $Z_n \to Z_0$ (resp. $Im(Z_n) \to Z'$). Then it can be shown Z_0 and Z' are in the same connected component of FP'. Thus if π is a motion in FP that passes through only a finite number of discontinuities of Im then $Im(\pi)$ is a piecewise connected path in FP'. But the well-behaved property above ensures that we can patch up $Im(\pi)$ into a motion in FP'.

Since FP' (called a "generalized" Voronoi diagram in Ó'Dúnlaing, Sharir, & Yap, 1983) is a two-dimensional manifold we must further reduce the motion planning problem to a lower dimensional space $FP'' \subseteq FP'$. See the original paper for how this is done.

We now give some general conditions for the retraction approach to work. We are basically interested in finite CW-complexes (Cooke & Finney, 1967; Schwartz & Sharir, 1983b).

Definition. A *cell complex* K of a compact topological space S is a partition of S into finitely many subsets (called *cells*) such that there is a finite sequence

$$\varnothing = K_{-1} \subseteq K_0 \subseteq K_1 \subseteq \cdots \subseteq K_n = K$$

which satisfies:

(1) For each $d = 0, \ldots, n$, the union $\cup K_d \subseteq S$ of the cells in K_d is a closed subset of S. We write $\|K_d\|$ for $\cup K_d$, so $S = \|K_n\|$.

(2) For $d \geq 0$, each cell $c \in K_d \sim K_{d-1}$ is relatively open in $\|K_d\|$. We call c a d-cell.

(3) For each d-cell c, there exists a continuous map f from the closed unit ball B of dimension d to the closure $\bar{c}$ of c where f restricted to the interior of B is a homeomorphism onto c. (Note that $\bar{c}$ need not be homeomorphic to B.)

Clearly a cell complex is also a cell decomposition. The 0-cells and 1-cells of K are called *vertices* and *edges*. If for each c in (3), there is a homeomorphism from B onto $\bar{c}$ then we say K is *regular*. A simple example of a cell complex that is not regular is a closed annular region that is partitioned into 6 cells as follows:

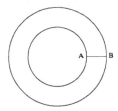

Here there are two vertices (A, B), three edges (AA, BB, AB) and one 2-cell. (Which of the cells fails the test of regularity?)

Typically, the topological spaces we consider are not compact. In order to impose a cell complex structure on the space, we can usually make the space compact by intersecting it with a sufficient large compact set. (This was implicitly done in section 5, for instance, though for a different reason.) "Dimension" of a cell complex K is the largest d such that K has a d-cell. The dimension of a space S is d if it has a cell complex of dimension d: it follows from standard topological facts that this definition of dimension does not depend on the choice of cell complex. (Of course we shall only be concerned with spaces that admits a cell complex structure.) A space of dimension 1 is called a *network*. It is easy to see that the Voronoi diagram in section 5 is a network.

Definition. A function $Im : FP \rightarrow FP'$ is *retraction-like* if

   (a)   *Im* is continuous except on a lower dimensional subset $V \subseteq FP$.

   (b)   *Im* preserves connectivity in FP, viz., $Im(Z)$ and Z belong to the same path-component of FP.

   (c) FP' is a faithful representation of FP, viz., each path-component of FP contains exactly one path component of FP'.

Retraction-like functions reduce motion planning in FP to motion planning in FP' in the following sense.

Lemma 9. (*retraction method*) Suppose there is a retraction-like function Im from FP to FP'. Then there is a motion π from Z to Z' in FP iff there is a motion π' in FP' from $Im(Z)$ to $Im(Z')$.

To see this, suppose we are given π. From (a) and standard arguments in differential topology (Sard's lemma), we may assume that π avoids the discontinuities of Im except for a finite number of instances. Then $Im(\pi)$ is a piecewise continuous path in FP'. By (c), we can patch $Im(\pi)$ into the required π'. Conversely, if π' is given, then we can extend π' to the required π using (b).

To ultimately reduce motion planning to a network, we must compose retraction-like functions:

Lemma 9. Retraction-like functions are closed under function composition.

Suppose that we manage to define a retraction-like function Im from FP to a network $N \subseteq FP$. To make the retraction method effective, we need (a) an algorithm to construct a graphical representation $G(N)$ of the network N, (b) an algorithm to compute Im, or more precisely, to compute for any placement Z the node or edge of $G(N)$ that contains $Im(Z)$. It is important to note that we do not need to have an algorithm to actually 'patch up' the discontinuities of Im.

8. SOLUTION TO THE GENERAL MOTION PLANNING PROBLEM

We describe a major result due to Schwartz and Sharir (1983b), namely, that the general motion planning problem can be reduced to path finding in a suitable submanifold of some multidimensional Euclidean manifold and this latter is solvable. This result is obtained by the decomposition approach.

A set $S \subseteq \mathscr{E}^k$ (for some $k \geq 1$) is *semi-algebraic* if
$$S = \{x \in \mathscr{E}^k : \phi(x) \text{ holds}\}$$
where $x = (x_1, \ldots, x_k)$, $\phi(x)$ is a *polynomial expression*, i.e., a Boolean combination of inequalities of the form $p(x) \geq 0$ with $p(x)$ a multivariate polynomial with integer coefficients. We say $\phi(x)$ is a *defining formula* for S and $\phi(x)$ *defines S*. Note that if $\phi(x)$ does not actually contain any of the variables in x then $\phi(x)$ defines either $\mathscr{E}^k$ or $\varnothing$. A polynomial expression is *satisfiable* if it defines a non-empty S, and two expressions are *equivalent* if they define the same set.

Recall the scheme (P1) of the general motion planning problem in section 3. To obtain the desired result we require that the input to the problem be finitely specifiable via semialgebraic sets. This amounts to three requirements:

(i)   The robot components B_i and the environment E are semialgebraic sets.

(ii)   The set $\mathscr{A}$ of admissible placements of the robot system B is semialgebraic.

(iii)   Given that the motion objective O is specified by two sets O_1 and O_2 of placements, indicating that a motion from any placement in O_1 to any placement in O_2 is desired, the sets O_1, O_2 are semialgebraic.

We shall ignore the optimality criteria C in (P1) although some suitable semialgebraic criteria can be admitted. Hence, an input to the general motion planning problem is a quadruple (B, E, O_1, O_2), where each semialgebraic set is represented by some polynomial expression. In the following let n denote the "size" of this input where we choose some standard convention for measuring size (e.g., the size of an integer m is $\log(2 + |m|)$ and the size of a polynomial of degree d, with b non-zero coefficients each of size $\leq c$ is $bd\log c$, etc). The complexity of the motion planning algorithm is a function of n. Our looseness in defining 'size' here is justified by the fact that the results we shall state are insensitive to any reasonable choice. For simplicity we consider the decision version of (P1), although it is not hard to modify the method to derive a path.

If we prefix a polynomial expression with a sequence of quantifiers of the form $(\forall x_i)$ or $(\exists x_i)$, we obtain a (prenex form) first-order formula whose interpretation is the standard one with variables ranging over the real numbers.[13] For example, $(\exists x)(\forall y)[(x^2 - y^2 \geq 0)\lor(3x + yz = 0)]$ is a formula with variables x, y, z (but only z is free). Such expressions will be called *Tarski formulas*, after Alfred Tarski (1951) who first showed that the satisfiability of such expressions can be decided. Indeed, Tarski showed that the quantifiers in the above expressions can be eliminated in the sense that for each Tarski formula, we can effectively find an equivalent polynomial expression; the satisfiability of polynomial expressions can be shown to be decidable. Below we indicate how this is done. Hence we conclude that a subset of $\mathscr{E}^k$ is semialgebraic if it can be described by a Tarski formula.

---

[13]The interested reader may refer to any standard textbook on logic.

For concreteness, the following discussion assumes $S = \mathscr{E}^3$ and the robot is $B = (B_1, \ldots, B_m)$ for some $m \geq 1$. Let R be a 3×3 rotation matrix (see example 1). If we regard R as a point in $\mathscr{E}^9$, then R must satisfy the following 7 equations:[14]

   (1) $R_i^T R_i = 1$, $i = 1, \ldots, 3$, where R_i is the ith column of R and R_i^T is the transpose.
   (2) $R_i^T R_j = 0$, $1 \leq i < j \leq 3$.
   (3) $det(R) = 1$.

This shows that the set of rotations is semialgebraic. Since a rigid transformation is represented as a pair (t, R) where t is a translation, we conclude that $\mathfrak{R}$ is a semialgebraic subset of $\mathscr{E}^{12}$. Therefore[15] the set $\mathfrak{R}^m$ of placements of B is a semialgebraic subset (in fact, a smooth $6m$-dimensional submanifold) of $\mathscr{E}^{12m}$. The requirement (ii) above that the set of admissible placements $\mathscr{A}$ be semialgebraic is reasonable if we grant that $\mathscr{A}$ is specified by couplings as illustrated in example 4: all the holonomic (resp. nonholonomic) couplings correspond to additional contraints of the form $p(x) = 0$ (resp. $p(x) \geq 0$) for suitable polynomials $p(x)$.

The requirement (i) above ensures that the set of free placements $FP \subseteq \mathscr{A}$ is also semialgebraic: we may write $FP = \mathscr{A} - \mathscr{F}$ where $\mathscr{F}$ represents the "forbidden placements," i.e., those placements Z such that $B[Z] \cap E \neq \varnothing$. Thus $Z \in \mathscr{F}$ satisfies the informal formula

$$(\exists x \in \mathscr{E}^3)[x \in B[Z] \text{ and } x \text{ is in the environment } E].$$

For each Z, clearly there is a polynomial expression $\phi(Z, x, y)$ that says that x is the Z-transformation of y. Also, let $\psi_B(y)$ and $\chi_E(x)$ be the polynomial expressions that say that '$y \in B$' and '$x \in E$' (respectively). Hence

$$\mathscr{F} = \{Z \in \mathscr{E}^{12m} : (\exists x)(\exists y)[\phi(Z, x, y) \wedge \psi_B(y) \wedge \chi_E(x)]\}.$$

Since this is a Tarski formula, we have established that $\mathscr{F}$, and hence FP, is semialgebraic.

To continue, we must now refer to an important result due to Collins (1975). A very accessible exposition of this result is given by Schwartz and Sharir (1983b); the paper also contains a wealth of information on various computational details.

---

[14]The first six equations reduce the degrees of freedom to 3, illustrating the well-known fact that the set of rotations constitute a smooth 3-dimensional submanifold of $\mathscr{E}^9$. The last equation equation does not reduce the dimensionality of the set of rotations since the first six already restrict $det(R)$ to ± 1: rather it selects one component $SO(3)$ of the manifold of the orthogonal group $O(3)$.

[15]The above semialgebraic realization of the Euclidean group in $\mathscr{E}^{12}$ is not the most efficient. Using quaternions, the rotation group is embedded in $\mathscr{E}^4$ so that $\mathfrak{R}$ can be embedded in $\mathscr{E}^{7m}$.

Definition. An *algebraic cell decomposition* K of $\mathscr{E}^k$ is a cell decomposition of $\mathscr{E}^k$ where each cell is either a singleton point whose coordinates are algebraic numbers, or a semi-algebraic set that is homeomorphic to $\mathscr{E}^d$ for some $d = 1, \ldots, k$. A *sample* for K is a function $f\colon K \to \mathscr{E}^k$ such that for all $c \in K$, $f(c) \in c$ and each coordinate of $f(c)$ is an algebraic number. Let P be a set of polynomials over k variables. The decomposition K is *P-invariant* if for each $p \in P$ and for each cell $c \in K$, the sign $signum(p(x))$ is constant as x ranges over c. Here $signum(r)$ is -1, 0 or $+1$ according as $r < 0$, $r = 0$ or $r > 0$ (resp.).

For the purposes of quantifier elimination and decision procedures, though not for motion planning applications, we need a special type of algebraic cell decomposition:

Definition. The *projection* of a point $p \in \mathscr{E}^k$ is a point $p' \in \mathscr{E}^{k-1}$ obtained by omitting the kth component of p. The projection of a subset of $\mathscr{E}^k$ is the set of projections of its members. A *cylindrical decomposition* K of $\mathscr{E}^k$ is an algebraic cell decomposition with the following properties: if $k > 1$ there exists (recursively) a cylindrical decomposition K' of $\mathscr{E}^{k-1}$, called the *base decomposition* of K, with the following property: for each $c \in K$ there is a $c' \in K'$ such that the projection of c is equal to c'. We say c is *based* on c'. A *cylindrical sample* of K is a sample $f\colon K \to \mathscr{E}^k$ such that if c_1, $c_2 \in K$ have a common base then the projections of $f(c_1)$ and $f(c_2)$ agree. (Thus f induces a cylindrical sample f' of K'.)

We can now state the result of G. E. Collins.

Theorem 10. There is a double-exponential time algorithm which, given any finite set P of polynomials with integer coefficients over k variables, constructs a *P-invariant* cylindrical decomposition K of $\mathscr{E}^k$ together with a cylindrical sample for K. More precisely, the algorithm takes time $O(m^{2^{O(k)}})$ where $m = \max \{d, s, p\}$, where d is the maximum degree of the polynomials in P, s is the maximum size of integer coefficients in P and p is the number of polynomials in P.

Some clarifications should be made about the preceding result:

(i)   The algebraic cell decomposition K is represented by a set of polynomial expressions that form the defining formulas for each of the cells of K. Also, the algorithm produces the base decomposition K' of K, the base decomposition of K', etc.

(ii)   All operations in the algorithm are ultimately reduced to arithmetic operations on rational numbers and are "exact" in a certain sense: each (real) algebraic number χ is represented "exactly" by a pair $(\phi(x), I)$ where $\phi(x)$ is an equation with rational coefficients and I is an interval with rational endpoints that contains a unique root of $\phi(x)$ equal to χ. We refer to [Schwartz and Sharir (1983a)] for the various techniques available.

(iii) The complexity estimate in the theorem allows fairly liberal interpretations. From (ii), it is enough to show how rational arithmetic operations are charged: for instance, we just require that the rational operations $x + y$ and $x \times y$ are given a cost of some polynomial in n where the rational numbers x

and y have binary representations of n bits each. Typically, addition is charged $O(n)$ and multiplication is charged $O(n^2)$ or $O(n\log n\log\log n)$.[16]

Example 8. We illustrate how the preceding theorem results in a decision procedure for the truth of Tarski sentences.[17] Suppose we are given a Tarski sentence $(\forall x)(\exists y)[\phi(x, y)]$ where ϕ is a polynomial expression. Let P be the set of polynomials occurring in $\phi(x, y)$. Let K be a P-invariant cylindrical decomposition for $\mathscr{C}^2$, K' a base decomposition of K and $f: K \rightarrow S$ a cylindrical sample, as computed by Collins' algorithm on P. [We must take care to order the coordinates of the space $\mathscr{C}^2$ such that the last coordinate (the one to be omitted in projecting from K to K') corresponds to variable x rather than to y. More generally, if P has variables $x_1, \ldots, x_k$ then we ensure that the quantifier prefix of the formula ϕ introduces these variables in reverse order: $x_k, x_{k-1}, \ldots, x_1$.] The given sentence is true if and only if the following holds for each cell $c' \in K'$: if c_1, $\ldots, c_m$ are the cells with base c' and $f(c_i)$ is (α, β_i) ($\alpha \in c'$) then $\phi(\alpha, \beta_i)$ is true for some $i = 1, \ldots, m$. The reader should have no difficulty generalizing this to a general decision method for Tarski sentences. •

Example 9. As another example, suppose we want to eliminate the quantifier $(\exists y)$ from some Tarski formula $(\exists y)[\phi(x, y)]$. This means we want another polynomial expression $\psi(x)$ that defines the same semi-algebraic set as $(\exists y)[\phi(x, y)]$. Let K, K' and f be as in the last example. For any cell c in $K \cup K'$, denote its defining formula by ψ_c: Collins' procedure can give us these formulas. Then $\psi(x)$ is the disjunction of all those $\psi_{c'}$ (for $c' \in K'$) where for some $c \in K$, c' is the base cell for c, $f(c) = (\alpha, \beta)$ and $\phi(\alpha, \beta)$ holds. Again, it is easy to generalize this to a general quantifier-elimination procedure. •

To return to the problem of motion planning, it should now be plain that we have in hand the basic tool to decompose the set FP of free placements into cells: it is evident that with suitable conventions, we can construct a suitable polynomial expression Φ defining FP. If P is the set of polynomials occurring in Φ, we can construct a P-invariant cylindrical decomposition K. Clearly FP is a union of some set of cells of K. For each cell $c \in K$, we check if $c \subseteq FP$ by testing if a sample point in c satisfies Φ.

Finally, we must form the adjacency graph $G(K)$. Arnon (1979) has noted that testing adjacency is a decidable problem. To see this, we simply have to show a Tarski formula saying that $b \cap \bar{c} \neq \varnothing$ where b, c are two cells:

$$(\exists x)(\forall \epsilon)(\exists y)[(\epsilon > 0) \wedge \phi_b(x) \cdot \Rightarrow \cdot (d(x, y) < \epsilon) \wedge \phi_c(y)] \qquad (*)$$

Thus we can compute the adjacency graph $G(K)$. We also need to compute the cell c that contains a given placement Z: this is simply a matter of finding the cell

---

[16]This is because the known upper bound for multiplying two n-bit numbers on a Turing machine is $O(n\log n\log\log n)$.

[17]Sentences are formulas with no free variables. A sentence is true if it is satisfiable as a formula.

c such that $\phi_c(Z)$ is true. We have thus shown the following important result of Schwartz and Sharir:

Theorem 11. The general motion planning problem is decidable.

With a careful accounting of the complexity, the time complexity can be shown to be doubly exponential (in the dimension k). It is not hard to see that any improvement of Collins' algorithm would lead to a corresponding improvement for motion planning. The converse implication is not known. Recently, Ben-Or, Kozen, and Reif (1984) has shown that the decision problem for Tarski sentences can be solved in a single-exponential space (which immediately implies a double exponential running time). However their result is not based on computing a cell decomposition, and hence does not have an immediate impact on the complexity of the general motion planning problem. This situation has been remedied by a forthcoming paper (Kozen & Yap, 1985). The new algorithm not only retains the single-exponential space bound of Ben-Or, Kozen, and Reif (1984) and also computes all the adjacency relations between cells, something which was not done in the original Collins' procedure.

The basic planning problem is tractable. Recall that the basic planning problem arises when we restrict the inputs of the general planning problem to some fixed robot system B. Since this is an important case in practice, it is of interest to show a more efficient bound here. Refering to Collins' theorem, we note that of the four complexity parameters k, d, s, p in the decomposition algorithm, there is an exponential dependence on k only. Thus with k fixed the complexity is polynomial time. We claim that the value of k is constant in the basic planning problem; indeed, $k \leq 12m$ where m is the number of components in the robot system. With reasonable assumptions, the values of the other parameters d, s, p used when calling the decomposition algorithm has a (fixed, usually small) polynomial dependence on the input size n. Hence constructing K can be done in time polynomial in n. This also implies that K has a polynomial number of cells.

What remains to be shown is that the adjacency relationship can also be computed in polynomial time. Since there are a polynomial number of cells we only need to show that testing adjacency for any pair of cells is polynomial in n. Note that the size of the polynomial expressions ψ_c that define each cell c is polynomial in n since the algorithm to construct these expressions is polynomial in n. Using Arnon's adjacency formula (*) above, we can eliminate the three quantifiers and obtain another polynomial size formula ϕ. Checking satisfiability of the final formula ϕ is polynomial in n.[18] We conclude that:

Theorem 12. The basic planning problem is tractable.

Further refinements. In Schwartz and Sharir (1983b) a more sophisticated test for adjacency is outlined; unfortunately it is not known if the method there is polynomial. On the other hand, they showed that, under certain additional hypotheses:

---

[18]The observation that this adjacency test is polynomial time seems to have been missed in the paper of Schwartz and Sharir.

(a) Motion planning can be confined to those cells in K of full dimension or codimension 1.

(b) The test for adjacency between cells c, c' where one is of full dimension and the other of codimension 1 can be done in polynomial time.

Of course (b) already follows from above but a particularly simple test is shown by Schwartz and Sharir for this case. One hypothesis we need is that the cylindrical decomposition K of $\mathscr{C}^k$ is "well-based": We will not define this notion but simply note that the well-basedness assumption leads to cell decompositions that are regular in the sense previously defined. Since well-based P-invariant cylindrical decompositions for an arbitrary set P of polynomials may not exist, Schwartz and Sharir (1983b) suggests that a random rotation of the coordinate axes would solve this with high probability.[19] The justification of (a) depends on the well-basedness of K and the mild assumption that the space of admissible placements $\mathscr{A}$ can be represented by a Cartesian product of $\mathscr{C}^k$ (for some k) with some copies of the rotation groups S^i of dimension $i = 1, 2, 3$. By omitting some exceptional placements whose removal does not increase the number of components of $\mathscr{A}$, this representation can be simplified to the product of $\mathscr{C}^k$ with several copies of S^1. This form can be "covered" by several copies of $\mathscr{C}^k$ for suitable k. Now we may decompose each $\mathscr{C}^k$ using Collins' procedure and also make identifications of cells across the copies. We refer the interested reader to the original paper. However, we note that much of this complicated development is rendered unnecessary with the new algorithm for cell-decomposition in Kozen and Yap (1985).

9. RETRACTION VIA CELL COMPLEXES

The previous section shows that the decomposition approach is a "universal" method. This section shows that the retraction method is similarly universal. To illustrate these ideas, we then prove a result similar to the main result in Hopcroft and Wilfong (1984b).

Lemma 13. Let K be a cell complex for ε^k. For $d = 2, \ldots, k$ there exists a retraction-like function:

$$Im_d : \|K_d\| \to \|K_{d-1}\|.$$

Proof. It is enough to show a retraction-like function f from the closed unit d-ball D to its boundary ∂D. We can take f to be the function that takes each point $(x_1, \ldots, x_{d-1}, x_d) \in D$ to

$$\left(x_1, \ldots, x_{d-1}, \pm \left(1 - \sum_{i=1}^{d-1} x_i^2 \right)^{1/2} \right).$$

----

[19]This shows that a good direction can be found in random polynomial time, but the deterministic complexity of this procedure is not clear.

where the sign of the square-root is that of x_d; if $x_d = 0$ then the positive square-root is meant. This function is continuous on D except at points on the hyper-plane $x_d = 0$. This clearly satisfies conditions (a) and (b) of the definition of retraction-like. Condition (c) follows from the well-known fact that the boundary of a d-cell is connected if $d \geq 2$.∎

As noted by P. Kahn (see Arnon, 1979), the cylindrical decomposition obtained by Collins' procedure in the previous section is a cell complex (though not necessarily regular [Schwartz & Sharir, 1983b]). We exploit this fact to obtain the next result.

Theorem 14. Under the same hypotheses as theorem 11, there is an algebraic 1-dimensional subspace $N \subseteq FP$ together with a retraction-like function $Im:FP \rightarrow N$.

Proof. By introducing artificial constraints that do not affect the existence of motions, we may further assume that $FP \subseteq \mathscr{E}^k$ is compact. Then the cell decomposition K of FP obtained in the proof of theorem 11 is actually a cell complex. Let Im_d $(d = 2, \ldots, k)$ be the retraction-like functions of the previous lemma. The theorem follows by choosing $N = \|K_1\|$ and $Im:FP \rightarrow N$ be the function composition of $Im_k, Im_{k-1}, \ldots, Im_2$, so $Im(x) = Im_2(Im_3(\ldots (Im_k(x)) \ldots))$.
∎

As noted earlier, the effectivity of this result depends on the existence to two algorithms: (a) One to construct the graphical representation of the network N. Since N is the network $\|K_1\|$, we can employ the algorithm of Kozen and Yap (1985) for computing Collins' cell decomposition together with adjacency information. (b) We also need another algorithm to compute the retraction Im. We claim the functions Im_i can be defined in such a way that they are effective in the sense that for any algebraic $Z \in \mathscr{E}^k$, we can compute an algebraic representation of $Im(Z)$ and furthermore, this computation is relatively efficient. A careful analysis of this will be presented elsewhere.

On motion in contact. In the rest of this section we given an application of the above ideas to a problem of Hopcroft and Wilfong. Consider a *loosely coupled* planar robot system $B = (B_1, \ldots, B_m)$; by this we mean that a placement Z of B will be considered admissible if

$$B_i[Z] \cap B_j[Z] = \varnothing$$

for all $i \neq j$.[20] We shall be interested in *semiadmissible* placements, ie, where components can touch but not overlap. A semiadmissible placement Z is *connected* if $B[Z]$ forms a connected set of points in the underlying physical space $S = \mathscr{E}^2$. Let *CONN* denote the set of connected semiadmissible placements. Fol-

---

[20]Hopcroft and Wilfong (1984a) considers a more general situation where the components are further partitioned into 'composite objects' where a placement is admissible if (i) the union of the bodies of a composite object forms a connected set and (ii) components from distinct composite objects do not overlap. Our proofs easily extend to handling composite objects.

lowing Hopcroft and Wilfong, we further assume the following restrictions: (i) each component B_i is a polygon; (ii) the placements of each component have a fixed orientation (ie. rotation is disallowed); and (iii) the environment E is empty. The last restriction implies that the set of semifree placements SFP for B is equal to the set of semiadmissible placements.

In the context of retraction algorithms, it is convenient to say that two motions π, π' are *correspondent* if $\pi(0) = \pi'(0)$ and $\pi(1) = \pi'(1)$. The main result in Hopcroft and Wilfong, 1984a) is the following:

Theorem 15. Given a loosely coupled robot system $B = \langle B_0, B_1, \ldots, B_m \rangle$ and a pair Z_1, Z_2 of placements in *CONN*. Then there is a semiadmissible motion in *SFP* from Z_1 to Z_2 if and only if there is a correspondent motion in *CONN*.

The proof uses the Mayer-Vietoris theorem. In a companion paper (Hopcroft & Wilfong, 1984b), it was further shown that, with B as before, there exists a network $N \subseteq CONN$ such that for any Z_1, $Z_2 \in N$, there is a semiadmissible motion in *CONN* from Z_1 to Z_2 if and only if there is a correspondent motion in N. Of course, this result is trivial if N is unrestricted (for example, choose N to consist of a single point); also there ought to be a reasonable retraction-like function from *CONN* to N. To state the result precisely, we next develop the appropriate machinery.

We may as well regard B_0 to be immobilized, so a placement Z of B can be specified by a point in $\mathscr{C}^{2m}$ (giving the placements of $B_1, \ldots, B_m$ relative to B_0). Note that each B_i has only two translational degrees of freedom since we do not allow rotation. We will partition the boundary of each B_i into k *corners* and k *sides* if B_i is a k-gon. We call the corners and sides of B_i the *features* of B_i. For placement Z and feature s of B_i, let $s[Z]$ denote the subset of the physical space S occupied by s in placement Z. Thus $s[Z]$ is a singleton point or an open line segment, according to whether s is a corner or a side.

A *clause* is a pair $C = (b, c)$ of features such that (i) b and c come from distinct components of B and (ii) b is a side and c a corner. For each clause C, we have an associated linear form ϕ_C in the $2m$ variables Z given by

$$\phi_C(Z) = L_C \cdot Z - u_C^*$$

where $L_C \cdot Z$ denotes the dot-product of Z (regarded as a $2m$-vector) with a constant $2m$-vector L_C, and u_C is a constant. Intuitively, $\phi c(Z) = 0$ hold for the clause $c = (b, c)$ if the point c [z] lies in the line through b [z]. Let D be the set of all clauses. A *sign-assignment* σ is a function from D to $\{-1, 0, +1\}$. We say Z *satisfies* σ if for each $C \in D$, $\phi_C(Z)$ is positive, zero or negative according as $\sigma(C)$ is -1, 0 or $+1$. Of course, some sign-assignments may be inconsistent in the sense that there are no placements Z satisfying them. We denote the unique sign assignment satisfied by Z as $\sigma(Z)$. Corresponding to σ we may associate two matrix inequalities:

$$M_\sigma \cdot Z = u_\sigma \text{ and } N_\sigma \cdot Z > v_\sigma$$

where M_σ is the matrix whose rows correspond to the vectors L_C in (*) for which $\sigma(C) = 0$, and the corresponding entry in u_σ is the u_C in (*); N_σ and v are similarly defined from those clauses C for which $\sigma(C) = \pm 1$. Let H_σ denote the set of placements Z, admissible or not, that satisfies the matrix equation $M_\sigma \cdot Z = u_\sigma$. The *rank* of a sign-assignment σ is the matrix rank of M_σ. Clearly H_σ is a linear subspace of dimension equal to $2m - r$ where r is the rank of σ. The set of placements satisfying a fixed sign-assignment σ is called a *cell* and denoted C_σ. It is not hard to show that (i) *CONN* is a union of cells and (ii) *CONN* is compact. In fact we have:

Lemma 16. The collection K of cells in *CONN* forms a regular cell complex.

Proof. Define K_d to consist of those cells in *CONN* with rank $\geq 2m - d$. To show that $\|K_d\|$ is closed we use induction on d. The result is clear for $d \leq 0$. Suppose $c = C_\sigma$ is a d-cell for $d \geq 1$. Thus c consists of those placements Z that satisfies $M_\sigma Z = u_\sigma$ and $N_\sigma Z > v_\sigma$. It is easy to see

Claim. The closure $\bar{c}$ consists of those placements Z that satisfies $M_\sigma Z = u_\sigma$ and $N_\sigma Z \geq v_\sigma$.

Thus if $Z \in \bar{c} \sim c$ then at least one of the inequalities of $N_\sigma Z \geq v_\sigma$ is actually an equality. It follows that $Z \in \|K_{d-1}\| \subseteq \|K_d\|$. Hence $\|K_d\|$ is closed. Next, to show that a d-cell c is relatively open in $\|K_d\|$, let $Z \in c$. Then it is easy to see that there is a neighborhood U of placements of Z such that $U \cap (\|K_d\|)$ is contained in c [choose U such that $Z' \in U$ implies if $\sigma = \sigma(Z)$, $\sigma' = \sigma(Z')$ then for each clause C, $\sigma'(C) = 0$ implies $\sigma(C) = 0$]. Finally, for each d-cell c, we must show a homeomorphism from the unit closed ball B of dimension d onto the closure $\bar{c}$ of c. This follows easily from the above claim which shows that each d-cell is homeomorphic to a closed convex polyhedron in $\mathscr{E}^d$. ∎

We may conclude:

Theorem 17. Let $N = \|K_1\|$. There is a retraction-like function from *CONN* to N. In particular, if $Z_1, Z_2 \in N$, then there is a semi-admissible motion in *CONN* from Z_1 to Z_2 if and only if there is a correspondent motion in N.

Note that our result is not identical to Hopcroft and Wilfong (1984b) in that our network N is defined slightly differently.

10. LOWER BOUNDS

A main goal of complexity theory is the classification of the inherent complexity of computational problems. Certain classes of roblems (called *complexity class-es*) are recognized to be 'natural': this is an informal notion but one criterion for naturalness is that membership in the class should be invariant under a variety of definitions. Among the most well-known of these are

$$L, P, NP, PSPACE. \tag{**}$$

The class P (resp. NP) corresponds to those problems that are computable by deterministic (resp. non-deterministic) machines in polynomial time. The class L

(resp. *PSPACE*) corresponds to problems computable by deterministic machines in logarithmic (resp. polynomial) space. Note that *P* gives precisely the problems we have called 'tractable'. There is a linear hierarchy of difficulty among the classes in (**):

$$L \subseteq P \subseteq NP \subseteq PSPACE.$$

Actually, except for the proper inclusion of *L* in *PSPACE*, it is not known if this hierarchy is proper: our present knowledge does not rule out the possibility that *P* = *PSPACE* or *L* = *NP*. However, there is strong circumstantial evidence that this hierarchy is proper: for instance if *P* = *NP* then a whole array of computational problems currently not known to be tractable would become tractable and many other unlikely consequences follow (e.g., see Garey & Johnson, 1979).

Since the "precise" classification of problems is typically very difficult (perhaps not even well-defined), we are usually contented if a problem can be precisely located in some hierarchy of classes. The classifications of this section will be relative to the hierarchy (**) above. We illustrate the basic technique for classifying problems by considering the following motion planning problem:

(Q1) *Moving many rectangles:* Given a set $B = (B_1, \ldots, B_m)$ of rectangles, a polygonal environment *E*, initial and final placements Z_1, Z_2 of *B*, to decide if there exists a semifree motion of *B* from Z_1 to Z_2.

Observe that (Q1), like all the problems considered in this section, is a decision problem. Furthermore, we will be considering semifree motion because it simplifies our constructions: the proofs can be modified to handle the (free) motion we usually treat. We will show that (Q1) is *at least as hard as* as the following problem:

(Q2) *Set partition:* Given a set $X = \{a_1, \ldots, a_n\}$ of positive integers, does there exist a subset $Y \subseteq X$ such that

$$\sum_{a \in Y} a = \sum_{b \in X - Y} b \ ?$$

The later problem is known to be *NP-complete* (Gary & Johnson, 1979) meaning that it (i) is in *NP* and (ii) is at least as hard as any problem in *NP*. The technique for showing that (Q1) is at least as hard as (Q2) is to 'reduce' (Q2) to (Q1): given an input *X* for (Q2) we transform it to an input $t(X)$ for (Q1) so that a (Q1)-answer for $t(X)$ is also a (Q2)-answer for *X*. Hence the existence of an efficient algorithm for (Q1) implies the existence of an efficient algorithm for (Q2), provided $t(X)$ can be computed at least as efficiently. For instance, if (Q1) is tractable and $t(X)$ can be computable in polynomial time then (Q2) is tractable. Technically, the term for a problem that is at least as hard as any problem in *NP* is *NP-hard*.

We now show that (Q1) is *NP*-hard. Suppose $X = \{a_1, \ldots, a_n\}$ is an input to
(Q2). Let $a_0 = \max \{a_i : i = 1, \ldots, n\}$, $2A = \sum_{i=1}^{n} \Sigma a_i$ (we may assume A is
an integer for otherwise there is no solution for X). We construct an input $t(X)$ for
(Q1): the robot system is $B = (B_0, B_1, \ldots, B_n)$ where the dimensions of the
rectangles are:

B_0: 2 by n, and B_i: 1 by a_i ($i = 1, \ldots, n$).

The environment consists of a north-south rectangular hall of dimensions n by 1
$+ 4a_0$, and an east-west transept of dimensions $2A + n$ by 1, symmetrically
arranged as shown in Fig. 3.10. The rectangle B_0 is initially placed squarely in
the northern end of the hall and the rest $B_1, \ldots, B_n$ are arranged vertically at the
southern end. The final placement is the reverse (B_0 at the southern end, etc).
This completes our description of the input $t(X)$ corresponding to X. It should be
clear that there exists a (semifree) motion satisfying $t(X)$ iff there is a (Q2)-
solution to X. A little thought shows that this input $t(X)$ can be computed rather
efficiently (quadratic time at most). The reduction of (Q2) to (Q1) is now done.

We have thus shown an *NP*-hardness lower bound on moving rectangles. It
turns out that the problem is really *PSPACE*-hard (i.e., at least as hard as any
problem in *PSPACE*). The proof is considerably more difficult, and we refer to
Hopcroft, Schwartz, and Sharir (1984). What about upper bounds? If we can
show that the problem is actually in *PSPACE* then our classification would be
tight relative to the hierarchy (**). This remains open. A partial result in this
direction (Hopcroft and Wilfong, 1984b) is that if we restrict the placements of
rectangles such that their sides are always parallel to the coordinate axes (hence
no rotation is allowed) then the problem can be shown to be in *PSPACE*. This
follows essentially from the results of the previous section (that such motion can
be restricted to the vertices and edges of the cell complex of *FP*).

Next consider the *moving many discs* problem which is a variant of (Q1)
where we have discs instead of rectangles. It is not so easy to do a similar

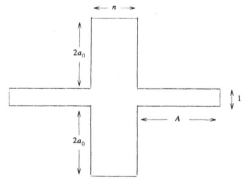

FIG. 3.10.   Reducing (Q2) to (Q1).

reduction from (Q2) to this problem. One reason why such a reduction may be more difficult is that moving discs is (despite some possible initial impressions) a simpler problem than moving rectangles:

(i)   Discs have only two degrees of freedom of motion but rectangles have three.
(ii) Discs are single parameter objects while rectangles are two parameter objects.

(Note that (i) and (ii) are independent considerations: for instance a rod has three degrees of freedom but one parameter.) With more care, it can still be shown that moving many discs is *NP*-hard (in fact 'strongly' *NP*-hard) (Spirakis & Yap, 1984).

Another problem that can be shown to be *NP*-hard is the following due to Hopcroft, Joseph, and Whitesides (1982):

(Q3) *Reachability for a many-linked arm:* Given an arm $B = (B_1, \ldots, B_m)$ in an initial position Z, a polygonal environment E and a point p, is there a (semifree) motion that moves the tip of the arm to position p?

The examples above all show lower bounds for problems with arbitrarily many degrees of freedom for the robots. Indeed, we already know (from section 8) that problems with fixed degrees of freedom are in P and hence cannot be *NP*-hard (unless $P = NP$). For lower bounds on the basic planning problem, we next show that certain basic planning problems are *L*-hard (i.e., at least as hard as any problem computable in logarithmic space). Consider the following planning for a unit cube B in $\mathscr{E}^3$:

(Q4) *Reachability for a cube:* Given a polyhedral environment E, a pair (Z_1, Z_2) of initial and final placements of the cube, does there exist a (semifree) motion from Z_1 to Z_2?

Unlike the problems considered until now, we must be careful about how the input for problems with low complexity is encoded: we will assume that E is represented by a set of disjoint polyhedra, each polyhedron is represented by two planar graphs that are duals of each other (one graph corresponding to the vertices and edges of the polyhedron, the other corresponds to the edges and faces of the polyhedron). It is assumed that the input is correct (i.e., the polyhedra are indeed disjoint and Z_1, Z_2 are free) since it is unclear that we can afford to check these in logarithmic space. We will reduce the following problem, known to be *L*-hard, to (Q4):

(Q5) *Undirected graph accessibility:* Given an undirected graph G with nodes labelled consecutively from 1 to n (for some n) does there exist a path from 1 to n?

Graphs in (Q4) and (Q5) are represented by a set of the nodes and a set of its edges. We remark that the problem (Q5) is not known to be in L but is complete for the class SL of problems computable in *symmetric logarithmic space* (see Lewis & Papadimitriou, 1980). SL lies between L and another well-known class NL of nondeterministic logarithmic space problems. The class SL comes from a restriction of nondeterministic machines which makes them ''symmetric'' in the sense that the transitions of such machines allow them to run ''forward'' or ''backward.'' Intuitively, the reason that symmetric computations correspond closely to motion planning is that motions can always be reversed.

Suppose G is the input to (Q5) where G has nodes $1, \ldots n$. We will construct a corresponding input to (Q4), but it is convenient to first describe a graph G' embedded in the 3-dimensional integer lattice with lattice points (x_1, x_2, x_3). For each node i, $(i = 1, \ldots ,n)$, of G we introduce the following three sets of nodes (lattice points) of G':

$$U(i) = \{(i, j, 0): j = 0, 1, \ldots ,i\};$$
$$V(i) = \{(i, i, k): k = 1, 2, \ldots ,n\};$$
$$W(i) = \{(i, k, i): k = 1, 2, \ldots ,n\}.$$

There are edges $[(i, j-1, 0), (i, j, 0)]$ for $j = 1, \ldots ,i$; and edges $[(i, i, k-1), (i, i, k)]$ for $k = 1, \ldots ,n$. Note that these nodes and edges are in the $x_1 = i$ plane. Figure 3.11 illustrates the case $n = 5$, $i = 3$.

If (i, j) is an edge in G, we introduce the following edges in the $x_1 = i$ plane: horizontal edges connecting (i, i, j) to (i, j, j), i.e.,

$$[(i, k, j), (i, k + 1, j)], k = i, \ldots ,j-1$$

and the vertical edges connecting (i, j, j) to (i, j, i), i.e.,

$$[(i, j, k), (i, j, k + 1)], k = i, \ldots ,j-1.$$

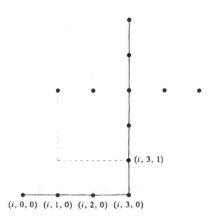

$(i, 3, 1)$

$(i, 0, 0)$ $(i, 1, 0)$ $(i, 2, 0)$ $(i, 3, 0)$

FIG. 3.11. Nodes and edges in the $x_1 = 3$ plane for $n = 5$.

The dashed edges in Fig. 3.11 illustrates these additional edges when $(i, j) = (3, 1)$ is an edge of G. If $i < j$ then we also connect the $x_1 = i$ and $x_1 = j$ planes using the edges:

$$[(k, j, i), (k + 1, j, i)], k = i, \ldots ,j - 1.$$

It is now easy to construct a polygonal environment E that has passage-ways with unit square crosssections where the axes of the passage-ways correspond to the edges of G'. Furthermore, the nodes of G' correspond to empty spaces of E of unit cube sizes. Then there is a semifree motion of the unit cube in E from the the position corresponding to $(1, 0, 0) \in G'$ to $(n, 0, 0) \in G'$ iff there is a path in G from node 1 to node n. This completes the reduction of (Q5) to (Q4).

11. SUMMARY: NEW DIRECTIONS AND OPEN PROBLEMS

The following includes both specific technical questions arising out of previous work as well as challenging new areas of motion planning. Indeed, most of the specific questions mentioned here are probably "within reach."

(1) *The basic planning problems:* We now know that both decomposition and retraction are universal approaches to motion planning, and are even tractable for the (any) basic planning problems. Unfortunately, the general techniques are too inefficient, leaving considerable room for inventiveness in producing computationally attractive algorithms. For example, using Voronoi diagrams as retracts has attraction because of their nice theoretical properties but they are not always computationally amenable (cf. Ó'Dúnlaing, Sharir, & Yap, 1984b). Hence we can expect the design of increasingly efficient algorithms for specific robot systems to continue to be an area of practical as well as theoretical interest. In particular, problems in 3-dimensions are mostly open.

(2) *Complexity of the general motion planning problem:* What is the precise complexity of this decision problem? Is it equivalent to deciding the truth of Tarski sentences? Note that presently, the best known lower bound for the latter problem is a single exponential time (see Ben-Or, Kozen, & Reif, 1984; Berman, 1980).[21] On the other hand, the best lower bound for the general motion planning problem is polynomial space.

(3) *Kinematic motion planning:* One interesting new direction is the kinematic problem (P3) described in section 4. It is necessary to impose some optimality

---

[21]Note that the claim of exponential space lower bound in that paper is erroneous.

criteria in this problem for otherwise it degenerates to the usual motion planning. Unfortunately is appears that such problems are quite difficult even under simple criteria such as minimizing time, or minimizing the maximum acceleration subject to a time deadline. To make the problem more tractable, we might ask for the fastest trajectory that avoids a set of obstacles and which is composed of "splines" from some class. We refer to a forthcoming paper of Reif and Sharir (1985).

(4) *Dynamic motion planning:* An even less understood issue is to generalize kinematic motion planning by explicitly taking into account the driving forces and torques. Questions (5), (6) and (9) below fall under this general heading.

(5) *Grasping:* This is essentially a local planning problem. Suppose the robot has to pick up an object and move it to another location. Planning the trajectory is not at issue here, but rather: how to best pick up the object for stability and for nonself-obstruction? Some work on stability of grasping has been done (see Baker, Fortune, & Grosse, 1984, and references therein). The consideration of nonobstruction arises because a particular grasp may obstruct the carrying out of certain tasks. E.g. the grasp holds on to a protruding part to be fitted into a hole.

(6) *Manipulation of objects:* At the robotics laboratory of the Courant Institute a four-finger (planar) hand has been constructed for the study of control algorithms. Suppose one is given a convex polygon P and such a four-finger hand. A grasp of P is defined by (at most) four points (in some circular order) corresponding to the fingers in contact with the boundary of the polygon, together with the applied force vectors. To make the problem interesting, we assume that friction is present and that (i) there is an upper bound on the magnitude of individual forces (so the finger does not puncture the polygon); (ii) there is a lower bounds on the sum of the magnitude of the normal forces. Here we imagine the polygon is a crosssection of a prism so that the frictional forces can support the prism against gravity.

Given initial and final stable grasps, can one move from one grasp to the other?

Here we define a motion as a finite sequence of "elementary moves," where an elementary move is one which changes the force vectors continuously until the force at one finger reaches zero, at which moment that finger can be "lifted" and placed at another contact point (and the forces again change continuously to the desired final configuration). Given a polygon, does there exist a motion between any two stable grasps? We have some initial results in this direction. Some new properties hitherto unobserved seem to arise: if we allow sliding motion in the definition of elementary moves above then motion becomes in general an irreversible phenomenon (i.e., when motion is discretized to a finite graph, the edges of the graph are directed)!

Another interesting question is this: Is four (or some fixed number of) fingers "universal" in the sense that for any polygon, for any pair G, G' of stable grasps of the polygon, there exists a motion from G to G'. If the four finger hand is universal, is there an a priori bound (independent of the polygon) on the number of elementary moves?

(7) *Motion coordination:* what is the complexity of coordinating the motion of k discs amidst polygonal obstacles? For $k = 2, 3$, Yap (1984a) has shown upper bounds of $O(n^k)$. Is the complexity $O(n^k)$ for all k? What about the coordination of two robot arms in some assembly task?

(8) *Movable obstacles:* Closely related to motion coordination are the problems where the environment has some inert obstacles that can be moved. If the obstacles can be moved without the direct intervention of the robot system then this can be reduced to the previous coordination problem: simply treat each movable obstacle as another robot component except that the final placements of these obstacles are "don't cares." But suppose that a movable obstacle M can move only when the roving robot B is in contact with M (with B pushing M, as it were). Then we have:

We are given a rigid robot B, a fixed environment E, a sequence of movable objects $(M_1, \ldots, M_m)$, initial placements of $(B, M_1, \ldots, M_m)$ and a final placement Z of B. Is there a motion that takes B to its final placement, possibly by shuffling the placements of the movable objects?

(9) *Independently moving obstacles:* This differs from the previous case in that some of the obstacles comes with a fixed known motion: We want to do motion planning which avoids these self-moving obstacles. (This recalls some well-known arcade games of trying to avoid asteroids.) This problem might arise if we want to introduce an additional robot to a manufacturing environment that already has other moving robots. Initial results along this line have been reported in Reif & Sharir (1985).

(10) *Compliant motion:* One of the main impediments to more widespread application of robots is the inability of current robots to control motion by force compliance rather than by position. A typical task requiring such a feature is the problem of inserting a pig into a hole: dead reckoning by position is undesirable if not impossible. Rather, the motion planning should be guided by force feedback in directions normal to the axis of the hole. Actually, this may require a hybrid-control since motion in the direction of the hole axis is possibly position-

controlled. Motion planning is now done in a higher dimensional space that includes the force dimensions.

(11) *Realtime or on-line motion planning:* For such problems, assume the algorithm is equiped with sensory primitives which allows it to interact with its partially known environment. To have a true motion planning problem, we allow the algorithm to plan motion (both for the purposes of discovering more of the environment as well as trying to attain its motion planning goal) in the midst of the partial information. What seems particularly challenging here is that it is not clear how we ought to pose a mathematically meaningful question in general. In Cole & Yap (1983) a limited problem is posed and solved: They considered the problem where we are given a convex polygon P and we want to determine its exact shape and location. We are allowed to probe P which is modeled by a directed infinite line; if the line intersects P, then we are told the first point of intersection. They showed that if we are given one point in the interior of P (otherwise there is no finitary solution, trivially) then $3n$ probes suffices and $3n - 1$ probes are necessary where P is an n-gon.

12.   CONCLUSION

We conclude with two major challenges, one theoretical and one more pragmatic.

(a)   We may consider motion planning problems in three increasing levels of difficulty: (a) physical path planning, (b) kinematic motion planning and (c) dynamics motion planning. This paper is essentially about (a). We can say that the fundamental issues in path planning are well-understood at this point. It is also clear from the above list of open problems that very little is known about kinematic and dynamic motion planning. This is the main theoretical challenge for future algorithmic research.

(b)   The other challenge is this: how can the theoretical insights already gained be put to use in real implementations? We do not suggest that this is a challenge solely for implementors to make their algorithms approximate the theoretical ones. A rather more fruitful mode of operation is to initiate a dialogue between the two groups so that continuing refinements of the theoretical models can be made to suit practical constraints, which can then have a greater chance of being implemented.

ACKNOWLEDGMENTS

I am extremely grateful for the comments and invaluable suggestions of various readers of the original draft: Brenda Baker, Micha Sharir, Daniel Leven, Gordon Wilfong, Colm Ó'Dúnlaing, John Hopcroft, Dean Krafft, Ellen Silverberg, Jack Schwartz.

REFERENCES

Akers, S. (1967). A modification of Lee's path connection algorithms. *IEEE Trans. Electronic Computers, EC-16,* 97–98.

Arnon, D. S. (1979). *A cellular decomposition algorithm for semi-algebraic sets.* (Comp. Sci. Tech. Report No. 353). Univ. of Wisconsin.

Baker, B. S., (1985, February). Shortest paths with unit clearance among polygonal obstacles. Submitted to *J. Algorithms,*

Baker, B. S., Fortune, S. J. & Grosse, E. H., (1984, November). Stable prehension with three fingers. *Proc. 17th ACM Symp. Theory of Comp. Sci.*

Ben-Or, M., Kozen, D., & Reif, J. (1984). The complexity of elementary algebra and geometry. *STOC, pp. 457–464.*

Berman, L. (1980). The complexity of logical theories. *Theoretical Comp. Sci, 11,* 71–77.

Blum, M., & Kozen, D. (1978). On the power of the compass (or why mazes are easier to search than graphs. *Proc. 19th IEEE Symp. Found. Comput. Sci.,* 132–142.

Brady, M., Hollerbach, J., Johnson, T., Lozano-Pérez, T., & Mason, M. (1983). (Eds.), *Robot motion: Planning and control,* Cambridge, MA: MIT Press.

Brooks, R. (1982). Solving the find-path problem by representing free space as generalized cones. *IEEE Trans. Systems, Man and Cybernetics SMC-13*(2), 190–197.

Brooks, R. A. (1983). Planning collision-free motions for pick-and-place operations. *The International J. of Robotics Research,* 2(4), 19–44.

Cole, R., & Yap, C. (1983, December). *Shape from probing.* NYU-Courant Institute, Robotics Lab. No. 15. (to appear *J. Algorithms*)

Collins, G. E. (1975). Quantifier elimination for real closed fields by cylindrical algebraic decomposition (pp. 134–183). *2nd GI conference on automata theory and formal languages, 33,* New York: Springer-Verlag.

Cooke, G. E., & Finney, R. L. (1967). *Homology of cell complexes.* New Jersey: Princeton University Press and University of Tokyo Press.

Ernst, H. A. (1961). *MH-1 A computer-operated mechanical hand.* Sc. D. Thesis, M.I.T.

Garey, M. R., & Johnson, D. S. (1979). *Computers and intractability: A guide to the theory of NP-completeness.* San Francisco: W. H. Freeman.

Hopcroft, J. E., Joseph, D. A., & Whitesides, S. H. (1982). On the movement of robot arms in 2-dimensional bounded regions. *23rd FOCS,* 280–289.

Hopcroft, J. E., Schwartz, J. T., & Sharir, M. (1984, February). On the complexity of motion planning for multiple independent objects; PSPACE hardness of the "warehouseman's problem," (NYU-Courant Institute, Robotics Lab. Report No. 14).

Hopcroft, J. E., Wilfong, G. (1984a, May). *On the motion of objects in contact.* (Comp. Sci. Dept. TR 84-602). Cornell University.

Hopcroft, J. E., & Wilfong, G. (1984b, July). *Reducing multiple object motion planning to graph searching.* (Comp. Sci. Dept. TR 84-616). Cornell University.

Kantabutra, V., & Kosaraju, S. R. (1984). Algorithms for robot arm movement. *Proc. Conf. on Information Sciences and Systems, Princeton University.*

Kirkpatrick, D. G. (1979). Efficient computation of continuous skeletons. *20th FOCS,* 18–27.

Kozen, D., & Yap, C. K. (1985). Algebraic cell decomposition in *NC. Proc. 26th IEEE Symp. Found. Comput. Sci.,* 515–521.

Lee, C. (1961). An algorithm for path connections and its applications. *IEEE Trans. Electronic Computers, EC-10,* 346–465.

Lee, D. T., & Drysdale, R. L. III (1981). Generalization of Voronoi diagrams in the plane. *SIAM J. Comp., 10,* 73–87.

Leven, D., & Sharir, M. (1985). An efficient and simple motion planning algorithm for a ladder moving in two-dimensional space amidst polygonal barriers. *Proc. first ACM Symp. on Computational Geometry*, pp. 221–227.

Lewis, H. R., & Papadimitriou, C. H. (1980). Symmetric space bounded computations. *ICALP, 80*.

Lozano-Pérez, (1981). Automatic planning of manipulator transfer movements. *IEEE Trans. Systems, Man and Cybernetics, SMC-11*(10), 681–689.

Lozano-Pérez, T. (1983). Spatial planning: A configuration space approach. *IEEE Trans. Computers, C-32*(2).

Lozano-Pérez, T. & Wesley, M. (1979). An algorithm for planning collision-free paths among polyhedral obstacles. *C.ACM, 22*, 560–570.

Moore, E. (1959). Shortest path through a maze. *Annals of the Computation Lab. of Harvard Univ.*, *30*, 285–292.

Moravec, H. P. (1981). *Robot Rover Visual Navigation*. Doctoral thesis, Stanford, UMI Research Press.

Nilsson, N. J. (1969). A mobile automaton: An application of artificial intelligence techniques. *Proc. IJCAI-69*, 509–520.

Ó'Dúnlaing, C., Sharir, M., & Yap, C. K. (1983). Retraction: A new approach to motion planning. *15th FOCS*, 207–220.

Ó'Dúnlaing, C., Sharir, M., & Yap, C. K. (1984a). Generalized Voronoi diagrams for moving a ladder: I. topological analysis. (Robotics Lab. Tech. Report No. 32). NYU-Courant Institute. To appear, *Comm. Pure and Applied Math.)*

Ó'Dúnlaing, C., Sharir, M., & Yap, C. K. (1984b). Generalized Voronoi diagrams for moving a ladder: II. Efficient construction of the diagram. (Robotics Lab Tech. Report No. 33). (To appear, *Algorithmica*)

Ó'Dúnlaing, C., & Yap, C. K. (1982). A retraction method for planning the motion of a disc. *J. Algorithms, 6*, 104–111.

Paul, R. (1972). *Modelling trajectory calculation and servoing of a computer controlled arm*. Doctoral *thesis, Stanford University*.

Paul, R. (1981). *Robot manipulators: Mathematics, programming and control*. Cambridge, MA: MIT Press.

Pieper, D. (1968). The kinematics of manipulators under computer control. *ARPA Order No. 957*, Stanford University.

Rajan, V. T. (1985). Minimum time trajectory planning. *IEEE International Conf. on Robotics and Automation*, 759–764.

Reif, J. H. (1979). Complexity of the mover's problem and generalizations. *Proc. 20th IEEE Symp. Found. Comput. Sci.*, 421–427.

Reif, J. H., & Sharir, M. (1985). Motion planning in the presence of moving obstacles. *Proc. 26th IEEE Symp. Found. Comput. Sci.*, pp. 144–154.

Rowat, P. F. (1979). *Representing spatial experience and solving spatial problems in a simulated robot environment*. Doctoral thesis, University of British Columbia.

Schwartz, J. T., & Sharir, M. (1983a). On the piano movers' problem: I. The special case of a rigid polygonal body moving amidst polygonal barriers. *Advances in Appl. Math., XXXVI*, 345–398.

Schwartz, J. T., & Sharir, M. (1983b). On the piano movers' problem: II. General techniques for computing topological properties of real algebraic manifolds. *Advances in Applied Math, 4*, 298–351.

Schwartz, J. T., & Sharir, M. (1983c). On the piano movers' problem: III. Coordinating the motion of several independent bodies: The special case of circular bodies moving amidst polygonal barriers. *The International J. of Robotics Research, 2*(3), 46–75.

Schwartz, J. T., & Sharir, M. (1984). On the piano movers' problem: V. The case of a rod moving in three-dimensional space amidst polyhedral obstacles. *Comm. Pure and Appl. Math, XXXVII*, 815–848.

Shamos, M. I., & Hoey, D. (1975). Closest point problems. *Proc. 16th IEEE Symp. Found. Comput. Sci.*, 151–162.

Shamos, M. I., & Hoey, D. (1976). Geometric intersection problems. *Proc. 17th IEEE Symp. Found. Comput. Sci.*, 208–215.

Sharir, M. (1983). *Intersection and closest-pair problems for a set of circular discs.* (Courant Institute—NYU Technical Report No. 67).

Sharir, M., & Ariel-Sheffi, E. (1983, February). *On the piano movers' problem: IV. Various decomposable two-dimensional motion planning problems.* NYU-Courant Institute No. 58.

Sharir, M., & Schorr, A. (1984). On shortest paths in polyhedral spaces. *Proc. 16th AMC Symp. Theory of Comp.*, pp. 144–153.

Shimano, B. E. (1978). *The kinematic design and force control of computer controlled manipulators.* Doctoral thesis, Stanford University. (Computer Sci. Report No. STAN-CS-78-660).

Spirakis, P., & Yap, C. K. (1984). Strong *NP*-hardness of moving many discs. *Information Processing Letters, 19,* 55–59.

Sutherland, I. (1969). A method for solving arbitrary-wall mazes by computer. *IEEE Trans. Computers, C-18*(12), 1092–1097.

Tarski, A. (1951). A decision method for elementary algebra and geometry. Berkeley: University of California Press. (2nd ed. rev.)

Udupa, S. M. (1977). Collision detection and avoidance in computer controlled manipulators. *Proc. IJCAI-5, MIT*, pp. 737–748.

Widdoes, C. (1974). *A heuristic collision avoider for the stanford robot arm.* C.S. Memo 227, Stanford University.

Yap, C. K., (1984a, February). Coordinating the motion of several discs. (NYU-Courant Institute, Robotics Lab. No. 16).

Yap, C. K., (1984b, December). How to move a chair through a door. (NYU-Courant Institute, Robotics Lab).

Yap, C. K., (1985, May). An $O(n\log n)$ algorithm for the Voronoi diagram of a set of simple curve segments. (NYU-Courant Institute, Robotics Lab. Report No. 43).

4 Approximation and Decomposition of Shapes

Bernard Chazelle
Princeton University

ABSTRACT

This paper reviews some of the techniques developed recently for approximating or decomposing geometric shapes in two and three dimensions. The relevance of these techniques to robotics is compelling. As is well-known in the area of computer graphics, replacing complex objects by approximations or decompositions into simple components can greatly facilitate parts-manipulation (e.g. detection of interference, hidden-line elimination, shadowing). For this reason, as well as for the theoretical challenges raised by these problems, considerable work has been devoted on this subject in recent years. This paper attempts to survey some of the major advances in the area of approximation and decomposition of shapes. Methods of particular interest for their practical significance or their theoretical import are reviewed in detail.

1. INTRODUCTION

Planning collision-free motion and recognizing shapes are central tasks in robotics and automated assembly systems, in particular. In both cases, the process is complicated by the necessity to operate in real-time and the occasional presence of fairly complex objects. To cope with the complexity of the geometric problems at hand, two lines of attack can be contemplated: one involves simplifying the shapes by computing approximations of them; the other calls for rewriting the objects as combinations of simple parts. Both of these approaches have met with considerable success in computer graphics and automated design, and their relevance to robotics as a whole is evident. It is the aim of this paper to present a brief survey of the various methods and techniques known today in this area.

2.   APPROXIMATION OF SHAPES

The goal is to replace a complex shape S by a simpler figure F that captures morphological features of S. Since F will often be used to approximate the clearance of S amidst obstacles, it will be assumed to enclose S, so as to provide a conservative approximation. We review some of the most common approximation schemes previously devised.

2.1. Convex Hulls

The *convex hull* of a set S of n points is the intersection of all convex sets containing S. We omit the proof that the convex hull of S, denoted C, is a convex polygon whose vertices (the *extreme* points) are points of S. A simple characterization of a vertex of the convex hull states that a point is extreme if and only if it does not lie strictly inside any triangle formed by any three points. This leads to a trivial $O(n^4)$ time algorithm. For each such triangle, eliminate each point lying inside. The remaining points will be extreme. A number of more efficient algorithms have been (only recently) discovered. We propose here to review some of the most important, practical, and/or original methods. The honor of discovering the first *optimal* convex hull algorithm goes to R. Graham (1972).

Graham Scan: The method involves sorting the points in a preliminary stage and then retrieving the convex hull in linear time via a procedure traditionally known as a *Graham scan*. Let $\{p_1, \ldots, p_n\}$ be the points of S sorted angularly clockwise around p_1, the point of S with largest y-coordinate (take the one with largest x-coordinate to break ties, if necessary). Computing the list of p_i's can be done in $O(n\log n)$ time.

Before proceeding any further, we should make an important observation concerning the computation of angles in particular and the design of geometric algorithms in general. Although it is very helpful and intuitive to think in terms of angles in the design part of a geometric algorithm, computation of angles and of their natural functions (such as trigonometric functions) tends to be computationally expensive. There are usually ways around this difficulty, however. Most often, we find all that is needed is a primitive operation to determine the relative order of two vectors. This operation can be implemented with only a few multiplies and subtracts. Figure 4.1 illustrates this concept. Let $\theta = \angle(ab, ac)$ be the angle from ab to ac, measured in counterclockwise order. We may assume $-\pi < \theta \le \pi$. As a matter of terminology, we will often refer to the expression "*b, a, c form a right (resp. left) turn*" to mean that $\theta > 0$ (resp. <0). If p_x, p_y denote respectively the x and y coordinates of point p, we easily show that

$$\theta > 0 \leftrightarrow (b_x - a_x)(c_y - a_y) - (c_x - a_x)(b_y - a_y) > 0.$$

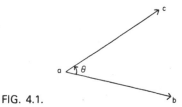

FIG. 4.1.

From a programming perspective, this equivalence allows us to compare, merge, and sort angles without having to compute them explicitly. For example, let a, $a_1, \ldots, a_p$ be $p + 1$ distinct points in E^2. To sort the points $a_1, \ldots, a_p$ angularly around a (clockwise, starting at a_1), one can simply use one's favorite sorting routine and re-implement the primitives $<$, $=$, $>$ used by the algorithm. Partition the original set of points into $L = \{a_i \mid \angle(aa_i, aa_1) > 0\}$, $M = \{a_i \mid \angle(aa_i, aa_1) = 0\}$ and $H = \{a_i \mid \angle(aa_i, aa_1) < 0\}$, and concatenate L, M, and H, after having sorted the sets separately. To do so, any comparison of the form $A_i < A_j$ in the algorithm should be replaced by $\angle(aa_j, aa_i) > 0$ and implemented with the formula given above. What we have here is an instance of *abstract data types*, a notion of crucial importance in geometric algorithm design. It is often possible and advantageous to specify geometric algorithms solely in terms of generic operations and treat the implementation of types separately. The reason is that many of these algorithms are extensions of well-known combinatorial algorithms couched in geometric terms. Identifying and separating their geometric and combinatorial components often simplifies the entire process of design, analysis, and implementation. With this observation made, we can now return to the convex hull problem.

Let px designate the horizontal ray emanating from p in the right direction and assume that $\angle(p_1p_2, p_1x)$ has the smallest value of any $\angle(p_1p_j, p_1x)$. Imagine that a rubber band is attached to p_1 at one end, and that the other end is taken to $p_2, p_3, \ldots, p_n$, in turn (Fig. 4.2). At the end of this process, the rubber band will be shaped exactly as the convex hull of S. The Graham scan is essentially a computer simulation of this process. Conceptually, it is easiest to describe this process as a stack manipulation algorithm. Recall that a stack is an *abstract data type* (i.e., a set and some operations defined on it), which behaves much like a pile of trays in a dining-hall. The only operations allowed are: *PUSH(e)* (add e to the

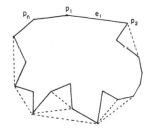

FIG. 4.2.

top); *POP* (remove top of stack); the top of the stack is designated *TOP*. The algorithm is based on the notion of *left* and *right turns*. At any stage, the stack contains the edges which should be on the convex hull were the algorithm to terminate at that instant. If the next vertex encountered witnesses a right turn, a new edge is added to the stack, otherwise, edges are popped off until the right turn condition is satisfied. The top of the stack is an edge *TOP*, whose endpoints are denoted t_1 and t_2, in clockwise order around the boundary.

```
begin
PUSH (p₁p₂)
for i = 3, . . . ,n
    begin
        while (t₁, t₂, pᵢ) turns left
            begin POP end
        PUSH (t₂, pᵢ)
    end
end
```

In an actual implementation, it might be more convenient (although conceptually more complicated) to simulate the stack with an array where each record holds a vertex and not an edge. The problem with representing edges is that information is essentially duplicated. See Sedgewick (1983, pp. 329) for an example of working code. A simple examination of the pseudo-code shows that the algorithm runs in linear time (the stack is *PUSH*ed at most once per point in *S*).

A Lower Bound: We easily see why Graham's $O(n\log n)$ time algorithm is optimal. We can sort n numbers $a_1, . . . ,a_n$ by computing the convex hull of the n points $\{(a_1, a_1^2), . . . ,(a_n, a_n^2)\}$ and reading off the vertices of the hull in clockwise order. Of course, this reduction assumes that by "computing the convex hull" we mean determining the order of the extreme points around the boundary. What if we only require the extreme points without reference to their order? Using a fairly technical argument, which we will not develop here, Yao has proven that even in this relaxed instance the convex hull problem requires $\Omega(n\log n)$ operations (Yao, 1981). Yao's model of computation is limited to decision trees with quadratic polynomial evaluations at the nodes. This assumption is actually quite realistic since all known methods seem to fall squarely in this model of computation. At any rate, calling upon deep results of algebraic geometry, Ben-Or has generalized Yao's lower bound to any decision tree with fixed-degree polynomial evaluation at the nodes (Ben-Or, 1983).

Gift-Wrapping: One drawback of the Graham scan is that it has the same complexity regardless of the size of the output. For example, if the convex hull happens to have only a small number of vertices, it is conceivable that a faster algorithm could be used. The *gift-wrapping* method, also known as the *Jarvis march* (Jarvis, 1973), provides a (partial) response to this concern. Once again, the algorithm is the simulation of a very simple physical process. Let p_1 be as usual the point of S with largest y-coordinate. Attach a long rope to p_1 and wrap it around S one step at a time. A *step* is the discovery of a new contact with the rope (i.e., a new extreme point). A simple angle "calculation" (see remark above concerning angles) allows us to implement every step in linear time. This leads to an $O(nh)$ time algorithm, where h is the number of extreme points. How will this method fare against the Graham scan on the average? Let $h^*(n)$ be the expected number of extreme points in a set of n points chosen uniformly and independently inside a bounded region R. It has been shown by Renyi and Sulanke (1963) that if R is a square or any convex k-gon for a fixed k, $h^*(n) = O(\log n)$, and that if R is a circle, $h^*(n) = O(n^{1/3})$ (Shamos, 1978). This shows that for the case of a uniform distribution in a k-gon the Jarvis march will asymptotically match the Graham scan in the average case.

Divide-and-Conquer: A very good algorithm in the sense of expected behavior was developed by Bentley and Shamos (1978). Assume that the points of S are stored in an array $A[1, n]$. Recursively, the algorithm computes the convex hull P of $A[1, \lfloor n/2 \rfloor]$ and the convex hull Q of $A[\lfloor n/2 \rfloor + 1, n]$; then in time proportional to the added size of P and Q it computes the convex hull of S. The expected running time of the algorithm $T(n)$ follows the recurrence relation, $T(n) = 2T(n/2) + O(h^*(n))$. It follows that $T(n) = O(n)$ in the case of a uniform distribution inside a k-gon or a circle. Two important facts to observe:

1. The recursive calls involve passing indices or pointers and *not* the entire arrays. The latter solution would always entail an $O(n\log n)$ run time;
2. The uniformity of the distribution is preserved through recursive calls since the subsets handled are defined independently of the value of their elements.

To complete the demonstration of the efficiency of Bentley and Shamos' method, we must describe a linear method for "merging" two convex polygons. Let P and Q be two convex polygons with respectively p and q vertices. We describe a method for computing the convex hull of $P \cup Q$ in $O(p + q)$ operations. Let $\{v_1, \ldots, v_p\}$ and $\{w_1, \ldots, w_q\}$ be a clockwise vertex-list of P and Q, respectively. Since P is convex the sequence of vertices $\{v_1, \ldots, v_p\}$ is angularly sorted around v_1. Is this also true of Q? If v_1 happens to lie inside Q, it is certainly so. Otherwise, the sequence of angles is *bimodal*, i.e. unimodal up to a circular permutation. By checking each vertex of Q, compute the two *segments of support* $v_1 w_i$ and $v_1 w_j$: these segments are such that v_1, w_i, w_{i-1} and v_1, w_i, w_{i+1} form right turns, and v_1, w_j, w_{j-1} and v_1, w_j, w_{j+1} form left turns (Fig. 4.3) — arithmetic on indices is done mod q. Note that these two conditions will be satisfied if and only if $v_1 \notin Q$, so whether $v_1 \in Q$ or $v_1 \notin Q$ need not be decided beforehand. Every vertex from w_{j+1} to w_{i-1} clockwise (if any) can be discarded and Q can be redefined as $\{w_i, w_{i+1}, \ldots, w_j\}$. In all cases, Q now forms a monotone sequence of angles with respect to v_1, so we can merge the vertices of P and Q together in linear time, and then apply the Graham scan on the resulting

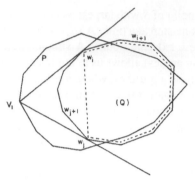

FIG. 4.3.

list of vertices. This will provide us with the convex hull of $P \cup Q$ in $O(p + q)$ time, which completes the description of Bentley and Shamos' algorithm. Despite the linear expected running time of the algorithm, one will notice the similarity between this method and *mergesort* (Aho, Hopcroft, & Ullman, 1974; Knuth, 1973; Sedgewick, 1983).

Divide-and-Conquer with Sorting: If we apply the Bentley-Shamos algorithm after having sorted the points of S by x-coordinate, the merge part can be easily implemented without resorting to a Graham scan (Preparata & Hong, 1977) (of course, by doing so we forsake any hope of breaking the $O(n\log n)$ barrier, as it is well known that sorting takes on the order of $n\log n$ operations, even on the average). Let P and Q be, as above, the two convex polygons to be *merged*. We assume that P lies totally to the left of Q. The convex hull of $P \cup Q$ is obtained by computing the *upper* and *lower bridges* of P and Q, i.e., the two unique segments joining a vertex of P and a vertex of Q with both polygons lying entirely on the same side of the infinite lines passing through the bridges (Fig. 4.4). Whether a segment is a bridge can be checked locally in constant time since P and Q are convex. For example, the upper bridge $v_k w_l$ is such that all four turns (v_k, w_l, w_{l-1}), (v_k, w_l, w_{l+1}), (v_{k-1}, v_k, w_l), (v_{k+1}, v_k, w_l) are right. To compute the bridges of P and Q, pick the leftmost vertex of P, say, v_1, and compute the segments of support, $v_1 w_i$ and $v_1 w_j$, as described earlier. Then proceed to *roll* the line U (resp. L) passing through $v_1 w_i$ (resp. $v_1 w_j$) clockwise (resp. counterclockwise) around Q, until the line contains the upper (resp. lower) bridge (Fig. 4.5).

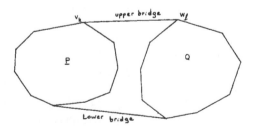

FIG. 4.4.

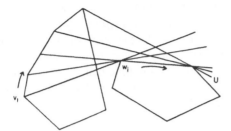

FIG. 4.5.

Rolling the line U passing through v_1w_i can be done by maintaining the two current contact-points, v_a and w_b, and deciding whether v_{a+1} or w_{b+1} should be the next one. Once again, this procedure can be understood as a two-way merge (i.e., merge of two sorted lists), with the primitives $<, =, >$ re-implemented as *left turn, no turn, right turn* conditions. The only major difference with merging is in the way the algorithm terminates. Rather than scanning the two lists through the end, the algorithm must keep on checking whether the bridge has been found, and stop as soon as it has. Unlike the previous one, the running time of this method is inherently $O(n\log n)$ because of the preliminary sorting. Why then did we bother mentioning it at all? The truth is that the algorithm serves as the ideal stepping-stone for the next method, the first one known to be optimal in both input and output size.

Marriage-Before-Conquest: Aside from the fact that the method we last described entails a sort on n numbers, a simple look at Fig. 4.4 reveals its main shortcoming. Once the bridges have been found, many edges of P and Q are bound to become irrelevant, therefore all the work done to obtain these edges will turn out to be useless. It would be nice in some sense to be able to run the algorithm *backwards*. We would first compute the bridges and then discard all the points lying directly below or above them, at which stage we would simply iterate recursively on the two sets of remaining points. The advantage of such a scheme would be to guarantee that every bridge computation is effective; in other words, no bridge computed would ever be thrown away. The bottleneck caused by the preliminary sorting can in turn be alleviated by resorting to a linear-time median algorithm (Aho et al., 1974; Knuth, 1973). This will allow us to partition the input set into two roughly equal-sized sets at a linear cost. This approach was taken by Kirkpatrick and Seidel and led to the first optimal algorithm in the sense of input *and* output size. More specifically, if h denotes as usual the number of extreme points in S, Kirkpatrick and Seidel's method runs in time $O(n\log h)$. Optimality follows from an information-theoretic argument given in Kirkpatrick and Seidel (1983).

We briefly review the main facets of the algorithm, referring the reader to Kirkpatrick and Seidel (1983) for further details. To begin with, we may restrict ourselves to the *upper hull*, i.e., the chain of the convex hull that runs clockwise from the leftmost to the rightmost point. The lower hull will be computed in a similar fashion, from which the convex hull will follow directly. Let $\{p_1,$ . . . $,p_n\}$ be the points of S in arbitrary order, with $x(p_i)$ denoting the x-coordinate of p_i. Let a be a real number and P (resp. Q) be the convex hull of $\{p \in S \mid x(p) \leq a\}$ (resp. $\{p \in S \mid x(p) > a\}$). The expression "*upper bridge over a*" refers to the upper bridge of P and Q. Let's assume for the time being that we know how to compute the *upper bridge over a* in a linear number of steps. The upper hull of S is computed by calling $HULL(1,n,S)$.

```
HULL(k,l,S)
Find median a of x-coordinates in S
Find upper bridge p_i p_j of S over a, with x(p_i) ≤ x(p_j)
S_1 ← {p ∈ S | x(p) ≤ x(p_i)}
S_2 ← {p ∈ S | x(p) ≥ x(p_j)}

"note that p_i ∈ S_1 and p_j ∈ S_2"

if i = k
      then print (i)
      else HULL(k, i,S_1)
if j = l
      then print (j)
      else HULL(j, l, S_2)
```

It is not difficult to see that in the worst case, one bridge computation will require $O(n)$ steps, two will require $O(n/2)$ steps each . . . , and more generally 2^i bridge computations will require $O(n/2^i)$ steps each. In other words, there is some integer k such that $\Sigma_{i \leq k} 2^i \gtrsim h$ and the running time is $O(\Sigma_{i \leq k} 2^i(n/2^i))$. This leads to the optimal $O(n \log h)$ time complexity announced earlier. To complete the demonstration, we need to describe a linear time algorithm for computing the upper bridge of P and Q. In a linear number of operations, we wish to be able to discard at least a fixed fraction of the points of S. Let α be an arbitrary slope $\in (-\infty, +\infty)$ and let L_α be the line of the form $Y = \alpha X + \beta$, with largest β such that there exists at least one index i with $p_i \in L_\alpha$. Assume that $x(p_i) < a$. The key observation is that for any pair (p_u, p_v) such that $x(p_u) < x(p_v)$ one may discard the point p_u provided that the slope of $p_u p_v$ exceeds α (Fig. 4.6). Indeed, because of the relative positions of p_i and p_v, the point p_u cannot be an endpoint of the upper bridge. Symmetrically, if we have $x(p_i) > a$, we will be able to discard p_v every time slope$(p_u p_v) < \alpha$. In order to balance the odds, we will pair up (p_1, p_2), (p_3, p_4), etc. and compute the median slope α of the $\lfloor n/2 \rfloor$ segments $p_{2j-1} p_{2j}$. Let $p_i \in L_\alpha$; once we have identified whether $x(p_i) < a$ or not, as many as $\lfloor n/4 \rfloor$ points will immediately fall out of contention. Iterating on this process will eventually produce the upper bridge. The time $T(n)$ taken by the algorithm satisfies the recurrence relation $T(n) = T(n - \lfloor n/4 \rfloor) + O(n)$, which gives $T(n) = O(n)$. Our exposition has left a number of special cases conveniently hidden (e.g.

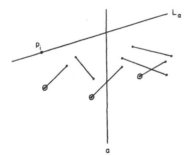

FIG. 4.6.

segments with infinite slopes, several segments with slope α, several points with x-coordinate a, etc.). All these difficulties can be easily handled with extra care, however, so we simply refer the reader to Kirkpatrick and Seidel (1983) for details.

Quickhull: Eddy and Floyd independently developed a convex hull algorithm, which although quadratic in the worst case, behaves remarkably well in practice (Eddy, 1977). This is not without resembling the case of *quicksort,* a sorting method which fares better than almost any other in practice albeit vulnerable to extremely poor behavior in the worst case (Aho et al., 1974; Knuth, 1973; Sedgewick, 1983). For the reader familiar with *quicksort* the resemblance will shortly appear much deeper than this, which is the reason why we christened the algorithm *quickhull.* Let A and B denote respectively the highest and lowest point in S. Compute the two points C, D of S on the leftmost and rightmost lines parallel to AB. Eliminate the points inside the quadrilateral $ABCD$. Sweep lines parallel to AC, CB, BD, DA, in turn, *outward* from the quadrilateral, and eliminate the points from inside the four triangles thus created (Fig. 4.7). Each of these triangles is formed by the segment from which the sweep starts and the last point swept. The algorithm iterates on this process. Under some reasonable assumptions concerning the distribution of the points in S, it is possible to show that the algorithm will run in linear expected time. In practice, this algorithm could be used as a preprocessor to Graham's algorithm, for example. The idea is that after a small number of passes, only very few points should be left, so a sort

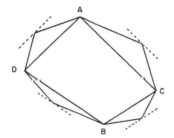

FIG. 4.7.

is unlikely to be time-consuming. Similar to *quicksort*, quickhull is a prime candidate for a simple, practical, and efficient implementation.

Approximation Method: On the practical side, it is worthwhile to mention a very fast convex hull algorithm due to Bentley, Faust, and Preparata (1982). The algorithm produces an approximation of the convex hull, that is to say, it might fail to report the *exact* convex hull, but can always be guaranteed to produce a convex hull arbitrarily *close* to the exact one (for some realistic measure). This might often be acceptable in situations where only a rough approximation of the morphology of the point-set is desired.

Higher Dimensions: In three dimensions, the situation is (ironically) much "clearer" than in E^2 because of the paucity of efficient algorithms. Preparata and Hong (1977) have devised an optimal $O(n\log n)$ time method which constitutes an extension of the divide-and-conquer method with sorting given for E^2. For higher dimensions ($d > 3$), algorithms are given in Chand and Kapur (1970), and Seidel (1981). Chand and Kapur (1970) extend the gift-wrapping method to arbitrary dimensions, while Seidel (1981) gives a general algorithm which is optimal when d is even. The running time of his algorithm is $O(n\log n + n\lfloor\frac{d+1}{2}\rfloor)$.

Dynamic Convex Hulls and Convex Layers: Preparata (1979) has described an optimal algorithm for inserting new points into a two-dimensional convex hull. If the convex hull has n vertices, any new point can be added to it in $O(\log n)$ time. One drawback of this method is that points which either fail to be on the convex hull, or stop being on the hull as a result of an insertion, are lost, thereby making deletions impossible. Overmars and van Leeuwen (1981) have shown that by sacrificing a little in time it is possible to accommodate deletions (see chapter by Dobkin and Souvaine in this volume). Their algorithm handles any update of this nature (insertion or deletion) in $O(\log^2 n)$ time. Chazelle has shown, on the other hand, that if only deletions from the convex hull are allowed, then optimal (amortized) time can be achieved (Chazelle, 1985). This, in particular, allows for the computation of the *convex layers* of an n-point set in $O(n\log n)$ time. These are the various polygons obtained by computing the convex hull of the point set, and then removing all the vertices from the set and iterating on this process until all the points are gone. This collection of polygons turns out to be very useful in statistical analysis (Shamos, 1978). It also allows for efficient range searching, as has been shown in Chazelle, Guibas, and Lee, 1983.

Convex Hull of Polygon: An important class of convex hull applications involves a set of points joined together by a simple curve. Recall that a curve is simple if it is not self-intersecting. Let P be a simple polygon with n vertices. This exist several algorithms for computing the convex hull of P in $O(n)$ time (Bhattacharya & El Gindy, 1984; Graham & Yao, 1983; Guibas, Ramshaw, & Stolfi, 1983; Lee, 1983; McCallum & Avis, 1979). Most of these algorithms can be viewed as generalizations of the Graham scan. A pointer scans the boundary of the polygon, while one or two stacks keep track of the "current" convex hull.

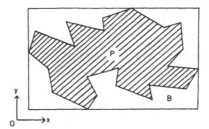

FIG. 4.8.   o └───→x

2.2. Rotating Calipers

A simple, albeit rough, method for approximating the shape of an object is to place it in a *bounding box*. In Fig. 4.8, object P has been tightly enclosed by a rectangle B whose sides are parallel to the axes. The term *axes* refers here to the two directions of an orthogonal system of coordinates (Ox, Oy). This method is very popular in computer graphics (Newman & Sproull, 1979), because it greatly simplifies the task of removing hidden parts from a scene, testing intersection between several objects, etc. In raster graphics, the horizontal and vertical directions are privileged, so it is natural to orient bounding boxes along them. Assume that P is represented as a polygon with vertices $p_1, \ldots, p_n$ in clockwise order. We easily compute B in $O(n)$ time by determining the maximum and minimum x and y coordinates among the p_i's.

Sometimes, insisting on a specific orientation of the bounding box can cause considerable degradation in the quality of the approximation. Figure 4.9 suggests that the proper measure of the *quality* of the bounding box should be the area that it occupies, regardless of its orientation. By rotating the box B counterclockwise by approximately 45 degrees, one reaches another enclosing box of much higher quality by this measure. The question that leaps to mind is then:

> *What is the complexity of computing the smallest-area rectangle enclosing the polygon P?*

If P is highly symmetric, it is easy to see that its smallest-area enclosing rectangle may not be necessarily unique. For this reason, we limit our search to *any* such rectangle, which we generically denote $\mathcal{A}(P)$. Toussaint (1983) has

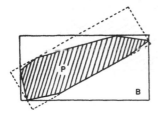

FIG. 4.9.

shown that finding $\mathcal{A}(P)$ can be done in $O(n)$ time. His method relies on a number of geometric observations. To begin with, since $\mathcal{A}(P)$ contains the convex hull C of P, we can advantageously drop P from consideration and restrict our investigation to C. As has been mentioned earlier, C can be computed in $O(n)$ time. The next crucial fact is a theorem by Freeman and Shapira (see Toussaint, 1983) which states that $\mathcal{A}(P)$ always has one side containing an edge of C. This suggests an $O(n^2)$ time algorithm based on the previous bounding box method. Make the x-axis parallel to each edge of C in turn, and for each orientation obtained solve the *bounding box* problem in $O(n)$ time. Toussaint observed that there is no need to repeat each computation from scratch for every position of the x axis. Instead, one should try to *batch* computations together by trying to derive the next answer directly from the previous one.

This intuition is materialized in the notion of *rotating calipers*. Let c_1, $\ldots, c_p$ $(p \leq n)$ be the vertices of C given in clockwise order. Consider the bounding box *anchored* at $c_{i-1}c_i$, i.e., the smallest-area enclosing rectangle with one side containing $c_{i-1}c_i$. This rectangle is formed by four infinite straight lines L_i, L_j, L_k, L_l (Fig. 4.10). Let c_i, c_j, c_k, c_l be the vertices in contact with the boundary of the rectangle, with $c_i \in L_i$, $c_j \in L_j$, $c_k \in L_k$, $c_l \in L_l$ (whenever a whole edge is in contact with the rectangle, we choose its *last* vertex clockwise as its representative). We look at L_i, L_j, L_k, L_l as calipers rotating clockwise around C in such a way that they preserve the integrity of the rectangle they form. Because of Freeman and Shapira's result, we may restrict our attention to the positions of the calipers for which one line contains an edge of C. To compute the next position from the current one, it suffices to compare the four angles, $\alpha_i, \alpha_j, \alpha_k, \alpha_l$ formed respectively by L_i and c_i, c_{i+1}, L_j and c_j, c_{j+1}, L_k and c_k, c_{k+1}, and L_l and c_l, c_{l+1}. The smallest of them (α_k in Fig. 4.10) determines the next position of the calipers. The rotation will continue until the calipers are back to their initial position. Because of the obvious symmetry of the calipers, the rotation can be limited to 90 degrees.

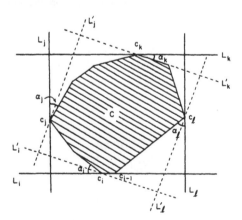

FIG. 4.10.

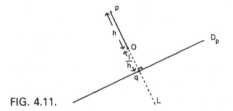

FIG. 4.11.

The algorithm is implemented by maintaining four pointers to keep track of c_i, c_j, c_k, and c_l. It strongly resembles the traditional method for merging four sorted lists into one, the so-called 4-way merge (Knuth, 1973). In particular, its complexity is clearly $O(n)$ in both time and space. The similarity with merging can be better seen if we map the polygon C into a different space. We resort to a technique of very common use in computational geometry: a *dual transform*.

The idea is (in this case) to put points and lines in one-to-one correspondence so as to be able to rephrase problems on lines as problems on points and vice-versa. Point $p = (a,b) \neq (0,0)$ is (bijectively) mapped to the line $D_p : aX + bY + 1 = 0$, and conversely D_p is mapped to p. Intuitively, this mapping is effected by considering the line L passing through both p and the origin O, and determining the point q of L at a distance $1/|Op|$ from p, with O on the segment pq (Fig. 4.11). The line D_p is obtained by taking the normal to L passing through q. It is easy to see that the transformation D preserves incidence relations. For this reason, it is tempting to consider each of the lines bounding C and map them via the dual transform D. Let L_i be the line containing the edge $c_i c_{i+1}$ and let $v_i = D_{L_i}$ denote its dual point ($1 \leq i \leq p$; index arithmetic taken mod p). Assume that the origin has been chosen to be *inside* the polygon C. It is easy to see that the sequence $\{L_1, \ldots, L_p\}$ is mapped into a sequence of points $\{v_1, \ldots, v_p\}$, which form a clockwise vertex-list of a convex polygon V.

A mechanical analogy of this correspondence is to roll a line (a caliper) around C clockwise and follow the motion of its dual point. The trajectory of this point will be precisely the convex polygon V. Define a *cross* to be a set of two lines passing through O normal to each other. A cross intersects V in four points, called *cross-points*. Our original set of four calipers is mapped into a set of four cross-points. The analogy is now complete: rotating the calipers corresponds to moving the cross clockwise. If L is a list of numbers, let $L + x$ be the list obtained by adding the number x to each element in L, and let θ_i be the slope of Ov_i ($1 \leq i \leq p$). Let L be the sorted list of angles $\{\theta_1, \ldots, \theta_p\}$. The 4-way merge of lists to which we evasively referred as the backbone of the rotating algorithm is now taking shape. A moment's reflection shows that the algorithm for computing the smallest-area rectangle all but boils down to merging together the four lists L, $L + 90$, $L + 180$, and $L + 270$.

This example and the little excursion into dual space which followed tell us something. First, a fairly specific geometric problem is solved in a very simple way by use of a general, unifying technique: the rotating calipers. Second, this technique is nothing but a geometric instantiation of a fundamental programming concept: k-way merging (Knuth, 1973). Put in this light, it is no surprise that the rotating calipers should find many other applications in computational geometry. For example, it is possible to find the *diameter* of a set of n points in E^2 in $O(n\log n)$ time, i.e., the largest distance between any two points (Shamos & Hoey, 1975; Toussaint, 1983). We can do so by first computing the convex hull of the point-set in $O(n\log n)$ time, and then rotating two parallel calipers around the hull in $O(n)$ steps. The two points realizing the maximum distance must appear as contact-points of the calipers in some position. Of more direct relevance to the subject of shape-approximation, we pose the following problem.

What is the complexity of computing the smallest-area triangle enclosing a convex polygon?

Note that solving this problem will also allow us to compute the smallest-area triangle enclosing an arbitrary set of points. We do so by taking the convex hull of the points as a preprocessing step, and thus reduce the problem to the one posed. Klee and Laskowski (1985) have proposed an $O(n\log^2 n)$ time algorithm for computing the smallest-area triangle of a convex polygon with n vertices (or any of them if there are several). Their algorithm was recently improved to optimal $O(n)$ time in O'Rourke, Aggarwal, Maddila, and Baldwin (1984). To quote O'Rourke et al., "the strength of their (Klee and Laskowski's) paper lies in establishing an elegant geometric characterization of these (local) minima, which permits them to avoid brute-force optimization." The strength of O'Rourke et al.'s paper, on the other hand, is to prove an *interspersing lemma*, which enables them to use rotating calipers as a guiding hand. Three calipers are used to represent tentative enclosing triangles: the only conceptual difference with the rectangular case is that (1) angles between calipers are not fixed; (2) the three calipers do not have a uniform behavior. Nevertheless, the algorithm decides on the basis of a constant-time test which caliper should be moved next and by how much. Since calipers move in the same direction, the running time is trivially linear. Once again, we can interpret the basic method as a k-way merge ($k = 3$), although the non-uniformity of the ranking criterion makes the analogy slightly less compelling.

2.3. Circle and Ellipse Enclosure

We leave aside polygonal shapes temporarily and turn our attention to smoother curves like circles and ellipses. Conic sections or for that matter any polynomial curves of reasonably small degree have the advantage of being space-effective: just a few coefficients are necessary to represent them. Also being *smoother* than polygons, these curves will often provide finer approximations than, say, triangles, rectangles, or more generally polygons with a small number of vertices. To begin with, we ask the following question:

What is the complexity of computing the smallest-area circle enclosing a set of points?

2.3.1. Circle Enclosure.   The case of circles is particularly interesting from an algorithmic standpoint. The problem can be traced as far back as Sylvester (1857) and has gone through numerous developments (see Shamos, 1978; Megiddo, 1983 for a short bibliography). Until recently the best solution known was due to Shamos and Hoey (1975) and was based on a geometric construction, called the *Voronoi diagram*. Its complexity of $O(n\log n)$ was subsequently improved to optimal $O(n)$ by Megiddo (1983) in a seminal paper on linear programming in $\mathfrak{R}^3$. The two methods are radically different and deserve separate treatments.

Shamos and Hoey's method: This is the simpler of the two. It rests crucially on a geometric construction, called the *farthest-point Voronoi diagram*. Let $S = \{p_1, \ldots, p_n\}$ be a set of n points in the Euclidean plane. Partition the plane into regions with common *farthest neighbors*. More precisely, for any point p, let $f(p)$ be the index i such that for each $j \neq i$, the distance $|pp_i|$ exceeds the distance $|pp_j|$. Whenever more than one index i can be found, $f(p)$ stands for the set of such indices. The function f partitions E^2 into faces, edges, and vertices. Let V_i denote the face associated with p_i. V_i is the intersection of each half-plane that is delimited by the bisector of (p_i, p_j) $(1 \leq j \neq i \leq n)$ and contains p_j. For this reason, either V_i is empty or it is a convex polygon (in the latter case, it is easy to show that V_i is unbounded). Furthermore, V_i is not empty if and only if p_i is a vertex of the convex hull of S. This leads to the fact that the faces V_i can be ordered cyclically around the boundary of the convex hull (Fig. 4.12). Since the set of edges forms a free tree (connected acyclic graph) it immediately follows that the subdivision contains at most $F - 2$ vertices, where F is the number of faces. Since $F \leq n$, we conclude that the farthest-point Voronoi diagram of n points in the plane has at most $n - 2$ vertices. Shamos and Hoey (1975) have

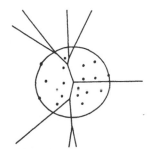

FIG. 4.12.

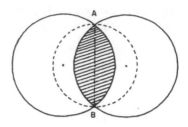

FIG. 4.13.

described how to compute this diagram in $O(n\log n)$ time, using a divide-and-conquer strategy.

We next show how to compute the smallest circle using this diagram. Let C be the smallest-area enclosing circle of S. To see that C is uniquely defined, assume the existence of two such circles C and C', and let A, B be their two intersection-points (note that the two circles *must* intersect). Every point of S lies in the shaded area in Fig. 4.13, therefore the circle of diameter AB contains S and is smaller than C and C', a contradiction. Let p_i, p_j be the pair of points in S whose distance to each other is maximum. This distance, called the diameter of S, can be obtained in $O(n)$ time using the farthest-point Voronoi diagram. To do so, determine the farthest neighbor of each point in S. Once the diameter is available, check whether every point of S lies inside the circle with diameter p_ip_j. If yes, this circle is clearly C; otherwise, C passes through at least three points p_u, p_v, p_w of S. Obviously the center of C is then the vertex at the intersection of V_u, V_v, V_w. Trying out all vertices of the farthest-point diagram and keeping the one whose associated circle has maximum radius will give us C. To conclude, we have described an $O(n\log n)$ time method for computing the smallest-area circle enclosing n points.

Megiddo's Method: Recently, Megiddo proved the surprising result that C can actually be computed in only a linear number of operations. His method is based on a general technique used primarily in the context of linear programming. It is this technique which inspired Kirkpatrick and Seidel to design their ingenious marriage-before-conquest convex hull algorithm (see section 2.1). Since the entire algorithm can be explained reasonably simply in a self-contained manner, we cannot resist the temptation of trying. As the reader will undoubtedly realize, the method contains a treasure of algorithmic insights. In the following, the pair (a_i, b_i) will denote the coordinates of p_i ($1 \leq i \leq n$). To begin with, we solve a simpler problem. Let L be an arbitrary line; a circle is said to be *centered* at L if its center lies on L. The question we pose is the following:

I. *What is the smallest-area enclosing circle centered at L?*

The same argument used above shows that this circle, denoted C_L, is uniquely defined. For the sake of convenience, we think of L as the x-axis, so we can represent the center of C_L by its x-coordinate, x^*. Megiddo's approach is similar in spirit to the well-known linear-time median algorithm (Aho et al., 1974; Knuth,1973). The idea is to identify and *discard* points that are irrelevant to the problem. A point p is irrelevant if it can be determined that C_L is the smallest-area circle enclosing $S - \{p\}$ centered at L. We will show a method for discarding at least a fraction α of the input set in linear time. This will allow us to zoom in on the solution in time $O(n + (1 - \alpha)n + (1 - \alpha)^2 n + \ldots) = O(n)$. To do so, we need to describe a number of primitive operations.

1. To begin with, it would be nice to be able to determine quickly whether $x^* \le x$ for an arbitrary value x. We can answer this question in linear time by computing the maximum distance d from $(x, 0)$ to any point in S. Let I be the set of points that realize this distance, i.e. $I = \{p_i \mid (a_i - x)^2 + b^2 = d^2\}$. If every point in I is to the left (right) of x, so must be x^*. More specifically, we have the following implications: if for each $p_i \in I$, $a_i > x$ then $x^* > x$; if for each $p_i \in I$, $a_i < x$ then $x^* < x$. Otherwise, we have $x = x^*$.

2. Secondly, given two points p_i, p_j ($a_i < a_j$), we would like to decide whether p_i or p_j can be discarded for good. Let z be the x-coordinate of the intersection of L with the bisector of (p_i, p_j) (we leave it to the reader to see what to do should this intersection be undefined). If it is known that $z < x^*$ (resp. $x^* < z$), then clearly p_j (resp. p_i) can be discarded.

Next we will see that with these two primitives it is possible to discard a fraction of the points in $O(n)$ operations. For each even value of i ($2 \le i \le n$), compute z_i, the x-coordinate of the intersection of L with the bisector of the pair (p_{i-1}, p_i). If $a_{i-1} = a_i$, it is legitimate to discard whichever of p_i or p_{i-1} is closer to L; we then leave z_i undefined. Next, compute the median element z_h of the list of z_i's thus obtained in $O(m)$ steps, where m is the number of remaining points. Using Primitive 1, determine in $O(m)$ time whether x^* lies strictly to the left or strictly to the right of z_h. If neither is the case, we have $x^* = z_h$ and we are finished. Otherwise, Primitive 2 allows us to immediately discard at least one point for each pair z_i falling on one side of z_h. This leaves us with roughly $\le 3m/4 \le 3n/4$ points, which leads to an $O(n)$ running time for computing x^*. Next we ask a slightly different type of question.

II. *Given a line L, does the center of the smallest-area enclosing circle lie on L: if yes, where? If not, on which side of L does it lie?*

Using our solution to Problem I, it suffices to determine the sign of y_C, the y-coordinate of the center of C. Without invoking convexity results explicitly, we will show by a simple geometric argument how to determine the sign of y_C. Let r be the radius of C_L and let J be the set of points of S at a distance r from $(x^*, 0)$, i.e. $J = \{p_i \mid (a_i - x^*)^2 + b_i^2 = r^2\}$. If J has a unique element, $J = \{p_i\}$, then $a_i = x^*$ and y_C has the sign of b_i (always assuming that L is the x-axis). If J has at least two elements, one of them must be to the left of x^* and the other to its right, otherwise it would be possible to "improve" x^*. Let p_i and p_j be such that p_i, p_j $\in J$, $a_i \leq x^* \leq a_j$, and let D_i (resp. D_j) be the disk centered at p_i (resp. p_j) of radius r. The center of χ of C must lie in the intersection $D_i \cap D_j$. It is immediate to see that $K = D_i \cap D_j$ is free of any point of L besides $(x^*, 0)$, therefore K lies totally on one side of L. Which one it lies on can be determined in constant time once p_i and p_j are available. This directly solves our problem since χ lies in K.

In some instances, it is possible to determine χ immediately. To identify these cases, we will compute $K^* = \cap p_i \in JD_i$ instead of simply K. To do so, we consider each $p_i \in J$ in turn (in any order) and update the current intersection. Since all the disks have a common point, the intersection either is a single point or consists of two circular arcs, so it can be computed in constant time per disk, i.e. in $O(n)$ time. If K^* happens to be reduced to a single point then $\chi = (x^*, 0)$ and we are finished. Otherwise, K^* lies entirely on one side of L, since we have (in particular) $K^* \subseteq K$. Next we solve the last subproblem of our series, from which we will be able to conclude directly.

III. *Eliminate a fraction of the points in S on the grounds that they are irrelevant to the definition of the smallest-area circle enclosing S.*

Using our algorithm for problem II, we can carry out a binary search over the Euclidean plane and thus zoom arbitrarily close towards the center χ. Unfortunately, this will not indicate the points contributing the center, however close we get to χ. Following a pattern by now familiar, we carry out the binary search not over E^2 but over a discrete set, namely S itself. To realize this objective, we must supplement our set of primitives with the following one: given a convex polygon (not necessarily bounded) in which χ is known to lie, and given the bisector of the pair (p_i, p_j), discard either p_i or p_j. This can be easily done as long as the bisector does *not* intersect the convex polygon. Indeed, in that case, one of the points, say p_i, lies on the same side of the bisector as every point in the polygon. This implies in particular that χ is closer to p_i than to p_j, and so p_i can be discarded. We are now in a position to present the algorithm in its entirety.

To begin with, pair up the points of S in an arbitrary fashion, e.g. (p_1, p_2), (p_3, p_4), . . . , and let L_1 be the bisector of (p_1, p_2), L_2 the bisector of (p_3, p_4), etc. Consider the slopes (in $[-\infty, +\infty)$) of the $\lfloor n/2 \rfloor$ bisectors $L_1, \ldots, L_{\lfloor n/2 \rfloor}$, and in $O(n)$ time, compute the median slope, α. Next, pair up each bisector L_i with slope$(L_i) \leq \alpha$ along with a distinct bisector L_j with slope $(L_j) \geq \alpha$. This gives us $\lfloor n/4 \rfloor$ pairs of the form (L_i, L_j). For convenience, let's re-orient the x-axis along the direction α; a line with slope α becomes a line with slope 0. For each pair (L_i, L_j), define the quantity y_{ij} as follows: if L_i and L_j have distinct slopes, let (x_{ij}, y_{ij}) be their intersection. Otherwise, let y_{ij} be the mean y-coordinate of L_i and L_j, i.e., the y-coordinate of the midpoint between $\{X = 0\} \cap L_i$ and $\{X = 0\} \cap L_j$. By convention, we let $y_{ij} = +\infty$ if both L_i and L_j are parallel to the

y-axis. At any rate, compute the median y of the $\lfloor n/4 \rfloor$ values of the form y_{ij} in $O(n)$ time. Using our solution to Problem II, we can decide in linear time whether χ lies on the line $Y = y$ (in which case we are finished), or lies above or below. Assume wlog that χ lies strictly below.

Next, consider all the $\lfloor n/8 \rfloor$ pairs (L_i, L_j) such that $y_{ij} \geq y$. If L_i and L_j are parallel, the intersection with $X = 0$ of one of them, say L_i, must lie strictly above $(0, y)$. Since L_i is then parallel to the x-axis, one of the two points defining the line can be discarded, namely, the one lying below L_i. Let n_1 be the number of points dropped this way and n_2 the number of remaining pairs $(n_1 + n_2 = \lfloor n/8 \rfloor)$. Consider now the set of remaining pairs (with $y_{ij} \geq y$) and compute the median x of the x_{ij}'s. Test on which side of the line $X = x$ the center χ must lie. Once again if χ lies on the line, we are able to solve the problem directly. Assume wlog that χ lies to strictly to the left of the line (Fig. 4.14). For each of the $\lfloor n_2/2 \rfloor$ pairs (L_i, L_j) to the right of $X = x$, one of the four points involved can be dropped from further consideration. Indeed, for each pair (L_i, L_j), one of the lines, say L_i, is oriented in such a way that it is possible to identify with certainty on which side of L_i the center lies. To see this, L_i can always be chosen with negative slope (by construction, either L_i or L_j satisfies this property). Let p_k and p_l be the two points of which L_i is the bisector, with p_k (resp. p_l) below (resp. above) L_i. Since χ must lie in the quadrant delineated in Fig. 4.14, whatever the exact location it will always be closer to p_k than p_l, therefore p_k can be discarded. This leads to the dismissal to $n_1 + \lfloor n_2/2 \rfloor \geq \lfloor n/16 \rfloor$ points from S, hence the linear running time of the overall algorithm.

Megiddo's method, however appealing, cannot be recommended for practical applications. For one thing, its reliance on linear-time median computation may critically hamper its performance when applied to problems of modest size. But since the algorithm does not need the exact median but any value *close enough*, computing the median out of a small sample of the input might perform just as well, and give a "somewhat" simpler algorithm.

2.3.2. Ellipse Enclosure.   After dealing with circles, the natural question is: how about ellipses? An ellipse is defined by five parameters whereas a circle

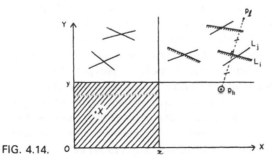

FIG. 4.14.

needs only three. These two additional degrees of freedom should in general allow for wrappings that fit more tightly around the set of points (Post, 1981; Silverman & Titterington, 1980).

What is the complexity of computing the smallest-area ellipse enclosing a set of points?

As is well-known, any ellipse in the plane centered at $u_0 = (x_0, y_0)$ can be analytically represented by an equation of the form $(u - u_0)^t A(u - u_0) = 1$, where A is a 2×2 positive-definite matrix. Since A is symmetric, five numbers suffice to define any ellipse and, as one should expect, at most five points uniquely define an ellipse. A simple algorithm consists of computing all candidate ellipses enclosing S. To so, we consider all ellipses passing through five points, and all smallest-area ellipses passing through four and three points respectively. In each case, we check whether the candidate ellipse contains every point of S, and throw it away if it does not. Elementary analysis shows that this algorithm will be extremely time-consuming, as it will require $O(n^6)$ operations.

Post (1981) has shown that it is possible to reduce the computation time to $O(n^2)$. The key to Post's method is to be able to throw away at least one point after a linear number of operations. The algorithm starts off with a spanning ellipse, checks whether it is the smallest, and if not, throws away at least one point and then shrinks the ellipse. Iteration takes over at this point after making sure that the new ellipse does enclose all the points of S. Additional work is necessary if this condition is not satisfied. The quadratic performance of the algorithm follows from the linearity of each pass.

2.4. Other Enclosure Problems

A large number of enclosure problems have been studied lately. We review some of the main results.

1. Given a set S of n points in the plane, consider all k-gons which can be formed with the points in S. Boyce, Dobkin, Drysdale, and Guibas (1982) have given an algorithm for finding the maximum perimeter triangle in $O(n\log n)$ time and the largest perimeter or area k-gon in $O(kn\log n + n\log^2 n)$ time for any k.

2. Given a simple n-gon P, find the largest-area convex polygon contained in P. This *potato-peeling* problem, as it is often called, has a number of variants, one of which calls for the maximum-perimeter enclosed polygon. Chang and Yap (1984) have proposed polynomial algorithms for all these problems.

3. Given a convex n-gon P, find a minimum-area convex k-gon enclosing P. Besides the aforementioned work of Klee and Laskowski (1985) and O'Rourke at al. (1984) for the case $k = 3$, a number of interesting results have been obtained. Using dynamic programming, Chang and Yap (1984) have described

an $O(n^3 \log k)$ time algorithm for the general problem. This bound has been improved to $O(n^2 \log n \log k)$ (See Aggarwal, Chang, & Yap, 1985). De Pano and Aggarwal (1984) have considered special cases of this problem: they give an $O(n \log k)$ (resp. $O(nk^2)$) time algorithm for the case where the enclosing polygon is equi-angular (resp. regular). See also O'Rourke (1984) for extensions of the smallest enclosing box in three dimensions.

Some problems in location theory are very similar in spirit to the type of questions we've been asking so far. *Given a set of points and disks in the plane, is it possible to arrange the disks so as to cover all the points?* Megiddo and Supowit (1984) have shown that the problem is NP-hard. However, the problem: *Given a set of n points in the plane, what is the smallest disk that covers at least k points?* can be solved in $O(k^2 n \log n)$ time, as demonstrated by Lee (1982). Also, the question: *Given n points in the plane, what is the largest number of points that can be covered by a single disk?* has been shown to be solvable in $O(n^2)$ time in Chazelle and Lee (1986).

A useful primitive to have in applications involving enclosure problems checks whether a polygon can fit into another. *Given two polygons P and Q, determine whether Q can contain P, if rotations and translations are allowed.* Chazelle (1983b) has given a general $O(p^3 q^3 (p+q))$ time algorithm for deciding on this question, where p (resp. q) denotes the number of vertices in P (resp. Q). If Q is convex, the complexity can be reduced to $O(pq^2)$. The algorithm essentially involves finding "all" possible locations where P could fit. The difficulty is to establish an upper bound on the number of locations. Despite its high complexity, the algorithm is optimal with respect to that measure, if p is a constant. This still leaves open the question of whether the decision problem can be solved more efficiently.

A few results are known in the special case where only translations are allowed. Letting $n = p + q$, the problem can be solved in $O(n)$ time if both polygons are convex (Chazelle (1983b). For the case where either the inside polygon is convex or both polygons are rectilinearly convex, Baker, Fortune, and Mahaney (1984) have given an $O(n^2 \log n)$ time algorithm.

3.   DECOMPOSITION OF SHAPES

3.1. Triangulations

Let P be a simple *n-gon*, ie., a simple polygon with n vertices. A *triangulation* of P is any partition of the polygon into disjoint triangles, all of whose vertices are vertices of the polygon. It is often useful to have a triangulation of P. For one thing, triangles are easy to handle, so a partition of this nature simplifies tasks such as testing for intersection, inclusion, etc. The idea is to perform computations iteratively on each part; for example, one can use this "decomposition" approach to detect interference or collision between a number of moving objects (see Ahuja, Chien, Yen, & Birdwell (1980) for example). Another benefit to gain from a triangulation is valuable information concerning topological properties of the polygon (Chazelle, 1982).

Triangulation problems are many and varied. In numerical analysis, for instance, a triangulation is often used to evaluate a function of two variables by interpolation or to integrate a function by the finite-element method. In these applications, the shape of the triangulation can be of great importance. In the context of the finite-element method, for example, it is likely that long, skinny triangles should be avoided because of the errors they tend to generate (see, for example, Baker, Gross, & Rafferty, 1985). We will not concern ourselves with such matters, however, since their relevance is relatively small if a triangulation is viewed solely as a tool for facilitating geometric computations. This is most often the case in robotics, computer graphics, computer animation, CAD/CAM, etc. As a starter, we pose the following problem:

What is the complexity of computing a triangulation of a simple n-gon?

Unfortunately, despite the large amount of work devoted to this problem, a definite answer still eludes us. There have been solid advances on the subject, however, both on the practical and theoretical sides. Garey, Johnson, Preparata, and Tarjan proposed the first $O(n\log n)$ time algorithm for triangulating P (Garey et al., 1978). Their algorithm works in two phases: to begin with, a *regularization* procedure is applied to P, which in $O(n\log n)$ time produces a partition of the polygon into *L-monotone* pieces, for some line L. A polygon is said to be *L-monotone* if any normal to L intersects the polygon in no more than one segment. With this decomposition in hand, Garey et al.'s algorithm completes the triangulation by cutting each monotone piece into triangles. This second (and last) pass takes linear time.

Chazelle (1982) has established a *polygon-cutting theorem*, which is useful for triangulation as well as other problems (e.g., visibility, internal distance). Assign a positive weight to each vertex of P in an arbitrary fashion, with the total weight not exceeding 1. Roughly speaking, the polygon-cutting theorem states that there exists two vertices a, b such that the segment ab lies entirely inside the polygon P and partitions it into two polygons, neither of whose weights exceeds $2/3$.

If we allow additional vertices to be introduced, then a successful approach to the triangulation problem is the *line-sweep* paradigm (Bentley & Ottmann, 1979; Shamos, 1978). Imagine a vertical line L sweeping the plane from left to right; the segments of the intersection $L \cap P$ are kept dynamically in a balanced search structure (e.g., AVL, Knuth, 1973), red-black, Sedgewick, 1983, trees). An *event* takes place every time the line encounters a new vertex. At that point, the search structure is updated to record the fact that a segment is about to vanish, to

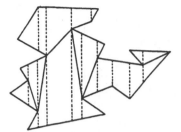

FIG. 4.15.

be created, or to have one of its endpoints switch to a different edge of P (Fig. 4.15). The net result of this line-sweep process is the so-called *vertical decomposition* of P, a set of vertical segments obtained by connecting every vertex of P to the edge immediately above or below it, in such a way that the segment created lies inside P (dashed lines in Fig. 4.15). It is easy to derive a triangulation of P from its vertical decomposition (note that the triangulation is not uniquely defined, but the vertical decomposition is). In a nutshell, the idea is to identify the *cusps* of P, i.e., the vertices that are locally in leftmost or rightmost position. For each trapezoid generated by the line-sweep procedure, connect each cusp to a vertex of the other vertical side (solid interior lines in Fig. 4.15). Removing all vertical segments produces a decomposition of P into monotone polygons (with respect to the horizontal direction). The linear time postprocessing of Garey et al.'s algorithm can then be applied to complete the triangulation.

Using a fairly similar method, Hertel and Mehlhorn (1983) succeeded in lowering the $O(n\log n)$ running time to $O(n + r\log r)$, where r is the number of reflex angles in P. Their method differs from the previous one in two important aspects. First, it performs the triangulation on the fly, i.e., it computes the triangles during the line-sweep process. Second, it skips over vertices that do not display reflex angles. The import of Hertel and Mehlhorn's result is to express the running time of the algorithm not only as a function of the input size but also as a function of a parameter which reflects a morphological property of the polygon.

In the same spirit, Chazelle and Incerpi (1984) developed a radically different algorithm based on divide-and-conquer. Their method requires $O(n\log s)$ time, with $s < n$. The quantity s measures the *sinuosity* of the polygon, that is, the number of times the boundary alternates between complete spirals of *opposite* orientation. The value of s is in practice a very small constant (e.g., $s = 2$ in the case of Fig. 4.15), even for extremely winding polygons. The running time of the algorithm depends primarily on the shape-complexity of the polygon. Informally, this notion of *shape-complexity* measures how *entangled* a polygon is, and is thus highly independent of the number of vertices. Aside from the notion of sinuosity, Chazelle and Incerpi have also characterized a large class of polygons for which the algorithm can be proven to run in $O(n\log\log n)$ time. Implementation of the algorithm has confirmed its theoretical claim to efficiency.

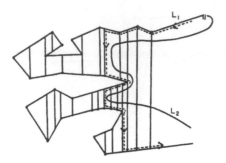

FIG. 4.16.

Briefly, the algorithm starts with the observation that vertical decompositions can be defined with respect to the boundary of P as opposed to the polygonal region P. This distinction enables a divide-and-conquer strategy involving (1) the computation of the vertical decompositions of $\{p_i, \ldots, p_{\lfloor \frac{i+j}{2} \rfloor}\}$ and $\{p_{\lfloor \frac{i+j}{2} \rfloor}, \ldots, p_j\}$; (2) the merge of these decompositions into the vertical decomposition of $\{p_i, \ldots, p_j\}$. The key to efficiency is to take *shortcuts* in the execution of the merge. Figure 4.16 illustrates this notion of shortcuts. L_1 and L_2 are the two polygonal lines whose vertical decompositions are being merged. The algorithm traverses the trapezoids of both decompositions, but visits *only* those that will be modified by the merge. For example, the trapezoids on the left-hand side of Fig. 4.16 will not be examined. The dashed line traces the steps of the merge (see Chazelle and Incerpi, 1985, for details).

3.2. Decompositions into Special Shapes

Among primitive shapes of particular interest, we also have quadrilaterals and convex or *star-shaped* polygons. We say that a polygon P is star-shaped if it contains at least one point p such that for every point q inside P the segment pq lies inside P. Sack and Toussaint (1981) have shown that any isothetic polygon with n vertices which is star-shaped can be partitioned into convex quadrilaterals in $O(n)$ time. Recall that a polygon is isothetic (also sometimes referred to as *rectilinear*) if its sides are parallel to two orthogonal axes. See Chazelle (1980, 1982), Chazelle and Dobkin (1985), Greene (1983), Hertel and Mehlhorn (1983) for examples of algorithms for decomposing a polygon into a number of convex polygons within a constant factor from the minimum number.

Kahn, Klawe, and Kleitman (1980) have shown that any isothetic polygon can always be partitioned into convex quadrilaterals. Building on this result, Sack (1982) has given an $O(n\log n)$ time algorithm for computing the decomposition. In a nutshell, the algorithm involves decomposing the polygon into monotone polygons, and then decomposing each of these polygons into quadrilaterals.

Avis and Toussaint (1981) have obtained similar results concerning star-shaped polygons. They have shown that any n-gon can be partitioned into star-shaped polygons in $O(n\log n)$ time.

3.3. Minimum Decompositions

> *Problem OCD (Optimal Convex Decomposition): Given a simple polygon P, which is the minimum number of convex polygons which form a partition of P?*

Let's briefly review the main results relating to this problem. Pioneering work appears in Feng and Pavlidis (1975) and Schachter (1978). with the design of heuristics for computing decompositions of shapes into a number (not necessarily minimum) of convex pieces.

Minimum Partitioning: One of the earliest results concerning the OCD problem per se was obtained by Chazelle and Dobkin (1979), who showed that the OCD problem was solvable in polynomial time (see an example of optimal decomposition in Fig. 4.18). Interestingly enough, this finding was followed by a stream of NP-hardness results for similar problems. For example, it was shown by Lingas (1982) that the presence of holes in the polygon P was sufficient to make the OCD problem NP-hard. Lingas also showed that a minimum decomposition into triangles was NP-hard. Other variants have been shown to be (most likely) intractable: Once again, assuming that P may have holes, Asano and Asano (1983) proved that decomposing P into a minimum number of trapezoids with two vertical sides is NP-hard. If holes are disallowed, however, they showed that the problem could be solved in $O(n^3)$ time, an upper bound later improved by Asano, Asano, and Imai (1984) to $O(n^2)$.

Minimum Covering: If now instead of partitioning the polygon, one would rather *cover* it with possibly overlapping convex pieces, it might be possible to save a few pieces. For example, a cross can be covered with two rectangles but cannot be partitioned into fewer than three convex polygons. Unfortunately, minimum coverings seem very difficult to compute. O'Rourke and Supowit (1983) have shown that this problem and a large number of its variants are NP-hard. A decision procedure for this problem is given in O'Rourke (1982). If the polygon is isothetic and the covering is made of rectangles, then a quadratic algorithm can be used (Franzblau & Kleitman, 1984).

No Steiner Points: Another class of decomposition problems stipulates that no new vertices should be introduced in the decomposition. These new vertices, called *Steiner points* are often useful—as we will shortly see—in reducing the number of pieces. Steiner points are sometimes undesirable, however. Once again, the presence of holes makes most of the problems NP-hard (see Keil, 1983; Lingas, 1982; O'Rourke & Supowit, 1983). If holes are disallowed, however, the situation is much brighter. Let c be the number of reflex angles in P; Greene (1983) has given an $O(n^2c^2)$ algorithm for decomposing P into a minimum number of convex pieces (not allowing Steiner points), and Keil (1983) has described an $O(c^2 n \log n)$ algorithm for the same problem.

Partitioning into Special Shapes: A decomposition problem of practical interest concerns the partition of an isothetic polygon into a minimum number of rectangles. Lipski (1983) and Imai and Asano (1983a, 1983b), have shown that the problem can be solved in $O(n^{3/2} \log n \log\log n)$ and $O(n^{3/2} \log n)$ time, respectively, by using a reduction to maximum matching of a particular bipartite intersection graph. In both cases, $O(n \log n)$ storage is required. Keil (1983) has shown that it is possible to partition any simple polygon into a minimum number of star-shaped polygons in $O(n^5 c^2 \log n)$ time, where c once again is the number of reflex angles in P. For other work, consult Ferrari, Sankar, Sklansky (1981), Keil and Sack (1985), Toussaint (1980b).

Minimum-Length Partitioning: Other objective functions besides the cardinality of the decomposition have been examined. Lingas (1981) and Lingas, Pinter, Rivest, and Shamir (1982) consider decomposition problems where the amount of "ink" used, i.e., the total length of the added edges is to be minimized.

Higher Dimensions: In three dimensions, the OCD problem becomes NP-hard (Lingas, 1982). A decision procedure based on a combinatorial description of the complex formed by a convex decomposition is given in Chazelle (1983a). Chazelle (1984) has established an $\Omega(n^2)$ lower bound on the time complexity of the problem. If n is the number of vertices in the polyhedron and c is the number of edges displaying reflex angles, in $O(nc^3)$ time it is possible to decompose the polyhedron into a number of pieces which, in the worst case, is minimal up to within a constant factor.

The OCD Problem: We now return to the original OCD problem: Given a simple n-gon P compute a minimum partition of P into convex pieces, using Steiner points if necessary. The remaining of this section is devoted to giving the technical basis of the polynomial time algorithms given in Chazelle (1980), Chazelle and Dobkin (1985). For details, we again refer the reader to the appropriate sources. In the following, a *notch* refers to a vertex of P with a reflex interior angle (e.g., a convex polygon has no notch). We let $\{p_1, \ldots, p_n\}$ be a vertex-list of P in clockwise order, and we assume that c of these n vertices are notches of the polygon. Chazelle and Dobkin have given several algorithms for computing an OCD (optimal convex decomposition) of P, the most efficient of which runs in $O(n + c^3)$ time (Chazelle, 1980; Chazelle & Dobkin, 1985). All the algorithms use two simple observations. First, each notch *can* be removed by the addition of a polygon to the decomposition. Second, *at most* two notches can be removed through the addition of a single polygon. It follows that the minimum number of convex parts always lies between $\lceil c/2 \rceil + 1$ and $c + 1$.

Below, we introduce the notion of *X-pattern*, the essential tool for generating minimum decompositions. An X_k-pattern is a particular interconnection of k notches which removes all reflex angles at the k notches and creates no new

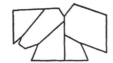 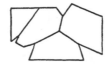

FIG. 4.17.

notches. A decomposition obtained by applying p patterns of type $X_{i_1}, \ldots X_{i_p}$, along with straight-line segments to remove the remaining notches can be shown to yield $c + 1 - p$ convex parts. It follows that the decompositions using the most X-patterns will at the same time minimize the number of convex pieces.

The next question to tackle is whether a given set of notches can be interconnected via an X-pattern. One solution around the multiplicity of combinatorially distinct X-patterns is to constrain the patterns in such a way that their detection becomes tractable. This leads to the introduction of Y-patterns, which we regard as X-patterns endowed with some structural property. A key property of minimum decompositions is that with the exception of X_4-patterns any X-pattern can be advantageously replaced by a Y-pattern. Since, as we will sketch out shortly, Y-patterns can be constructed in polynomial time via dynamic programming, we can prove in this way that the OCD problem is in the polynomial class. Further geometric analysis leads to substantial gains in the efficiency of the algorithm, but we will not elaborate on these improvements which are fairly intricate.

As a starter, we define the notion of *naive decomposition*. This is the decomposition obtained by removing each notch in turn by means of a simple line segment *naively* drawn from the notch. Figure 4.17 shows a naive decomposition of a nonconvex polygon next to an improved decomposition. To obtain a naive decomposition of P is easy: consider each notch $v_1, \ldots, v_c$ in turn, and extend a line segment from v_i in order to remove the reflex angle at the notch. Stop the line as soon as it hits another line already in the decomposition. Trivially, the naive decomposition of P produces exactly $c + 1$ convex parts. Next, we define a *nice* class of decompositions to which we may restrict our attention from then on. We say that a polygon is *interior* to P if it lies inside P and at most a finite number of its points lie on the boundary of P. An *X-decomposition* is then defined as any convex decomposition containing no interior polygons and such that no vertex is of degree greater than 3, except for the notches, which may be of degree at most 4. An important result, whose proof we will omit, states that the class of X-decompositions always contains an OCD.

We are now ready to introduce the notion of *X-pattern*. Consider the planar graph formed by the added edges of an X-decomposition. This graph is a forest of trees, where each node is of degree 1 or 3, except for notches which may be of degree 1 or 2. Note that it is a forest because we disallow interior polygons in the decompositions. A straightline embedding of a tree lying inside P (i.e., with no self-intersections) is called an X-pattern if

1. all vertices are of degree 1, 2 or 3.

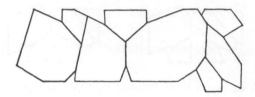

FIG. 4.18.

2. any vertex of degree 1 or 2 coincides with a notch of P, and its (1 or 2) adjacent edges remove its reflex angle.
3. none of the 3 angles around any vertex of degree 3 exceeds 180 degrees.

An X-pattern with k vertices of degree 1 or 2 is called an X_k-pattern. Informally, an X_k-pattern is an interconnection of k notches used to remove them while introducing $k - 1$ additional polygons into the decomposition. Figure 4.18 gives an example of an X-decomposition with one X_3-pattern and one X_4-pattern. Note that the leftmost tree is not an X-pattern because one of its vertices of degree 1 does not coincide with a notch of the polygon. In the following, the vertices of an X-pattern (regarded as tree vertices) of degree 1 (resp. 2,3) will be called N1-nodes (resp. N2, N3-nodes). We justify the introduction of X-patterns with the following observation. An X-decomposition with p X-patterns has at least $c + 1 - p$ convex parts. This suggests that using p X-patterns saves at most p polygons over the naive decomposition. This leads to the definition of *compatible X-patterns*. A set of X-patterns is said to be compatible, if no pair of edges taken from two distinct patterns intersect. It is not hard to show that any compatible set of p X-patterns can be used to produce an X-decomposition with exactly $c + 1 - p$ convex parts. The OCD problem can now be expressed as a generalized matching problem.

Let p be the maximum number of compatible X-patterns. An OCD can be obtained by first applying the p X-patterns and then applying the naive decomposition to any remaining non-convex polygon. Unfortunately the complexity of this approach is prohibitive, given the excessive number of candidates we might have to consider in the process. X-patterns allow Steiner points (i.e., vertices not on the boundary of P) to be adjacent. Looking at any X-pattern as a mechanical system of extendible arms and joints, it appears to be greatly underconstrained. We can show that X-patterns can be in general *reduced* to maximally constrained X-patterns, called Y-patterns. The next step is then to prove that Y-patterns can be computed in polynomial time. Roughly speaking, a Y_k-pattern is an X_k-pattern such that no edge joins two Steiner points. For example, the X_4-pattern of Fig. 4.18 violates this restriction. The utility of Y-patterns comes from the fact (which we will not prove) that all X_k-patterns ($k \neq 4$) can be transformed into Y-patterns. This allows us to limit attention to X-decompositions with only Y and X_4-patterns. The transformations, called *reductions*, involve the stretching, shrinking, and rotating of lines in the original pattern. The next question which we must be able to answer efficiently is:

Given k notches, does there exist an X-pattern connecting them?

Let v be an N3-node of a Y-pattern. Removing the three edges vv_i, vv_j, vv_k adjacent to v disconnects the Y-pattern into three subtrees. Note that the removed edges play the role of an X_3-pattern with respect to the three subtrees. This leads us to introduce the notion of *extended X-pattern*. An extended X_l-pattern ($l = 2$, 3) is either an X_l-pattern by itself or an X_l-pattern formed by a subgraph of an X_m-pattern ($m > l$). It is clear that an algorithm for testing the possibility of an X_l-pattern between a given set of notches can also be used to determine the possibility of an extended X_l-pattern, as long as the angles formed at the notches by the subtrees of the X_m-pattern are known in advance.

To see this, we must define the notion of *extended notch*. Let v_i be a notch of the extended X_l-pattern, and let W be any wedge centered at v_i. The extended notch at v_i with respect to W refers to v_i along with the angular specifications given by W. This means that the X-pattern edge emanating from the extended notch v_i must lie in the intersection of W with the visibility-polygon at v_i (El Gindy & Avis, 1981). In general the wedge W will be taken as the locus of rays (i.e., half-lines) which emanate from v_i and remove the reflex angle at v_i. When dealing with ordinary X-patterns, the wedge W is simply determined by the edges of P adjacent to v_i. In the case of extended patterns, however, the wedge W will take into account the other edges already adjacent to v_i, and will thus be wider than in the previous case. Sometimes, we will also consider cases where the wedge W is taken as the entire plane. This is done if we do not wish to remove a reflex angle at a particular notch. We can show that in all cases it is possible to check for the existence of an (extended) X_l-pattern between l given notches in polynomial time (for $l \leq 4$).

We can now turn to the decomposition algorithm, which as we mentioned earlier is based on dynamic programming. We rely on the observation that if a certain X_k-pattern belongs to an OCD of P, it decomposes P into k subpolygons, $P_1, \ldots, P_k$, so that finding an OCD for each P_i yields an OCD for P. We compute maximal compatible sets of patterns for each P_i. Since the notches of P_i are also notches of P, any X-pattern of P_i is also an X-pattern of P. Conversely, we want to show that any X-pattern of P involving only notches in P_i is also an X-pattern of P_i. This is crucial since dynamic programming proceeds bottom-up. Indeed, the algorithm will always compute maximal sets of compatible patterns involving notches of P_i before it even knows the shape of P_i, i.e. it will rely solely on the notches of P_i and not on its boundary.

To check this point, we define $V(i, j)$ as the set of notches between v_i and v_j in clockwise order. We have $V(i, j) = \{v_i, v_{i+1}, \ldots, v_j\}$, with index arithmetic taken modulo c. Let $z_1, \ldots, z_k$ be the notches of an X-pattern, T, given in clockwise order around the boundary of P, and let $V(i, j)$ be the notches of P between z_u and z_{u+1} in clockwise order ($z_u = v_{i-1}$, $z_{u+1} = v_{j+1}$). It is possible to prove that no X-pattern with its notches in $V(i, j)$ can intersect T. This independence result can be understood in combinatorial terms. It states that two X-patterns are intersection-free if and only if their notch sequences are not intermixed, i.e. one sequence falls completely between two consecutive elements of the other sequence.

Next, we define $S(i, j)$, for every pair of notches v_i, v_j, as a maximum compatible set of X_4 or Y-patterns in $V(i, j)$. The goal is to evaluate $S(1, c)$, which we achieve by computing all values $S(i, j)$ from $\{s(k, l) \mid V(k, l) \subset V(i, j)\}$. This can be done directly if v_i and v_j are not to be connected to the same pattern. For this case, we test all combinations $\{S(i, k), S(k + 1, j)\}$, for each $v_k \in V(i, j - 1)$. In the event where v_i and v_j should be connected together, we consider the possibility of an X_4 or a Y-pattern between the two notches.

To handle the latter case, we compute all candidate Y-patterns via dynamic programming. We compute Y-subtrees (i.e., subtrees of Y-patterns) as well as Y-patterns by patching Y-subtrees together. A Y-subtree is considered *not* to be a candidate if at the time it is computed we are ensured of the existence of at least one OCD which does not use this Y-subtree and satisfies previous constraints. As a shorthand we say that a pattern or a Y-subtree lies in $V(i, j)$ if all its notches do. We next give a brief description of the polynomial time algorithm.

Consider a Y-pattern of an OCD with at least one N2-node, v_i. This node splits the Y-pattern into two Y-subtrees, so there exists an index j such that

1. One of the Y-subtrees lies in $V(i, j)$ whereas the other lies in $V(j + 1, i)$.
2. All the other patterns in the OCD lie totally either in $V(i, j)$ or in $V(j + 1, i)$.

The idea is to examine the candidacy of the Y-subtree in $V(i, j)$ immediately after $S(i, j)$ has been computed. Let v_i, $v_{i_1}, \ldots, v_{i_m}$ be a clockwise-order list of the notches in the Y-subtree. The candidacy of this subtree may be rejected if the following equality is not satisfied:

$$|S(i,j)| = |S(i + 1, i_1 - 1)| + \ldots + |S(i_{m-1} + 1, i_m - 1)| + |S(i_m + 1, j)|. \tag{1}$$

$|S(k, l)|$ represents the number of patterns in $S(k, l)$, and the last term of the right-hand side is to be ignored if $i_m = j$. If Relation (1) is not satisfied, it is clear that the right-hand side is strictly smaller than $|S(i, j)|$. Since we only consider the presence of a single pattern connecting notches in both $V(i, j)$ and $V(j + 1, i)$, we can dismiss any Y-subtree that does not satisfy (1). Indeed, using the patterns of $S(i, j)$ would yield a smaller decomposition.

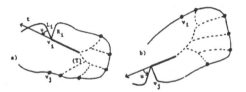

FIG. 4.19.

Another crucial fact will allow us to reduce the number of candidates to keep around. Let L_i (resp. R_i) denote the edge of P adjacent to v_i and preceding (resp. following) v_i in clockwise order. Both L_i and R_i are understood as *directed* outwards with respect to v_i. Let t be the edge adjacent to v_i in the Y-subtree lying in $V(i, j)$. We refer to t as the *arm* of the Y-subtree. When the arm of a Y-subtree enters the expression of an angle, we assume that it is directed towards the notch (here t is directed towards v_i). Among all the Y-subtrees in $V(i, j)$ for which v_i is an N2-node, (1) is true and $u = \angle(L_i, t) < 180$, we may keep the Y-subtree T which minimizes the angle u as the only candidate with respect to $V(i, j)$ (Fig. 4.19a). We then define $B(i, j)$ as a pointer to the arm of T. If no such subtree can be found, $B(i, j)$ is set to 0. The same reasoning applied counterclockwise with respect to $V(i, j)$ (v_i is now an N2-node), leads to $F(i, j)$, defined in a similar fashion (Fig. 4.19b).

We are now ready to present the decomposition algorithm. We define the function $\langle ARG \rangle$ for the purpose of assembling Y-subtrees in the computation of $S(i, j)$. The argument ARG is in general a pair of Y-subtrees in $B(u, v)$ or $F(u, v)$. If the two subtrees can be patched together and form a Y-pattern T, the function $\langle \rangle$ returns (C, T), where C is the maximum number of compatible patterns which can be applied in $V(i, j)$ including T. We discuss the implementation of this function after describing the main algorithm.

Initialization of the data structures is performed in STEP 1, followed by two nested loops setting up the framework for the dynamic programming scheme. Each stage corresponds to the computation of $S(i, j)$ for a given value of i and j. STEP 2 computes the best Y-pattern which connects v_i and v_j, by patching together precomputed candidate Y-subtrees. STEP 3 computes a maximum set of compatible patterns in $V(i, j)$, denoted L, assuming that v_i and v_j do not belong to the same pattern. It computes M, defined similarly, with the only difference that the presence of an X_4-pattern connecting v_i and v_j is now allowed. Finally the Y-pattern of STEP 2 (if any) is used to compute N, and the maximal set among L, M, N is chosen as $S(i, j)$. STEP 4 computes the Y-subtrees which lie in $V(i, j)$ and are considered candidates. These subtrees are to be used in further iterations through STEP 2. Once a maximum compatible set of patterns for P has been found, STEP 5 is able to complete the OCD with the naive decomposition.

NOTE: *The following pieces of pseudo-code are reproduced from Chazelle and Dobkin (1985) with the permission of the authors.*

Procedure *ConvDec(P)*

beginprocedure

STEP 1:

The preprocessing involves checking that P is simple and nonconvex. We make a list of the notches $v_1, \ldots, v_c$, and we initialize all $B(i, i)$ and $F(i, i)$ to 0.

> **for** $d = 1, \ldots, c - 1$
>   **for** $i = 1, \ldots, c$
> $j := i + d \, [\bmod \, c]$
>   **do STEPS 2, 3, 4**

STEP 2:

Compute the best Y-pattern connecting v_i and v_j as follows:
For each k; $v_k \in V(i + 1, j - 1)$, compute the set $Q = \cup_{1 \leq i \leq 4} Q_i$, where
$Q_1 = \{\langle F(i, k), B(k, j)\rangle\}$ /* N2-node on path */
$Q_2 = \{\langle B(i, k - 1), F(k, j)\rangle\} \cup \{\langle B(i, j - 1), F(j, j)\rangle\}$ /* no N2 or N3 nodes on path */
$Q_3 = \{\langle B(i, k - 1), B(k, j - 1)\rangle\}$ /* N3-node on path */
$Q_4 = \{\langle F(i + 1, k), F(k + 1, j)\rangle\}$ /* N3-node on path*/
The elements of Q are pairs of the form (C, T). Let T be the Y-pattern which has the maximum C value in Q.

STEP 3:

Let $S(i, j)$ be the maximum of L, M, N with respect to cardinality, where (max is taken with respect to cardinality)

$L = \max_{v_k \in V(i, j - 1)} \{S(i, k) \cup S(k + 1, j)\}$
/* corresponds to a patching together of Y-patterns */
$M = \max \{\{x_{i,a,b,j}\} \cup S(i + 1, a - 1) \cup S(a + 1, b - 1) \cup S(b + 1, j - 1)\}$
for all X_4-patterns $x_{i,a,b,j}$ connecting v_i, v_a, v_b, v_j, with $v_a, v_b \in V(i, j)$.
/*corresponds to the use of an X_4-pattern */
$N = \{$the Y-pattern T of STEP 2$\} \cup S(i + 1, i_1 - 1) \cup \ldots \cup S(i_{p-1} + 1, i_p - 1) \cup S(i_p + 1, j - 1)$
where $v_i, v_{i_1}, \ldots, v_{i_p}, v_j$ are the notches of T in clockwise order.
/* corresponds to the use of the Y-pattern T */

STEP 4:

Compute $B(i, j)$ and $F(i, j)$.

STEP 5:

Finish off the decomposition using the naive decomposition, i.e. adding one polygon for each remaining notch.

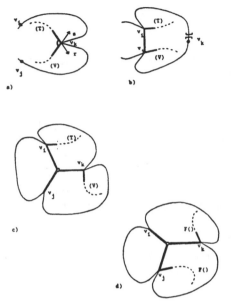

FIG. 4.20.

endprocedure

In the following we explore some of the primitive operations used by the algorithm in greater detail.

1. *Patching Y-subtrees* (STEP 2)

The function $\langle ARG \rangle$ takes two Y-subtrees and constructs a Y-pattern if these two subtrees can be patched together. ARG is any argument of the form: $(F(i, k), B(k, j))$, $(B(i, k - 1), F(k, j))$, $(B(i, k - 1), B(k, j - 1))$ or $(F(i + 1, k), F(k + 1, j))$, with v_i, v_k, v_j occurring in clockwise order.

> **Case 1.** $\langle F(i, k), B(k, j) \rangle$ (Fig. 4.20a).
> Let $F(i, k) = T$ and $B(k, j) = V$, with r and s their respective arms. If $\angle(r, s) < 180$ and $T \neq 0$ and $V \neq 0$, then set $\langle F(i, k), B(k, j) \rangle = (|S(i, k)| + |S(k, j)| + 1$, Y-pattern: $T \cup V)$, else $\langle F(i, k), B(k, j) \rangle = 0$.

> **Case 2.** $\langle B(i, k - 1), F(k, j) \rangle$ (Fig. 4.20b).
> Let $B(i, k - 1) = T$ and $F(k, j) = V$. If an extended X_2-pattern is possible between v_i and v_j then set $\langle B(i, k - 1), F(k, j) \rangle = (|S(i, k - 1)| + |S(k, j)| + 1$, Y-pattern: $\{v_i v_j\} \cup T \cup V)$, else $\langle B(i, k - 1), F(k, j) \rangle = 0$.

> **Case 3.** $\langle B(i, k - 1), B(k, j - 1) \rangle$ (Fig. 4.20c).
> Let $B(i, k - 1) = T$ and $B(k, j - 1) = V$. If an extended X_3-pattern W is possible between v_i, v_j, v_k, then
> $\langle B(i, k - 1), B(k, j - 1) \rangle = (|S(i, k - 1)| + |S(k, j - 1)| + 1$, $T \cup V \cup W)$, else $\langle B(i, k - 1), B(k, j - 1) \rangle = 0$.

Case 4. $\langle F(i + 1, k), F(k + 1, j)\rangle$ (Fig. 4.20d).

Let $F(i + 1, k) = T$ and $F(k + 1, j) = V$. If an extended X_3-pattern W is possible between v_i, v_j, v_k, then
$$\langle F(i + 1, k), F(k + 1, j)\rangle = (|S(i + 1, k)| + |S(k + 1, j)| + 1, T \cup V \cup W), \text{ else}$$
$$\langle F(i + 1, k), F(k + 1, j)\rangle = 0.$$

Because of Relation (1), it is clear that STEP 2 computes the Y-pattern connecting v_i and v_j (if any is to be found) such that the number of compatible patterns which can be applied in $V(i, j)$ is maximum. All we have to check is that all cases are indeed handled in STEP 2. Consider the path from v_i to v_j in any such Y-pattern. If it contains an N2-node, it will be detected in Q_1. Otherwise one N3-node may appear on this path and all these candidates will be reported in Q_3 and Q_4. The final case, handled by Q_2, assumes that the path from v_i to v_j is free of N2 and N3-nodes.

2. *Computing $S(i, j)$ (STEP 3)*

Assume by induction that $S(k, l)$ has been computed for all v_k, $v_l \in V(i, j)$ (except for $S(i, j)$). The algorithm investigates the three following cases in turn:

1. Disallow the presence of any pattern having both v_i and v_j as vertices.
2. Consider the possibility of an X_4-pattern connecting v_i and v_j.
3. Consider the possibility of a Y-pattern connecting v_i and v_j.

3. *Constructing Y-subtrees (STEP 4)*

We compute $B(i, j)$ and $F(i, j)$ by iteratively patching Y-subtrees together via two functions, $Y(i, ARG)$ and $Y'(i, ARG)$. ARG is an argument of the form $B(a, b)$ or $(B(a, b), B(c, d))$ (or the same with F). We describe these functions with respect to B's only, all other cases being similar.

Case 1. $Y(i, B(a, b))$ (Fig. 4.21a)

The vertices v_a, v_b, v_i occur in clockwise order. Let $T = B(a, b)$. Extend the notch at v_a to take into account the arm of T. Extend the notch at v_i by making its associated wedge be the entire plane. If an extended X_2-pattern is possible between v_a, v_i and if $w = \angle(R_i, v_iv_a) < 180$, set $Y(i, B(a, b)) = (Y\text{-subtree:} \{v_iv_a\} \cup T)$, else $Y(i, B(a, b)) = 0$.

Case 2. $Y(i, (B(a, b), B(c, d))$ (Fig. 4.21b)

The vertices v_a, v_b, v_c, v_d, v_i occur in clockwise order. Let $T = B(a, b)$ and $V = B(c, d)$. Extend the notch at v_a (resp. v_c) to take into account the arm of T (resp. V). Extend the notch at v_i by making its associated wedge be the entire plane. If an extended X_3-pattern is possible between v_a, v_c, v_i, compute the locus of its N3-node. Let S be the point in the locus which maximizes the angle $w = \angle(R_i, v_iS)$.

If $w \in 180$, set $Y(i, (B(a, b), B(c, d)) = (Y\text{-subtree:}\{Sv_i\} \cup \{Sv_a\} \cup \{Sv_c\} \cup T \cup V)$, else $Y(i, (B(a, b), B(c, d)) = 0$.

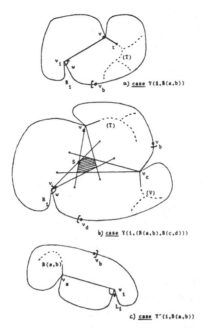

a) case Y(i,B(a,b))

b) case Y(i,(B(a,b),B(c,d)))

FIG. 4.21.

c) case Y'(i,B(a,b))

Case 3. $Y'(i, B(a, b))$ (Fig. 4.21c)

We define $Y'(i, B(a, b))$ in the same way as $Y(i, B(a, b))$, with $\angle(R_i, v_i v_a)$ replaced by $\angle(v_i v_a, L_i)$. $Y'(i, F(a, b))$ is defined similarly, so we omit the details.

We are now ready to implement STEP 4 of the decomposition algorithm. We will only describe the computation of $B(i, j)$, since the case of $F(i, j)$ is strictly similar. We begin by computing the four sets B_1, B_2, B_3, B_4. Let C be the value of $|S(i, j)|$ computed in Step 3.

$B_1 = \{$ Y-subtree of $B(i, k)\}$, for all $v_k \in V(i, j - 1)$ such that $|S(i, k)| + |S(k + 1, j)| = C$.
/*v_j is not a notch of the Y-subtree */
$B_2 = \{Y'(i, B(k, j))\}$, for all $v_k \in V(i + 1, j - 1)$ such that $|S(i + 1, k - 1)| + |S(k, j)| = C$.
/* v_i's neighbor is an N2-node */
$B_3 = \{Y'(i, F(i + 1, j))\}$, if $|S(i + 1, j)| = C$.
/*v_i's neighbor is an N2-node */
$B_4 = \{Y'(i,(F(i + 1, k), F(k + 1, j)))\}$, for all $v_k \in V(i + 1, j - 1)$ such that $|S(i + 1, k)| + |S(k + 1, j)| = C$.
/* v_i's neighbor is an N3-node */

Let T be the Y-subtree of $B_1 \cup B_2 \cup B_3 \cup B_4$ which maximizes the angle $u = \angle(t, L_i)$, where t is understood here as the arm of T directed outward from v_i (Fig. 4.22). We define $B(i, j)$ as a pointer to the arm of T (now understood to be directed towards v_i).

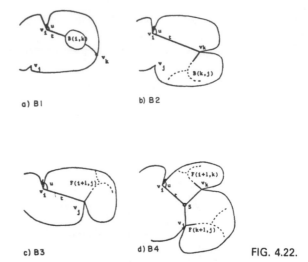

a) BI   b) B2

c) B3   d) B4   FIG. 4.22.

It follows from previous remarks that $B(i, j)$ can be computed in polynomial time. Y-subtrees can be merged in constant time by linking their respective arms together. B_1 through B_4 evaluate all candidate Y-subtrees adjacent to v_i and lying in $V(i, j)$, and keep a single candidate, i.e., the subtree which has maximum angle u. It is easy to show by induction that it is sufficient to consider only the Y-subtrees in the B's and F's. B_1 considers all subtrees which do not have both v_i and v_j as notches (Fig. 4.22a). B_2 and B_3 compute the subtrees whose vertex adjacent to v_i is an N2-node. Note that B_1 and B_2 may share common subtrees. The two possible configurations are illustrated in Fig. 4.22 b,c. Finally B_4 detects all candidate subtrees such that the vertex adjacent to v_i is an N3-node (Fig. 4.22d).

4. *Completing the OCD* (Step 5)

The last step of procedure *ConvDec* consists of removing the remaining notches with the naive decomposition. This can be done in polynomial time, leading to the main result: an optimal convex decomposition of a simple polygon can be computed in polynomial time.

A rough analysis of the algorithm's complexity shows that the exponent in the polynomial is prohibitively high. Cutting down the complexity to $O(n + c^3)$ is a long and complicated process which we will not address here. The reader can find the basic scheme in Chazelle and Dobkin (1985) and all details in Chazelle (1980).

4. EPILOGUE

Approximations and decompositions of two and three dimensional shapes are only beginning to be understood. Powerful algorithmic techniques are available, and problems once treated in an ad hoc fashion can now be solved with general, versatile methods. All of the algorithms reviewed here have been analyzed and many of them have been implemented. Their relevance to robotics is compelling, but more work needs to be done to adapt these methods to the context of specific applications. In particular, relatively little is known regarding problems in three dimensions.

The aim of this article has been to present state-of-the-art methods for approximating and decomposing geometric shapes. We have given emphasis to techniques with direct practical applications as well as methods of purely theoretical interest. Our rationale for including the latter has been that these methods often reveal enough insights to suggest simple, practical methods. For example, our long development on the OCD problem readily suggests efficient approximation methods (based on the naive decomposition and, say, X_2-patterns only). It is certainly counterproductive to dismiss complicated theoretical methods solely on the grounds of impracticality. Many recent advances in theoretical computer science might never come into practice as such but may very well, as has often happened in the past, trigger the making of practical breakthroughs. For example, the availability of practical methods for planar point location today (a central operation in many tasks—see chapter by Yap in this volume) owes a great deal to an earlier algorithm by Lipton and Tarjan, theoretically remarkable, but unfit for practical use.

The field of computational geometry is blossoming. One of its challenges is to span the entire spectrum from theory to practice, enabling powerful mathematical constructions to have an impact in practical domains. Robotics along with statistics, computer graphics, and a number of other applications areas serve as the prime providers of fascinating problems to researchers in computational geometry. In return, we believe that robotics is advised to keep an eye on an area that is blossoming today and is likely to come up with many of the algorithmic tools that it needs in the years to come.

ACKNOWLEDGMENTS

I wish to thank A. Aggarwal, J. Incerpi, C. K. Yap, and the referee for many helpful comments and suggestions.

The author was supported in part by NSF grants MCS 83-03925 and the Office of Naval Research and the Defense Advanced Research Projects Agency under contract N00014-83-K-0146 and ARPA Order No. 4786.

REFERENCES

Aggarwal, A., Chang, J. S., & Yap, C. K. (1985). Minimum area circumscribing polygons. *J. of Computer Graphics, Vision, and Image Processing.*

Aho, A. V., Hopcroft, J. E., & Ullman, J. D. (1974). *The design and analysis of computer algorithms.* Reading, MA: Addison-Wesley.

Ahuja, N., Chien, R. T., Yen, R., & Birdwell, N. (1980). Interference detection and collision avoidance among three dimensional objects. *Proc. 1st Annual Nat. Conf. on Artificial Intelligence*, Palo Alto, pp. 44–48.

Asano, T., & Asano, T. (1983). Minimum partition of polygonal regions into trapezoids. *Proc. 24th Annual FOCS Symp.*, pp. 233–241.

Asano, T., Asano, T., & Imai, H. (1984). *Partitioning a polygonal region into trapezoids*. (Res. Mem. RMI84-03). Dept. Math. Eng. & Instrumentation Physics, Univ. of Tokyo.

Avis, D., & Toussaint, G. T. (1981). An efficient algorithm for decomposing a polygon into star-shaped polygons. *Pattern Recognition*, *13*, 395–398.

Baker. B. S., Fortune, S. J., & Mahaney, S. R. (1984). Inspection by polygon containment. *Proc. 22nd Allerton Conf. on Comm. Control and Computing*, pp. 91–100.

Baker, B. S., Crosse, E., & Rafferty, C. S. (1985). *Non-obtuse triangulation of a polygon*. (Technical Report). Numerical Analysis Manuscript 84-4, AT&T.

Ben-Or, M. (1983). Lower bounds for algebraic computation trees. *Proc. 15th ACM Annual Symp. on Theory of Computing*, pp. 80–86.

Bentley, J. L., Faust, G. M., & Preparata, F. P. Approximation algorithms for convex hulls. *Comm. ACM*, *25*, pp. 64–68.

Bentley, J. L., & Ottmann, T. (1979, September). Algorithms for reporting and counting geometric intersections. *IEEE Trans. Comp. C-28, 9*, 643–647.

Bentley, J. L., & Shamos, M. I. (1978). Divide and conquer for linear expected time. *Info. Proc. Lett.*, *7*, 87–91.

Bhattacharya, B. K., & El Gindy, H. (1984). A new linear convex hull algorithm for simple polygons. *IEEE Trans. Infor. Theory, IT-30*, pp. 85–88.

Boyce, J. E., Dobkin, D. P., Drysdale, R. L., & Guibas, L. J. (1982). Finding extremal polygons. *Proc. 14th ACM Annual Symp. on Theory of Computing*, pp. 282–289.

Chand, D. R., & Kapur, S. S. (1970). An algorithm for convex polytopes, *J. ACM*, *17*(1), 78–86.

Chang, J. S., & Yap, C. K. (1984). A polynomial solution for potato-peeling and other polygon inclusion and enclosure problems. *Proc. 25th Annual FOCS Symp.*, pp. 408–416.

Chazelle, B. (1980). Computational geometry and convexity. Doctoral thesis, Yale University (Also available as Technical Report CMU-CS-80-150, Carnegie-Mellon University, Pittsburgh, PA).

Chazelle, B. (1982). A Theorem on polygon cutting with applications. *Proc. 23rd Annual FOCS Symp.*, pp. 339–349.

Chazelle, B. (1983a). A decision procedure for optimal polyhedron partitioning. *Info. Proc. Lett.*, *16*(2), 75–78.

Chazelle, B. (1983b), The polygon containment problem. *In advances in computing research* (pp. 1–33). Greenwich, CT: JAI Press.

Chazelle, B. (1984). Convex partitions of polyhedra: A lower bound and worst-case optimal algorithm. *SIAM J. on Comput.*, *13*(3), 488–507.

Chazelle, B. (1985). On the convex layers of a planar set. *IEEE Trans. Inform. Theory, IT-31*(4), 509–517.

Chazelle, B., & Dobkin, D. P. (1979). Decomposing a polygon into its convex parts. *Proc. 11th Annual ACM Symp. on Theory of Computing*, pp. 38–48.

Chazelle, B., & Dobkin, D. P. (1985). Optimal convex decompositions. *Computational geometry* (pp. 63–133), North-Holland.

Chazelle, B., Guibas, L. J., & Lee, D. T. (1983). The power of geometric duality. *Proc. 24th Annual FOCS Symp.*, pp. 217–225.

Chazelle, B., & Incerpi, J. (April 1984). Triangulation and shape-complexity. *ACM Trans. on Graphics*, *3*(2), 135–152.

Chazelle, B., & Lee, D. T. (1986). On a circle placement problem. *Computing, 36*, 1–16.

DePano, A., & Aggarwal, A. (1984). Finding restricted k-envelopes for convex polygons. *Proc. 22nd Allerton Conf. on Comm. Control, and Computing*, pp. 81–90.

Eddy, W. (1977). A new convex hull algorithm for planar sets. *ACM Trans. Math. Software, 3*(4), pp. 398–403.

El Gindy, H., & Avis, D. (1981). A linear algorithm for computing the visibility polygon from a point. *J. Algorithms, 2,* 186–197.

El Gindy, H., Avis, D., & Toussaint, G. T. (1983). Applications of a two dimensional hidden line algorithm to other geometric problems. *Computing, 31,* 191–202.

Feng, H., & Pavlidis, T. (1975). Decomposition of pollygons into simpler components: Feature generation for syntactic pattern recognition. *IEEE Trans. Comp., C-24,* 636–650.

Ferrari, L., Sankar, P. V., & Sklansky, J. (1981, December). Minimal rectangular partitions of digitized blobs. *Proc. 5th International Conference on Pattern Recognition,* Miami Beach, pp. 1040–1043.

Franzblau, D. S., & Kleitman, D. J. (1984). An algorithm for constructing regions with rectangles: Independence and minimum generating sets for collection of intervals. *Proc. 16th Annual ACM Symp. on Theory of Computing,* pp. 167–174.

Garey, M., Johnson, D. S., Preparata, F. P., & Tarjan, R. E. (1978). Triangulating a simple polygon. *Info. Proc. Lett., 7*(4), 175–180.

Graham, R. L. (1972). An efficient algorithm for determining the convex hull of a finite planar set. *Info. Proc. Lett., 1,* 132–133.

Graham, R. L., & Yao, F. F. (1983). Finding the convex hull of a simple polygon. *J. Algorithms, 4*(4), 324–331.

Greene, D. H. (1983). *The decomposition of polygons into convex parts.* In F. Preparata (Ed.), Advances in computing research (pp. 235–259). Greenwich, CT: JAI Press.

Guibas, L. J., Ramshaw, L., & Stolfi, J. (1983). A kinetic framework for computational geometry. *Proc. 24th Annual FOCS Symp.,* pp. 100–111.

Hertel, S., & Mehlhorn, K. (1983). Fast triangulation of simple polygons. *Proc. FCT'83,* Borgholm, LNCS Springer-Verlag, pp. 207–218.

Imai, H., & Asano, T. *An efficient algorithm for finding a maximum matching of an intersection graph of horizontal and vertical line segments.* (Papers of IECE Tech. Group on Circuits and Systems, CAS), 83–143.

Imai, H., & Asano, T. (1983b, October). *Efficient algorithms for geometric graph search problems.* (Res. Memo. RMI 83-05). Dept. Math. Eng. & Instrumentation Physics, Univ. of Tokyo.

Jarvis, R. A. (1973). On the identification of the convex hull of a finite set of points in the plane. *Info. Proc. Lett., 2,* 18–21.

Kahn, J., Klawe, M., & Kleitman, D. (1980). Traditional galleries require fewer watchmen. *SIAM J. Alg. Disc. Method, 4*(2), 194–206.

Keil, J. M. (1983). *Decomposing polygons into simpler components.* Doctoral thesis, University of Toronto.

Keil, J. M., & Sack, J. R. (1985). Minimum decompositions of polygonal objects. *Computational Geometry,* North-Holland.

Kirkpatrick, D. G., & Seidel, R. (1983, October). *The ultimate planar convex hull algorithm.* (Tech. Rep. No. 83-577). Cornell University. A preliminary version appeared in *Proc. 20th Annual Allerton Conf. on Comm., Control, and Comput.,* pp. 35–42.

Klee, V., & Laskowski, M. C. (1985). Finding the smallest triangles containing a given convex polygon. *J. Algorithms, 6*(3), 359–375.

Knuth, D. E. (1973). *The art of computer programming: Sorting and searching.* (Vol. 3). Reading, MA: Addison-Wesley.

Lee, D. T. (1982). On k-nearest neighbor Voronoi diagrams in the plane. *IEEE Trans. Comput., C-31,* No. 6, 478–487.

Lee, D. T. (1983). On finding the convex hull of a simple polygon. *Int'l J. Comput. and Infor. Sci., 12*(2), 87–98.

Lingas, A. (1981, November). *Heuristics for minimum edge length rectangular decomposition.* Unpub. Manuscript, MIT.

Lingas, A. (1982). The power of non-rectilinear holes. *Proc. 9th Colloquium on Automata, Languages and Programming,* Aarhus, LNCS Springer-Verlag, pp. 369–383.

Lingas, A., Pinter, R., Rivest, R., & Shamir, A. (1982, October). Minimum edge length partitioning of rectilinear polygons. *Proc. 20th Annual Allerton Conf. on Comm., Control and Comput.,* pp. 53–63.

Lipski, W., Jr. (1983). Finding a Manhattan path and related problems. *Networks, 13,* pp. 399–409.

McCallum, D., & Avis, D. (1979). A linear time algorithm for finding the convex hull of a simple polygon. *Info. Proc. Lett., 9,* 201–206.

Megiddo, N. (1983). Linear time algorithm for linear programming in $\Re^3$ and related problems. *SIAM J. Comput., 12*(4), 759–776.

Megiddo, N., & Supowit, K. J. (1984). On the complexity of some common geometric location problems. *SIAM J. Comput., 13*(1), 182–196.

Newman, W. M., & Sproull, R. F. (1979). Principles of interactive computer graphics. New York: McGraw-Hill.

O'Rourke, J. (1982). The complexity of computing minimum convex covers for polygons. *Proc. 20th Annual Allerton Conf. on Comm. Control and Comput.,* pp. 75–84.

O'Rourke, J. (1984). *Finding minimal enclosing tetrahedra.* The Johns Hopkins Univ., Techn. Report.

O'Rourke, J., Aggarwal, A., Maddila, S., & Baldwin, M. (1984). *An optimal algorithm for finding minimal enclosing triangles.* (Tech. Rep. JHU/EECS-84/08). Dept. of EE/CS. Johns Hopkins Univ.

O'Rourke, J., & Supowit, K. J. (1983). Some NP-hard polygon decomposition problems. *IEEE Trans. Inform. Theory, IT-29*(2), 181–190.

Overmars, M. H., & van Leeuwen, J. (1981). Maintenance of configurations in the plane. *JCSS, 23,* 166–204.

Post, M. J. (1982, May). *Computing minimum spanning ellipses.* (Tech. Rep. No. CS-82-16). Brown Univ. A preliminary version appeared in *Proc. 22nd Annual FOCS Symp.,* pp. 115–122.

Preparata, F. P. (1979). An optimal real-time algorithm for planar convex hulls. *C. ACM, 22,* 402–405.

Preparata, F. P., & Hong, S. J. (1977). Convex hulls of finite sets of points in two and three dimensions. *Comm. ACM, 20*(2), 87–93.

Roger, C. A. (1964). *Packing and covering.* Cambridge, England: Cambridge University Press (1964).

Sack, J. R. (1982). An $O(n\log n)$ algorithm for decomposing simple rectilinear polygons into convex quadrilaterals. *Proc. 20th Annual Allerton Conf. on Comm., Control, and Comput.,* pp. 64–74.

Sack, J. R., & Toussaint, G. T. (1981). A linear time algorithm for decomposing rectilinear star-shaped polygons into convex quadrilaterals. *Proc. 19th Annual Allerton Conf. on Comm., Control, and Comput.,* pp. 21–30.

Schachter, B. (1978). Decomposition of polygons into convex sets. *IEEE Trans. on Computers, C-27,* 1078–1082.

Schoone, A. A., & van Leeuwen, J. (1980). *Triangulating a star-shaped polygon.* (Tech. Rep. No. RUV-CS-80-3). Univ. of Utrecht.

Sedgewick, R. (1983). *Algorithms.* Reading, MA: Addison-Wesley.

Seidel, R. (1981). *A convex hull algorithm optimal for points in even dimensions.* M.S. Thesis, Tech. Rep. 81-14, Univ. British Columbia, Canada.

Shamos, M. I. (1978) *Computational geometry.* Doctoral thesis, Yale University.

Shamos, M. I., & Hoey, D. (1975). Closest-point problems. *Proc. 16th Annual FOCS Symp.,* pp. 151–162.

Silverman, B. W., & Titterington, D. M. (1980). Minimum covering ellipses. *SIAM J. Sci. Stat. Comput., 1*(4), 401–409.

Toussaint, G. T. (1980a). Pattern recognition and geometrical complexity. *Proc. 5th Int. Conf. on pattern recognition*, Miami Beach, pp. 1324–1347.

Toussaint, G. T. (1980b, October). Decomposing a simple polygon with the relative neighborhood graph. *Proc. 18th Annual Allerton Conf. on Comm., Control, and Comput.*

Toussaint, G. T. (1983, May). Solving geometric problems with the "rotating calipers". *Proc. IEEE MELECON'83*, Athens, Greece.

Toussaint, G. T., & Avis D. (1982). On a convex hull algorithm for polygons and its application to triangulation problems. *Pattern Recognition, 15*(1), 23–29.

Yao, A. C. (1981). A lower bound to finding convex hulls. *J. ACM, 28,*(4), 780–787.

5 Intersection and Proximity Problems and Voronoi Diagrams

Daniel Leven
Micha Sharir
Tel Aviv University

ABSTRACT

In this article we consider the problem of efficient detection of intersection or proximity in a collection of bodies in two or three dimensional space. In two-dimensional space we use (generalized) Voronoi diagrams as a tool for solving these problems. We review basic properties of these diagrams and describe various efficient techniques for their construction. In three dimensional space Voronoi diagrams are no longer adequate for efficient intersection detection, and we present other, though less powerful, techniques for restricted classes of objects.

1. INTRODUCTION

An important topic in planning and controlling the motion of robot systems is that of intersection and proximity detection. That is, given a specific configuration of the system subparts, and the position and orientation of all the obstacles present in the robot's workspace, we wish to determine whether at this configuration there occurs an intersection between a moving robot subpart and a stationary obstacle, or between two separately moving subparts of the robot. This particular form of the problem arises when checking a prescribed planned motion of the robot for collision avoidance, by dividing the path into short segments and by checking whether an intersection would occur for the position of the robot system at the beginning of each such segment.

There are however many variations of the problem that can also arise in practice. For example, instead of checking whether an intersection has occurred, we might wish to check whether any two objects (robot subparts or obstacles) have approached one another to a distance less than some "safety distance" Δ. Such a proximity detection would be useful when controlling an actual motion, so that the system motion can be abruptly stopped whenever such a dangerous proximity is detected, before an actual collision will have occurred.

Another variation is, instead of taking "snapshots" of the system position at frequent intervals and detecting intersection or proximity at each such snapshot, to perform in advance a single "global" intersection detection analysis, given the planned path of motion of the system.

If the shape of the obstacles or the system subparts has a too complex geometric description, we might wish to approximate the problem, by enclosing each such object inside another object of simpler shape and by analyzing for intersection or proximity of these new objects. When making such an approximation, and sometimes also when considering the original precise problem, we may have to allow certain pairs of the given objects to intersect (e.g., if they cover two adjacent subparts of the robot, such as two links of the same arm, etc.) so that such intersections are not considered harmful. In such cases it is helpful to assign to each object a certain "color," and then to check only for intersection or proximity between pairs of different colored objects. These considerations anticipate that intersection detection will become more difficult as the shape of the given objects becomes more complex. Not surprisingly, this is indeed the case, as is reflected in subsequent sections of this article.

It is clear that in most of the applications considered so far, efficiency of the corresponding algorithms is crucial. For example, in proximity detection during the control of an actual motion, it is essential that the time required by the algorithm will be small enough to enable the system to stop in time before a detected dangerous proximity results in an actual collision. Even in the case of "off-line" intersection testing for a planned system motion, efficient detection procedures would allow us to perform the test for a denser collection of motion instances, thereby gaining a higher level of reliability of the intersection detector.

Before continuing, it is worth making the following simple observation: If the number of objects is n, and if one can determine in constant time whether a given pair of these objects intersect (or even calculate in constant time the distance between two given objects), then detection of intersection or proximity within the given collection of objects can be accomplished in time $O(n^2)$ by the naive technique that considers all possible pairs of objects. This straightforward technique is also easy to implement, and for small n is probably the best choice in practice. However, if n is large we will want to use, whenever possible, techniques that are asymptotically faster than this simple approach. Thus the remainder of this article is devoted to the description of techniques that run indeed in faster than quadratic time. Typically, these techniques will run in time $O(n \log^k n)$ for $k = 1$ or 2 usually.

The preceding discussion thus motivates consideration of the following series of problems:

Let S be a set of n objects of relatively simple structure in two or three dimensional space.

I. Do any two objects in S intersect?

II. More generally, suppose that we assign a "color" to each object in S. Does there exist a pair of objects in S having different colors and intersecting each other?

III. If no two objects in S intersect, what is the smallest distance between any two objects in S (or, in the colored version, what is the smallest distance between two objects in S having different colors)? More generally, for each object B in S find the object in S (of different color) nearest to B.

IV. Preprocess S so that, given an arbitrary "query point" X, the object in S nearest to X can be found quickly.

V. Suppose that the position of each object B in S varies with time in a known prescribed (and relatively simple) manner. Is there a time t at which any two objects in S intersect? In the colored version, is there a time t at which two differently colored objects intersect? If no intersection occurs, what is the smallest distance ever attained between any two (differently colored) objects?

As noted above, efficient solution of these problems, especially in the three-dimensional case, would facilitate construction of an "off-line" debugging system for robot control programs to check whether collision occurs along a planned path of motion, and would also make it more feasible to check in real time whether a moving subpart of the system is getting dangerously close to another (moving or stationary) object.

In this chapter we describe various techniques for solving these problems, that have appeared in the literature in the past few years. In Section 2 we begin to discuss the two dimensional case. First we describe a technique for detecting intersections, that was initially used by Shamos (1978) for detecting intersections between line segments but that can be used for other objects as well (typically, circular objects and other convex objects of some simple form). Then, in Section 3, we introduce the notion of (generalized) Voronoi diagrams as the major tool for solving some of the more complex problems mentioned above, and describe their basic properties. In Section 4 we describe the application of Voronoi diagrams to the solutions of these problems, and also review some additional more classical applications of these diagrams. Section 5 covers efficient techniques for the construction of Voronoi diagrams, as developed by Shamos (1978), Lee and Drysdale (1981), Sharir (1985) and also some original extensions of these techniques developed by the authors. Section 6 briefly discusses the dynamic versions of the intersection and proximity detection problems, as posed in Problem V above. Finally, in Section 7 we turn to the three dimensional case; we show why in this case Voronoi diagrams are inadequate for our problems, and review alternative techniques that solve only the simplest intersection detection problem I, and only for collections of objects having some restricted simple shape, such as rectilinear boxes or spheres. These methods are due e.g., to Edelsbrunner (1980a, 1980b), to Edelsbrunner and Maurer (1981), and to Hopcroft, Schwartz, and Sharir (1983).

2. A SIMPLE INTERSECTION DETECTION ALGORITHM IN TWO-DIMENSIONAL SPACE

In this section we discuss the first intersection problem posed above for a set of planar objects such as line segments, circular objects or other objects of simple shape and show that it can be solved in time $O(n \log n)$. The algorithm is a straightforward modification of an algorithm due to Shamos (1978) that tests for intersection of straight line segments. We assume that the bodies in S are monotone in the x-direction (i.e., the boundary of each object $B \in S$ consists of an upper portion and a lower portion and both portions extend monotonously from left to right). Thus for each such B there exist exactly two vertical lines tangent to it, and any vertical line between these lines intersects B in a closed segment. The structure of the objects in S is assumed to be simple enough so that each of the following operations takes constant time:

(i) Check whether two specific bodies in S intersect each other.
(ii) For each $B \in S$, find the smallest and largest abscissae of points in B.
(iii) For each $B \in S$, find, for a given abscissa x, a point $(x, y) \in B$ (if such a point exists).

Let $B_1, \ldots, B_n$ be the objects in S. For each $j = 1, \ldots, n$ let a_j (resp. b_j) be the smallest (resp. largest) abscissa of a point in B_j. For simplicity we assume that the $2n$ numbers $a_j, b_j, j = 1, \ldots, n$, are all distinct (see Fig. 5.1).

Note that if we draw a vertical line $x = x_0$ it will cut (some of) the objects in S in straight segments which, if the objects do not intersect each other along that line, are disjoint from each other. Hence the objects in S that intersect the line $x = x_0$ can be linearly ordered in a list $L(x_0)$ in which an object B_i precedes another B_j if the segment cut off B_i by $x = x_0$ lies below the corresponding segment cut off B_j. Note that since the objects in S are monotone, the list $L(x_0)$ remains unchanged as x_0 increases until either the line $x = x_0$ meets a new object B_k (this will happen when $x_0 = a_k$), or when it stops making contact with some object B_k (just after $x_0 = b_k$), or when two of the objects in $L(x_0)$ intersect at $x = x_0$. Moreover, the leftmost intersection (if any exists) of any two objects $B, B' \in S$

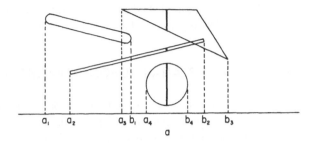

FIG. 5.1. An instance of the intersection problem.

will occur at some $x = x_0$ such that, for x slightly less than x_0, the list $L(x)$ contains B and B' as adjacent elements. (Here we ignore the special case in which B and B' intersect at a point that is leftmost in one of these objects, that requires a slightly different argument.)

In view of these observations, to detect the presence or absence of an intersection one simply has to check repeatedly whether any two adjacent elements in the list $L(x_0)$ intersect each other, and stop as soon as an intersection has been detected. Plainly, these checks have to be performed only at points where $L(x_0)$ changes (i.e., at the points $x_0 = a_j, b_j, j = 1, \ldots, n$), and one only needs to test newly adjacent pairs in $L(x_0)$. To facilitate execution of these operations, the list L can be maintained as a 2-3 tree, allowing all the required list-maintenance operations to be performed in time $O(\log n)$

Details are as follows. The algorithm begins by sorting the $2n$ numbers $a_j, b_j,$ $j = 1, \ldots, n$, in increasing order, and then processes them from left to right. Initially, the list L is empty. Suppose that the abscissa currently being processed is one of the a_j. Then L is updated by inserting the object B_j into L in its proper place, using a standard 2-3 tree search during which the comparison of any two objects B, B' is accomplished by comparing two representative points in the intersection of $x = a_j$ with B, B' respectively (we have assumed that this can be done in constant time). After insertion, the algorithm finds the two objects B, B' immediately preceding and succeeding B_j in L, and checks whether either B or B' intersects B_j.

Similarly, if the abscissa currently being processed is b_j, then the object B_j is deleted from L, using essentially the technique just outlined. After deletion, the algorithm finds the two objects in L that immediately preceded and followed B_j prior to its deletion (these will have become newly adjacent in L after deletion of B_j), and determines whether they intersect each other.

The algorithm halts whenever an intersection is detected, or, if no intersection has been detected, when all the abscissae a_j and b_j have been processed, in which case the algorithm reports that there is no intersection between the objects in S.

The correctness of the algorithm follows from the preceding observations. The time complexity of the algorithm is $O(n \log n)$ since processing of each of the $2n$ abscissae a_j, b_j can be accomplished in $O(\log n)$ time, using a 2–3 tree representation for the list L (cf. e.g., Aho, Hopcroft, & Ullman, 1974).

3.   GENERALIZED PLANAR VORONOI DIAGRAMS

The sweeping algorithm presented in the preceding section does not solve the more complicated "colored object" intersection problem (or any of the other intersection or proximity problems mentioned above). Indeed, the argument justifying the correctness of the algorithm breaks down as soon as an intersection is detected, so that if the first intersection detected is between two objects having the same color we can no longer use the procedure described to find additional intersection points. Even though this problem can be overcome by appropriate patching of the algorithm (see e.g., Bentley & Ottman, 1979), the algorithm may still have to process a large (possibly quadratic) number of "harmless" intersections before detecting the first intercolor intersection. We therefore need more sophisticated techniques and data structures to handle the various other proximity and intersection problems listed above, and propose to use to this end an approach based on generalized Voronoi diagrams. We first define these diagrams and analyze their basic properties; then in Section 4 we present some applications of these diagrams including the proximity problems mentioned in the introduction; finally, in section 5 we present techniques for the efficient construction of these diagrams.

3.1. Definitions and Basic Properties

Roughly speaking, the Voronoi diagram of a set S of objects is a partition of space into cells with disjoint interiors, having the property that each cell contains all points that are at least as near to one object of S as to any of the other objects of S. More precisely, let $S = \{s_1, s_2, \ldots, s_n\}$ be a set of n disjoint planar convex objects. The generalized *Voronoi diagram Vor(S)* associated with the set S is defined as follows. For each $i \neq j$ define

$$H(s_i, s_j) = \{y \in E^2 : d(s_i, y) \le d(s_j, y)\},$$

i.e., the set of all points whose distance from s_i is no greater than their distance from s_j. (The distance of y from s_i is taken to be zero if y lies inside s_i.) Then define the (closed) *Voronoi cell $V(s_i)$* associated with s_i to be

$$V(s_i) = \bigcap_{j \neq i} H(s_i, s_j).$$

i.e., the set of all points y whose distance from s_i is no greater than y's distance from any other element of S. Finally, the *Voronoi diagram Vor(S)* is defined to be the set of points that belong to more than one Voronoi cell. If the objects in S

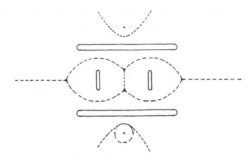

FIG. 5.2.   The Voronoi diagram of
a set of convex objects.

are disjoint from each other then *Vor(S)* is a collection of one-dimensional arcs
(see below); otherwise some modifications of this definition may be required in
order to preserve this property (cf. Kirkpatrick, 1979, for example). The edges of
this diagram are maximal connected arcs that belong to exactly two Voronoi cells
and the vertices are those points that belong to more than two cells. For simplicity we assume that no point in *Vor(S)* lies in more than three Voronoi cells or,
equivalently, that no more than three objects in S can be tangent to the same
circle and lie in its exterior. In the case where the objects of S are single points,
this assumption requires that no more than three of these points be cocircular.
Figure 5.2 shows an example of such a Voronoi diagram.

The above definition generalizes several more specific cases studied recently,
such as cases in which the objects of S are singleton points (Kirkpatrick, 1979;
Shamos, 1978), or are line segments and points (Kirkpatrick, 1979; Lee &
Drysdale, 1981), or are circular discs (Lee & Drysdale, 1981; Sharir, 1985); see
also Yap (1984). Most of the properties of Voronoi diagrams hold for general
convex objects, as will now be established.

The generalized Voronoi diagram just defined has the following properties
(cf. Sharir, 1985).

(1) The collection of Voronoi cells covers the whole plane.

Proof: obvious.

(2) Let y be a point in some cell $V(s_i)$. Then the line segment that joins y to the
closest point on s_i is wholly contained in $V(s_i)$, and in addition the interior of this
segment meets no other Voronoi cell. (We will capture this property by calling
$V(s_i)$ *star shaped* with respect to s_i.)

Proof: Let $y \in V(s_i)$, and let I be the segment connecting y to the closest point x
on s_i. Suppose there exists a point z in the interior of I that belongs to a second
cell $V(s_j)$ of an object $s_j \neq s_i$. Hence we have

$$d(s_j, z) \leq d(x, z).$$

and from this and the triangle inequality (making use of the fact that z lies between x and y on a straight line) we get

$$d(s_j, y) \leq d(s_j, z) + d(z, y) \leq d(x, z) + d(z, y) = d(s_i, y).$$

Thus we must have equality throughout (or else y could not belong to $V(s_i)$). This implies that the point on s_j nearest to z and to y must be x itself, that is impossible since we have assumed that s_i, s_j are disjoint.

Remark: If two objects s_i, s_j of S touch one another at some point x (e.g., as in the case of two line segments with a common endpoint), then the above argument breaks down. Nevertheless, it is generally easy to modify the definition of Voronoi diagrams, by adding to it extra edges connecting such points x to points on the Voronoi diagram, thereby obtaining a diagram with similar properties to those listed here. For more details see for example (Lee & Drysdale, 1981) or (Kirkpatrick, 1979).

(3) Suppose I_1, I_2 are two line segments in $V(s_i)$ that join two points x_1, x_2 in $V(s_i)$ with their respective closest points, y_1, y_2, on s_i. Then either I_1, I_2 do not intersect except possibly at their endpoints or else one of these lines is wholly contained in the other.

proof: Suppose I_1, I_2 intersect at a point z that is not one of their end points and that one segment is not wholly contained in the other. Suppose also, without loss of generality that $d(z, y_1) \geq d(z, y_2)$. Then from the triangle inequality we get

$$d(y_2, x_1) < d(y_2, z) + d(z, x_1) \leq d(y_1, z) + d(z, x_1) = d(y_1, x_1)$$

and this contradicts the fact that y_1 is the closest point on s_i to x_1.

(4) **Corrollary:** Suppose we are given k points $a_1, \ldots, a_k$ in a clockwise order on the boundary of some Voronoi cell $V(s_i)$. Then the points that are closest to them on the boundary of s_i also appear in clockwise order.

(5) The intersection I of three Voronoi cells $V(s_i)$, $V(s_j)$, $V(s_k)$ can consist of at most two points.

Proof: Let y be a point in I. By (2) the interior of the segment L_i (resp. L_j, L_k) connecting y with its nearest point on s_i (resp. s_j, s_k) is contained in $V(s_i)$ (resp. $V(s_j)$, $V(s_k)$) and in no other cell. Let y' be another point in I, and let L_i' (resp. L_j', L_k') be the segments connecting y' with its nearest point on s_i (resp. s_j, s_k). It is clear that no two of these six segments can intersect one another (except at an endpoint). It follows that I cannot contain a third point z, because at least one of the segments connecting z to the three closest points on s_i, s_j, s_k would have to intersect either one of the preceding six segments or one of these objects, which is impossible.

(6) Let a be a point on the locus C of points that are equidistant from two objects s_i, s_j and let y_i, y_j be the (unique) points that are closest to a on s_i, s_j respectively. Then the unique tangent line to C at a is the bisector line of $\angle y_i a y_j$.

Proof: Let b be a point close to a on C and denote $d(a, b)$ by t. Let T_i (resp. T_j) be the tangent line to s_i (resp. s_j) at y_i (resp. y_j) that is perpendicular to ay_i (resp. ay_j). Since s_i, s_j are convex and b is a point on C we have:

$$d(b,y_i)^2 - t^2 \leq d(b,T_i)^2 \leq d(b,s_i)^2 \leq d(b,y_i)^2$$
$$d(b,y_j)^2 - t^2 \leq d(b,T_j)^2 \leq d(b,s_j)^2 \leq d(b,y_j)^2$$
$$d(b,s_i) = d(b,s_j).$$

Hence, using the law of cosines for the triangles bay_i, bay_j, we obtain

$$\left| \cos(\angle bay_i) - \cos(\angle bay_j) \right| \leq \frac{t}{2d(a,y_i)}$$

The required result now follows by letting t approach zero.

(7) $Vor(S)$ consists of simple smooth arcs.

Proof: The fact that $Vor(S)$ consists of simple arcs follows from (4). The fact that the arcs are smooth follows from (6) which proves the uniqueness of the tangent line at any point of $Vor(S)$ that is not a vertex.

(8) Let C be the convex hull of the union of all objects in S. Note that the boundary of C consists of an alternating sequence of portions of the boundaries of the objects of S and of portions of the common exterior tangents to pairs of these objects. We will call two objects s_i, s_j of S *adjacent* along the boundary of C if this boundary contains a straight segment tangent to both s_i and s_j. Then the unbounded edges of $Vor(S)$ are precisely those edges that separate two cells $V(s_i)$ and $V(s_j)$ for which s_i and s_j are adjacent along the boundary of C. Moreover, each such unbounded edge approaches asymptotically the perpendicular bisector of the portion of the common tangent of s_i and s_j connecting the two points of tangency.

Proof: Let e be an unbounded edge of $Vor(S)$, common to two Voronoi cells $V(s_i)$ and $V(s_j)$. It is easy to check that the portions of e sufficiently far away from s_i, s_j must be contained in one of the two semi-infinite strips whose base segments are the two exterior common tangents to s_i and s_j, and whose interiors are disjoint to both s_i and s_j. Suppose without loss of generality that s_i and s_j both touch the x-axis, that they both lie in the lower half-plane bounded by the x-axis, that the origin is the midpoint of the segment connecting the two points of tangency, and that e is contained in the upper semi-infinite strip bounded from below by the portion of the x-axis between the two points of contact with s_i and s_j. Let $a = (u, t)$ be a point on e. For sufficiently large t and for any $k = 1, \ldots , n$, $d(a, s_k)$ behaves asymptotically as $t - \eta_k$, where η_k is the y-coordinate of the highest point on s_k. It follows that

$$0 = \eta_i = \eta_j \geq \eta_k,$$

for every $k = 1, \ldots , n$. This however is easily seen to imply that s_i and s_j are adjacent to each other along C, since all objects in S lie in the lower half-plane bounded by the x-axis, and thus the portion of the x-axis between its points of tangency with s_i and s_j belongs to the boundary of C. (An extreme case that we need consider is that in which the x-axis is also tangent to a third object s_k, at a point lying between its points of contact with s_i and s_j. However, if such a situation arises, it is easy to check that all the points on e sufficiently far away are nearer to s_k than to one of s_i, s_j, contradicting the definition of e.) The converse statement, namely that any pair of adjacent objects along C induces an unbounded Voronoi edge, can be proved using the above argument in reverse.

(9) $Vor(S)$ need not be connected. In fact, it can have up to $O(n)$ connected components. However, every connected component of $Vor(S)$ is unbounded.

Proof: Consider the following set S of circular discs, that consists of a unit disc D_1, and of k additional small disks $D_2, \ldots , D_{k+1}$ all of radius ρ, such that the centers of these discs are placed near to the boundary of D_1 at equally spaced positions. If ρ is chosen to be sufficiently small (e.g., of the order $O(1/k^2)$) then it is easily checked that for each $j = 2, \ldots , k + 1$ the discs D_1 and D_j are adjacent to each other along the boundary of the convex hull of the D_i's. Moreover, it can also be shown that for ρ sufficiently small each unbounded edge common to $V(1)$ and some other $V(j)$ is a full branch of the hyperbola defining the points that are equidistant from both D_1 and D_j and no two such edges intersect each other. This shows that $Vor(S)$ can have as many as $O(n)$ connected components.

(10) Adjacent edges I, J on the boundary of $V(s_i)$ cannot be tangent to each other. Moreover, the angle formed by the tangent lines to I, J at the joint vertex x, and contained in $V(s_i)$, is convex.

Proof: Suppose I, J separate $V(s_i)$ from $V(s_j)$, $V(s_k)$ respectively. Let y_i, y_j, y_k be points on s_i, s_j, s_k respectively that are closest to x. By (6) the tangent to I at x bisects $\angle y_i x y_j$ while the tangent to J at x bisects $\angle y_i x y_k$. This, together with the fact that y_i, y_j, y_k are equidistant from x and are all disjoint, proves that the two edges cannot be tangent to each other and also that the angle formed by I, J at x, and contained in $V(s_i)$, is convex.

(11) *Vor(S)* consists of at most $O(n)$ edges and vertices.

Proof: By (10) every vertex in *Vor(S)* has degree ≥ 3. Also we may assume that all the unbounded components of *Vor(S)* are connected at infinity and thus the planar graph formed by *Vor(S)* is connected. Counting the degrees of all vertices in *Vor(S)* we get

$$3|V| \leq \sum_{v \in V} \text{degree of } v \leq 2|E|$$

where V (resp. E) is the set of vertices (resp. edges) of *Vor(S)*. Substituting this inequality in Euler's formula for connected planar graphs, namely

$$|V| + |F| = |E| + 2$$

where F is the set of cells of *Vor(S)*, we get that $|E| \leq 3|F|$, $|V| \leq 2|F|$. Since each cell contains one object in S there are n cells and the required result follows.

Remarks: (1) More specific information on the structure of *Vor(S)* can easily be obtained for sets S whose elements have all some special shape. For example, if the elements of S are points, then the edges of *Vor(S)* are all straight segments, and each Voronoi cell is convex. If the elements of S are straight segments and points, the edges of *Vor(S)* are straight segments and parabolic arcs, and if S consists of circular discs, the edges of *Vor(S)* are straight segments and hyperbolic arcs.

(2) In this section we have ignored the shape of the objects of S. If these objects have too complex a shape, then the edges of the Voronoi diagram will also be relatively complex. In other words, the complexity of a discrete combinatorial description of the Voronoi diagram will grow with the complexity of such descriptions of the objects of S. In this article we will simply finesse this issue, by assuming each of the objects of S to be of relatively simple shape (e.g., as in the preceding remark). Consequently, we will regard each basic operation that we will need to apply to one or two objects in S (e.g., finding the leftmost or rightmost point on such an object, finding the locus of points equidistant from two such objects, etc.) as requiring constant amount of time. In general, however, the complexity bounds on the size of the Voronoi diagrams and on the performance of various algorithms for the construction and use of the diagrams, as given in this article, may have to be revised if the shape of the objects in S becomes too complex.

4. APPLICATIONS OF VORONOI DIAGRAMS

The reason why Voronoi diagrams are so useful for solving a variety of problems in computational geometry is that they provide global information about the nearness of objects in some given collection to one another. This basic property of Voronoi diagrams, plus the fact that their size (in the two-dimensional case) is linear in the number of objects involved, can be exploited to yield efficient algorithms that solve various problems in computational geometry (including our intersection and proximity problems), some of which are given in this section.

Consider first the problems of detecting intersections and proximity in a set S of n given closed convex objects. The problem of using the Voronoi diagram of S to detect intersections is somewhat complicated, because the definition of $Vor(S)$ given above assumes that the objects in S are disjoint. We will comment on this problem later in this section, but first discuss the easier case in which the objects in question are indeed disjoint, and our aim is to calculate the proximity of these objects to one another.

We begin with the following observation.

Lemma 4.1: Let $s_1, s_2 \in S$ such that s_2 is a nearest object to s_1 among all other objects in S (i.e. the (Euclidean) distance between s_1 and s_2 is not greater than the distance between s_1 and any other object in S). Then the shortest segment p joining a point on s_1 to a point on s_2 is completely contained in $V(s_1) \cup V(s_2)$.

Proof: Suppose to the contrary that p crosses into a third Voronoi cell $V(s_3)$. Then there exists a point $x \in p$ such that $d(x, s_3) < d(x, s_2)$, so that

$$d(s_1, s_3) \leq d(s_1, x) + d(x, s_3) < d(s_1, x) + d(x, s_2) = d(s_1, s_2)$$

which contradicts the fact that s_2 is the object of S nearest to s_1.■

Lemma 4.1 implies that for each object $s_1 \in S$ the object of S nearest to it can be found by examining all those objects whose Voronoi cells are adjacent to $V(s_1)$ in $Vor(S)$, and selecting from among them the object whose distance from s_1 is smallest. Moreover, by a single scan through all the $O(n)$ edges of $Vor(S)$ we can compute in this manner the nearest neighbor in S of each object of S. We therefore have the following theorem.

Theorem 4.1: The problem of finding for each object in S its nearest neighbor in S can be solved in $O(n)$ time, given the Voronoi diagram of S. In particular, the shortest distance between any two objects of S can also be calculated in $O(n)$ time from the diagram.

Consider next the colored version of the proximity problem; here we want to find the smallest distance between any two differently-colored objects of S.

Lemma 4.2: Suppose that a color is assigned to each object in S. Then the shortest distance between any two differently-colored objects of S is attained between two objects $s_1, s_2 \in S$ such that $V(s_1)$ and $V(s_2)$ have a common boundary edge.

Proof: Let s_1 and s_2 be the two nearest differently colored objects in S, and let p be a shortest segment joining a point on s_1 to a point on s_2. The midpoint x of p is equidistant from s_1 and s_2, and we claim that it must lie on $Vor(S)$ (i.e. on the edge separating $V(s_1)$ and $V(s_2)$). Suppose the contrary; then x must be contained in a third cell $V(s_3)$. It follows, by a similar argument to that in the proof of Lemma 4.1 that s_3 is closer to s_1 (resp. s_2) than s_2 (resp. s_1). But either s_1 or s_2 must be colored by a different color than that of s_3 (otherwise they would both have the same color), and this contradicts our choice of s_1 and s_2.■

Thus, using again the fact that the size of $Vor(S)$ is linear in n, we obtain

Theorem 4.2: The problem of finding the closest pair of differently-colored objects in S can be solved in $O(n)$ time, given the Voronoi diagram of S.

It is easily seen that the arguments in the proof of Lemma 4.2 also yield the following stronger result.

Theorem 4.3: For each color c let s_1 and s_2 be the two nearest objects in S such that one of them is colored c and the other is colored by a different color. Then $V(s_1)$ and $V(s_2)$ are adjacent Voronoi cells in $Vor(S)$. Consequently, the problem of computing for each color c the two nearest objects in S one of which is colored c and the other by a different color, can be solved in $O(n)$ time from the Voronoi diagram of S.

A related application of Voronoi diagrams is as follows. Let S be a given set of n convex objects. We wish to pre-process S so that given a query point x, we can easily find the object in S nearest to x. In terms of $Vor(S)$, this problem is equivalent to that of locating the Voronoi cell in $Vor(S)$ that contains x. If the edges of $Vor(S)$ are relatively simple, then this latter problem can be solved efficiently using e.g., the point-location algorithm of Kirkpatrick (1983), in which after $O(n \log n)$ pre-processing of $Vor(S)$, the location of the cell of $Vor(S)$ containing any given point x can be accomplished in $O(\log n)$ time.

Consider next the problem of detecting intersection between any two (differently colored) objects of S. If no colors are involved, then the simple sweeping algorithm described in Section 2 can be used to detect intersection among objects of S in time $O(n \log n)$. In the case where only inter-color intersections are to be detected, for Voronoi diagrams to be applicable, one would have to generalize them to sets of possibly intersecting objects. In general this is not possible without sacrificing the property of Voronoi diagrams that their size is linear in the number of objects. Indeed, one can easily construct an example of a set S of n line segments intersecting each other in $n^2/4$ points. Each such point ought to belong to $Vor(S)$ because it is simultaneously nearest to the two segments of S intersecting at it. Thus $Vor(S)$ will have at least $n^2/4$ distinct edges in this case.

For certain shapes of the objects of S, however, it is possible to obtain a generalized Voronoi diagram of S of linear size even though the objects in S might intersect one another (even quadratically many times). For example this is the case when S consists of circular discs, as has been demonstrated in Sharir (1985). We will omit here details about this generalization (that can be found in Sharir, 1985), but only comment that the idea used there is to extend the distance function $d(x, S)$ between a point x and a disc S so that when x lies inside S, $d(x, S)$ is negative and is equal in absolute value to the distance from x to the circumference of S. Then, if one defines $Vor(S)$ essentially as in Section 3 above, but taking into account also negative distances, one obtains a diagram with properties similar to those listed in Section 3. Then, generalizing the arguments in Lemmas 4.1 and 4.2, one can show:

(i) If there exists an intersection between some pair of discs in S, then there exists such an intersecting pair s_1, s_2 such that $V(s_1)$ and $V(s_2)$ are adjacent Voronoi cells.

(ii) If colors are assigned to the discs in S, and if two differently-colored discs intersect, then there exists a pair of intersecting and differently-colored discs whose Voronoi cells are adjacent to one another in $Vor(S)$.

Thus we can conclude the following result.

Theorem 4.4: The problem of detecting intersection between differently-colored discs in a set of n planar discs can be solved in $O(n)$ time, given the Voronoi diagram of S as defined in Sharir, 1985.

Remark: Voronoi diagrams are not the only available tool for efficient detection of intersections and proximity. There exist several efficient and more direct methods for calculating e.g. the shortest distance between any two points in a set S, or for finding all nearest neighbors in such a set; see Clarkson (1983), Bentley (1980), Bentley, Weide, and Yao (1982).

Voronoi diagrams have also many additional applications, many of which are described in Shamos' doctoral thesis (1978). Among these, we will mention the following ones:

Triangulation.   Let S be a set of n points in the plane. We want to construct a partition of the convex hull of S into triangles such that the vertices of the triangles are the points in S. This partition is called a *triangulation* of the convex hull. To do this we define the following planar graph V^*, called the *Voronoi dual*. The vertices of V^* are the points in S. The edges of V^* are straight line segments that connect two points in S if their respective cells in $Vor(S)$ have an edge in common (see Fig. 5.3). We have the following theorem.

Theorem 4.5: If no four points of S are cocircular then the Voronoi dual is a triangulation of the convex hull of S.

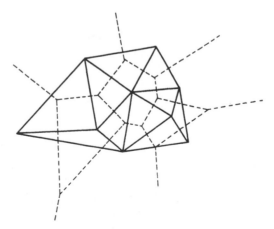

FIG. 5.3. The Voronoi dual (the Voronoi diagram is drawn in dashed lines).

Proof: We first argue that V^* is a planar graph. Indeed, suppose to the contrary that two of its edges AB, CD intersect, where the four points A, B, C, D are all distinct. Let u be a point on the Voronoi edge separating $V(A)$ and $V(B)$, and let v be a point on the Voronoi edge separating $V(C)$ and $V(D)$. Since the circle Γ of radius $|uA|$ about u and the circle Γ' of radius $|vC|$ about v contain no point of S in their interior, it follows that one of the segments uA, uB must intersect one of the segments vC, vD. Suppose for specificity that uA intersects vC. Then, using the triangle inequality twice, it follows that either $|uC| < |uA|$ or $|vA| < |vC|$; but if the first inequality holds, u cannot lie on the boundary of $V(A)$, and if the second one holds, v cannot lie on the boundary of $V(C)$, contradictions which imply that V^* is indeed planar. The assertion is now an immediate consequence of the fact that by our assumptions each vertex of $Vor(S)$ has degree 3, so that each face of V^* (except the outermost one) is a triangle. ∎

Even if S does not satisfy the assumption of Theorem 4.5, it is easy to obtain a triangulation of the required form from V^*. In this case some of the faces of V^* may be polygons having more than three sides, but then each of them can be triangulated in a straightforward manner.

Corollary: Given the Voronoi diagram of a set S of n points in the plane, one can compute in time $O(n)$ a triangulation of the convex hull of S so that the vertices of the triangles are the points in S.

Remark: The particular triangulation yielded by the Voronoi dual is also known as the Delaunay tesselation of S (see Shamos, 1978, for more details).

Euclidean Minimum Spanning Trees.   Let S be a set of n points in the plane. A Euclidean Minimum Spanning Tree T of S is a tree whose vertices are the points in S and whose edges are straight line segments which join pairs of points in S. We call T a Euclidian Minimum Spanning Tree (EMST in short) of S if the sum of its edges' lengths is the minimum among all spanning trees of S. We have the following property.

Lemma 4.3: The EMST of S is a subgraph of the Voronoi dual of S.

Proof: Let AB be an edge of the EMST. We claim that AB is completely contained in the union of the two cells $V(A)$ and $V(B)$. Suppose to the contrary that AB crosses into a cell of $C \in S$. By a similar argument to that of lemma 4.1 $|CA| < |AB|$, $|CB| < |AB|$. Suppose that the path in the EMST from C to A, B meets A first. Then by replacing AB by CB we get a spanning tree of lower weight (i.e. sum of the lengths of its edges) contrary to assumption. Hence $V(A)$ and $V(B)$ have a common boundary edge in $Vor(S)$ and therefore AB is an edge of the Voronoi dual. ∎

Now it has been shown in the proof of Theorem 4.5 that V^* is a planar graph. Since the minimum spanning tree of a planar graph can be computed in linear time (see for instance Cheriton and Tarjan, 1976), we have the following theorem.

Theorem 4.6: The EMST of a set S of n points in the plane can be computed in $O(n)$ time, given the Voronoi diagram of S.

5. EFFICIENT CONSTRUCTION OF VORONOI DIAGRAMS

Most of the known efficient techniques for the construction of Voronoi diagrams have the following schematic divide-and-conquer structure. They all begin by dividing S into two subsets R and L each containing at least some fixed constant fraction (usually half) of the objects in S. They next construct recursively the Voronoi diagrams of L and R, and then merge these two diagrams together to form the Voronoi diagram of S. (A recent alternative technique, based on line sweeping, has been developed by Fortune, 1985b.)

The main step of each of these algorithms is that which merges the diagrams $Vor(L)$ and $Vor(R)$ into the single diagram $Vor(S)$. For this, one must compute the set C of points y that are simultaneously nearest to an object $s_i \in L$ and to an object $s_j \in R$. Following Kirkpatrick (1979) we call C the *contour* separating R and L (see Fig. 5.4). It is easy to show that, once C has been computed, $Vor(S)$ can be obtained by taking the union of $Vor(R)$, $Vor(L)$, and C, and then by discarding (portions of) edges belonging to one of the partial diagrams $Vor(R)$, $Vor(L)$, whose points have become nearer to some object belonging to the other set (these portions will be delimited by the intersections of these edges with C).

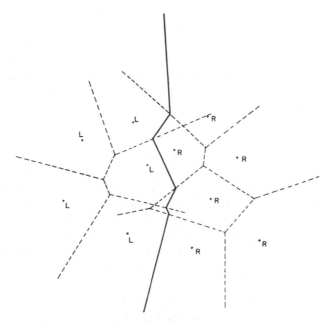

FIG. 5.4.   The contour C separating *Vor* (L) and *Vor* (R).

We first note that C is a subgraph of $Vor(S)$, as follows easily from its definition. In particular, each vertex of C is a vertex of $Vor(S)$ which, by assumption, is adjacent to exactly three cells $V(s_1)$, $V(s_2)$, $V(s_3)$ of that diagram. Hence of the three objects s_1, s_2, s_3 of S, two belong to one of the sets L, R, and one belongs to the other set. This implies that each vertex of C is adjacent to precisely two edges of C. Thus C is a union of disjoint simple curves (consisting of edges of $Vor(S)$) without endpoints. In other words, each connected component of C is either a closed simple curve, or else is an unbounded simple curve extending to infinity in both directions.

The contour C partitions the plane into disjoint connected open regions. Each such region Q either consists entirely of points each of which is nearer to some point of L than to any point in R (in which case we call Q an *L-region*), or consists entirely of points each nearer to some point in R than to any point in L (in which case we call Q an *R-region*).

The problem of tracing the contour may be divided into two subproblems. The first problem is that of finding a given point z_K on each connected component K of the contour (in general, the contour can have up to $O(n)$ components), and the second problem is that of tracing each contour component from the starting point found in the preceding step. These two steps are rather independent of one another, and we will consider each of them separately. Roughly speaking, the first step is generally harder than the second one, that can always be implemented in $O(n)$ time.

In this section we describe efficient implementation of the second step, that is general enough to be applicable to all algorithms to be described later in this article. In the following subsections we shall complete the discussion of each algorithm by describing first how the set S is split into the sets R and L and then how, during the merge phase, initial points on each contour component are located. The contour tracing procedure described below is that suggested in Kirkpatrick, 1979. To facilitate this tracing, we will find it convenient (following Kirkpatrick, 1979) to partition each Voronoi cell $V(s_i)$ (of either $Vor(L)$ or $Vor(R)$) into *subcells* by connecting each vertex v of $V(s_i)$ to s_i by a shortest straight segment (called a *spoke*, as in Kirkpatrick, 1979). Clearly each subcell is bounded by two spokes, one Voronoi edge and possibly a part of s_i. We will assume in what follows that the objects in S have relatively simple shape, so as to allow various geometric operations on these objects to be performed in constant time. In particular, we will assume that, given an arc e defined as the locus of points that are equidistant from two objects in S, its intersection points with the boundary of any Voronoi subcell (in $Vor(L)$ or $Vor(R)$) can be found in constant time, assuming that we use an appropriate representation of the corresponding diagram $Vor(L)$ or $Vor(R)$.

Suppose that we have somehow found a point $z \in C$ (but such that z is neither on $Vor(L)$ nor on $Vor(R)$), for which the two objects $s_i \in L$, $s_j \in R$ nearest to z are known, and suppose further that the two subcells $U(s_i)$ of $V(s_i)$ in $Vor(L)$ and $U(s_j)$ of $V(s_j)$ in $Vor(R)$ to which z belongs are also known. Then we can trace the component K of C containing z as follows. We first find the Voronoi edge e (in $Vor(S)$) containing z. Note that e is (contained in) the locus of points equidistant from s_i and s_j, and is an edge lying on K. We begin tracing K by following e from z in some direction, and by computing its intersection points with the boundaries of the two subcells $U(s_i)$ and $U(s_j)$. Suppose for specificity that the nearest of these points along e is the point z' at which e intersects the boundary of $U(s_i)$. If z' lies on a Voronoi edge α, then z' is a Voronoi vertex in $Vor(S)$ (since it is nearest to the three objects s_i, s_j, and the object s_k whose cell (in $Vor(L)$) lies on the other side of α). In this case the contour K crosses α at z' into a subcell of $V(s_k)$ in $Vor(L)$ (by assertion (10) of Section 3, two Voronoi edges that meet at a common vertex are never tangent to one another). In this case, K continues after z' along the Voronoi edge e' containing points equidistant from s_k and s_j. On the other hand, if z' lies on a spoke, K will continue after z' along the edge e, but will cross into another subcell of $V(s_i)$ (Note that the contour can never be tangent to a Voronoi spoke, by assertion (6) of section 3).

Tracing the contour in this way, we either come back to z, in which case the component K is a bounded component of the contour, or else we reach an unbounded edge of the contour, in which case we have to repeat the tracing procedure just outlined by starting from z in the other direction of the edge e in order to obtain the entire component K.

Next we show that if a point z_K is available on each of the components K of C, the total cost of constructing C is $O(n)$. The complexity of the tracing procedure just described is plainly $O(n_1^{(K)} + n_2^{(K)})$, where $n_1^{(K)}$ is the number of Voronoi edges in K, and where $n_2^{(K)}$ is the number of intersections of K with Voronoi spokes (in either $Vor(L)$ or $Vor(R)$). As in (Kirkpatrick, 1979), we can show that the sum over all of K of the quantities $n_1^{(K)} + n_2^{(K)}$ is $O(n)$. Namely we have

Lemma 5.1: Each Voronoi spoke (in either $Vor(L)$ or $Vor(R)$) is intersected by the contour C in at most one point. Moreover this remark also holds for any segment that is the locus of points nearest to some fixed point on an object in S.

Proof: Let e be a Voronoi spoke of a cell $V(s_i)$ in $Vor(L)$. Then each point z at which e and C intersect each other must lie on the boundary of the cell $V(s_i)$ in $Vor(S)$, and since (by assertion (2) of Section 3) the interior of the line joining z to s_i lies entirely in this cell it follows that at most one such intersection point can exist.∎

Corollary: The total time required to trace the contour, given a point z (and the two subcells in $Vor(L)$, $Vor(R)$ containing it) on each of its connected components, is $O(n)$.

Proof: The total number of edges on C is $O(n)$ because C is a subset of $Vor(S)$. The total number of intersections of C with Voronoi spokes is also $O(n)$, by Lemma 5.1. Hence the total complexity of the tracing procedure applied to each component K of C is

$$\sum_K (n_1^{(K)} + n_2^{(K)}) = O(n)$$

This completes our discussion of the contour tracing procedure. In cases where the preceding step, that locates a point on each connected component of C, can also be carried out in linear time, the overall time complexity $T(n)$ of the algorithm for constructing the Voronoi diagram of n objects will be governed by the recurrence formula.

$$T(n) = 2T\left(\frac{n}{2}\right) + O(n).$$

so that $T(n)$ is $O(n \log n)$. Although there are several important special cases in which the first point location step can also be accomplished in linear time (thus leading to $O(n \log n)$ algorithms), there is no known general method for achieving this time bound, so that there is no known $O(n \log n)$ algorithm for constructing the Voronoi diagram of n general convex objects.

5.1. Constructing the Voronoi diagram for a set of points

In this subsection we describe an algorithm for constructing the Voronoi diagram for a set S of n points in time $O(n \log n)$. The algorithm is due to Shamos (1978).

For simplicity, we will assume in what follows that no two points in S lie on the same vertical line. (If S does not have this property, then we can apply an infinitesimal rotation that will make the abcissae of all points in S distinct from each other.) The algorithm begins by splitting S into two sets L and R of equal size such that all points in L lie to the left of every point in R. This can be done by computing the median of the x-coordinate of these points in time $O(n)$ (see, for example, Aho et al., 1974). Having split the set S in this way we have the following lemma that simplifies considerably the process of locating initial points on each contour component.

Lemma 5.1.1: Under the preceding splitting of S, the contour consists of a single connected component that is monotonic in y.

Proof: Suppose first that a contour component is not monotonic in y. Then it is easily seen that, up to symmetric configurations, we can assume that there exist three consecutive vertices on the contour $z_i = (x_i, y_i)$, $i = 1, 2, 3$, such that the following conditions hold:

(a) $y_1, y_3 > y_2$;
(b) $x_1 < x_2 \leq x_3$; and
(c) the edge (z_1, z_2) separates the Voronoi cells (in $Vor(S)$) of a point $p_1 \in L$ and of another $p_2 \in R$ and the edge (z_2, z_3) separates the cell of p_2 from that of another point $p_3 \in L$ (see Fig. 5.5).

However, this is impossible since then $p_2 \in R$ would have to lie to the left of point $p_3 \in L$ (since the edge (z_2, z_3) is the perpendicular bisector of the line joining p_2 and p_3), contrary to the definition of L and R. It follows that each contour component is monotonic in y (and hence unbounded).

Suppose now that there exist at least two contour components C_1, C_2 that form respectively the left and right boundaries of an R-region Q. But then there would be an L-region lying to the right of Q on the other side of C_2, and hence at

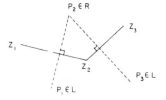

FIG. 5.5. Illustrating the proof of the contour monotonicity.

least one point in L would have to lie to the right of a point in R, again contrary to definition. Hence there exists exactly one contour component. ■

Since C consists of just one component, the problem of locating a point on this component in $O(n)$ time is rather straightforward. For example, we can proceed as follows. First we find the rightmost point $p \in L$ and determine the subcell in $Vor(R)$ in which it is located. Then we start moving from p to the right along a horizontal ray l, keeping track of the two points $q_L \in L$ and $q_R \in R$ whose subcells (in $Vor(L)$ and $Vor(R)$ respectively) are currently being crossed by l. Each time we cross into a new subcell we check whether l intersects the perpendicular bisector of the line joining q_L and q_R within their corresponding subcells. Eventually this condition must hold at the point of intersection of l with the contour. The whole process obviously takes $O(n)$ time, as shown by an analysis similar to that of the contour tracing process.

Together, the details described in this and the previous section yield the following schematic algorithm for constructing $Vor(S)$ for a set of points:

1. Split S into two equal size subsets L and R such that all points in L lie to the left of every point in R.
2. Compute $Vor(L)$ and $Vor(R)$ recursively.
3. Construct the contour C as follows:

   (a) Determine the rightmost point p in L and find the subcell in $Vor(R)$ containing it.
   (b) Find an initial point on the contour by moving horizontally from p to the right, in the manner described above.
   (c) Trace the contour in the manner described in the previous section.

4. Obtain the final diagram $Vor(S)$ by taking the union of $Vor(R)$ and of $Vor(L)$ and C, and then by discarding (portions of) edges of $Vor(R)$ and of $Vor(L)$ that are cut off from their cells by C.

The algorithm just sketched runs in time $O(n \log n)$ since the merge phase can obviously be accomplished in $O(n)$ time.

Remark: Step 4 of the algorithm can in fact be accomplished during execution of step 3(c). Indeed, each time the contour component being traced intersects a Voronoi edge e of either $Vor(L)$ or $Vor(R)$, thereby obtaining a new Voronoi vertex z of $Vor(S)$, we can update the data structure representing $Vor(S)$ right then, by making z the endpoint of the current contour edge, the edge e, and the new contour edge leaving z, and at the same time discarding the portion of e that has been cut off its cell at z. Then, after step 3(c) has terminated, an additional linear scan of the diagrams will remove from the data structure all portions of $Vor(L)$ and $Vor(R)$ that have been thus cut off by C, thereby obtaining $Vor(S)$.

(We have not described explicitly the data structure that we use to represent $Vor(S)$, and leave it to the reader to work out the rather straightforward representation details.)

5.2. Constructing the Voronoi Diagram for a Set of Convex Objects

In this section we describe an algorithm for constructing the Voronoi diagram of a set S of n convex objects in time $O(n \log^2 n)$. The algorithm is again based on the general scheme presented in the beginning of Section 5, in which the location of initial points on the contour components is accomplished by a slightly less efficient procedure than that sketched in the preceding subsection. This procedure is adapted from a similar one described by Sharir (1985), and is also similar to that used by Lee and Drysdale (1981).

In what follows we continue to assume that the shape of each of the objects in S is simple enough, and consequently regard each of the operations that we will need to apply to such an object, or to a pair of such objects, as requiring constant time.

Similar to preceding assumptions, we will assume in what follows that the leftmost points of any two distinct objects in S have distinct abscissae. (If not, we can apply an infinitesimal rotation that will make the abscissae of the left-most points of all objects in S distinct from each other.) The algorithm then splits S into two subsets R and L of equal size, such that the leftmost point w_i of each $s_i \in L$ lies to the left of the leftmost point w_j of every object $s_j \in R$. This partitioning of S can easily be done by finding the median of these leftmost points in time $O(n)$.

This particular partitioning of S is somewhat more advantageous than alternative partitionings, and has some useful properties stated below. However, unlike the case where S consists of single points, there is no general way to partition S into two equal size subsets so as to guarantee that the contour C will have a single component. In fact, C can in general have up to $O(n)$ components, so that substantially more efficient methods for locating initial points on its components than the method described in the preceding subsection have to be devised.

Before describing such an efficient technique, we first prove some properties of the contour C implied by the way in which L and R have been defined (see also Fig. 5.6).

Lemma 5.2.1: (a) Let $s_i \in L$ and let u be the horizontal half-line whose rightmost endpoint is the leftmost point of s_i. Then u does not intersect C (and consequently lies wholly inside an L-region).
(b) There exists precisely one L-region; all other components of the complement of C are R-regions.

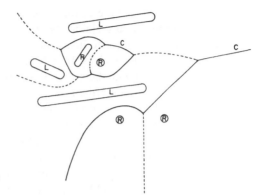

FIG. 5.6.   The contour in the case
of general convex objects.

(c) Each R-region has a connected boundary, consisting of a single (bounded or unbounded) component of C.

Proof: (a) Suppose the contrary, and let z be a contour point lying on u. Let $w \in u$ be the leftmost point of s_i. It follows that there exist objects $s_j \in R$ and $s_k \in L$ such that

$$d(z, s_j) = d(z, s_k) \leq d(z, s_i) = d(z, w)$$

But then some point on s_j (and in particular its leftmost point) must lie to the left of w, contradicting the way in which L and R have been defined.

(b) It suffices to show that the leftmost points w_i, w_j of each pair s_i, s_j of objects in L are connected to each other via a path that does not intersect the contour C. Let u_i (resp. u_j) be the horizontal half-line whose rightmost end-point is w_i (resp. w_j). For each t, let y_i (resp. y_j) be a point on u_i (resp. u_j) whose abscissa is t. We claim that if t is negative and has a large enough absolute value, the segment $e = y_i y_j$ does not intersect the contour, which, together with (a), implies that w_i and w_j are connected to each other via the polygonal path $w_i y_i y_j w_j$ that is wholly contained in a single L-region. To see this, suppose to the contrary that there exists a contour point $z \in e$, nearest to some $s_k \in R$ and to some $s_m \in L$. Then

$$d(z, s_k) = d(z, s_m) \leq d(z, s_i) = (d^2(y_i, w_i) + d^2(z, y_i))^{1/2} \approx d(y_i, w_i)$$

and similarly

$$d(z, s_k) \leq d(z, s_j) \approx d(y_j, w_j)$$

as $t \rightarrow \infty$. It follows that the leftmost point w_k of s_k does not lie to the right of w_i or of w_j. Since, by definition of L, R, w_k cannot lie to the left of w_i, w_j, it follows that the three points w_i, w_j, w_k all have the same abscissa, again contradicting the way in which L and R are defined. This proves (b).

(c) Suppose that M is an R-region whose boundary consists of at least two disjoint connected components K_1, K_2 of C. K_1 (resp. K_2) is, by the definition of

C, a topologically closed curve (either bounded or unbounded) and therefore partitions the plane into two disjoint regions, both bounded by K_1 (resp. K_2). Since K_1 and K_2 are disjoint, it follows that they collectively partition the plane into three regions, one of which is M, whereas the other two must contain L-regions. This, however, contradicts (b), thus proving our assertion.∎

We are now prepared to describe how points on each contour component are located. For this we observe that if we connect each object in R by a straight line segment e to an object in L these segments e must collectively intersect all the components of C. We will show that in total time $O(n \log n)$ we can find such segments e, each intersecting C just once, and that in additional $O(n)$ time we can trace these segments to find a point z on each component K of C. This will lead to an $O(n \log n)$ merging step, and hence to an $O(n \log^2 n)$ overall algorithm.

We iterate over all the cells $V(s_i)$ of $Vor(R)$ (and their corresponding objects s_i), proceeding as follows. First we pick one point, say w_i, on the boundary of each object $s_i \in R$. Let $s_j \in L$ be such that $w_i \in V(s_j)$ (in $Vor(L)$) so that s_j is the object in L nearest to w_i. The line segment e that joins w_i to its closest point on s_j obviously intersects C (since one of its endpoints lies on an object in L and the other lies on an object in R) and by Lemma 5.1 it intersects C exactly once (since it is wholly contained in a single subcell of $V(s_j)$).

We can proceed to find the unique intersection z of e with C using a technique quite similar to the contour-tracing procedure described at the beginning of Section 5. That is, we first find the Voronoi subcell (of $V(s_i)$) in $Vor(R)$ containing points on e near w_i (in the case where e penetrates from w_i into s_i, we simply find the next point of intersection of e with s_i and continue the tracing process from there). We then find the intersection of e with the Voronoi edge or one of the Voronoi spokes bounding that subcell, beyond which e crosses into another subcell of $Vor(R)$. (In the extreme case in which e coincides with a Voronoi spoke of $V(s_i)$, e will exit the two subcells of $V(s_i)$ in which it lies at the Voronoi vertex that is the other endpoint of the spoke; it is then a bit more complicated, but still straightforward, to determine the Voronoi subcell into which e enters past this vertex.) Continuing in this manner, we partition e into subsegments e_1, . . . ,e_m, each of which is contained in some Voronoi subcell of $Vor(R)$ or lies along a spoke of $Vor(R)$. As all this is done, we keep track of all the cells $V(s_k)$ of $Vor(R)$ that have already been encountered during such tracing of subsegments of previously traced segments. If e crosses into a cell that has already been encountered before, tracing of the sequence of edges e_1, . . . ,e_m stops immediately (this rule is justified by Lemma 5.2.2 below). This guarantees that the total cost of traversing subcells of $Vor(R)$ is bounded by the total number of such subcells, and hence by $O(n)$.

For each $t = 1$, . . . ,m let $V(s_{i_t})$ be the cell in $Vor(R)$ containing e_t. As tracing proceeds through the cell $V(s_{i_t})$, we check whether $V(s_{i_t})$ has been encountered before, and, if not, whether there exists $z \in e_t$ such that $d(z, s_j) = d(z, s_{i_t})$ (this can be done in constant time). As already shown, there will exist a

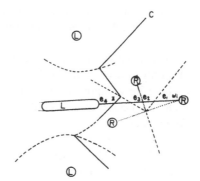

FIG. 5.7. Locating a point on a contour component.

unique point z on e having this property, and this z will be the required intersection point of e with C. Figure 5.7 illustrates this process. (Note that the algorithm will reach this z only if tracing is not abandoned earlier, because a previously encountered cell of $Vor(R)$ is encountered again.) Let e_k be the subsegment of e containing z. Note that z is found in time $O(p + k)$, where p is the number of subcells of $V(s_i)$ in $Vor(R)$.

To show that tracing of e can be abandoned as soon as any cell of $Vor(R)$ is encountered for the second time, we will use the following:

Lemma 5.2.2: Let k be as in the preceding paragraph. Then for each $t \leq k$, $Vor(S)$ contains a cell of s_{i_t}, and s_{i_t} lies in the same R-region as s_i.

Proof: Let M be the R-region containing $s_{i_1} = s_i$, and let $1 < t \leq k$. Pick any point $z_t \in e_t$ (but in case $t = k$, z_t must lie between w_i and z). Since $t \leq k$, the segment $w_i z_t$ does not intersect the contour, and therefore is wholly contained in M. We now argue that the line e' joining z_t to the closest point on s_{i_t} also wholly contained in M. This follows since e' is wholly contained in one cell of $Vor(R)$ and by Lemma 5.1 intersects C at most once but it cannot intersect C once since both of its endpoints are in an R-region.■

As we apply the procedure just described to each of the objects $s_i \in R$, one of the following two situations will arise: either

(a) While tracing subcells of $Vor(R)$ crossed by the segment e joining w_i to the closest point on s_j (where s_j is the object for which $w_i \in V(s_j)$), we encounter a subcell of some cell $V(s_r)$ whose R-region M in $Vor(S)$ was encountered before (along previously traced segments). In this case we conclude from Lemma 5.2.2 that s_r, as well as every other object of R whose cell in $Vor(R)$ has been crossed by e so far, lies in the R-region M. In this case we can stop tracing e and go on to process other objects of R, since we can be sure that the component of the contour C intersected by e has already been explored. The algorithm will also note that all cells of $Vor(R)$ crossed by e so far belong to M, to avoid repeated processing of these cells later on. (Note that this case will arise only when $V(s_i)$ is an inner cell in M, i.e. a cell not intersected by the contour); or

(b) The tracing procedure continues until an intersection z of e with the contour is found. In this case only new cells of $Vor(R)$ are being traced, and z will lie on a new component of C (this component must be new, because all the old components of C have already been traced, and all the R-subcells through which they pass have already been encountered and marked; hence before reaching any old component of the contour the scanning will stop by step (a) above). As in (a), we take note of the fact that all cells crossed by e during this tracing belong to the new R-region just found, to avoid repeated processing of these cells later on. These observations imply that we can find a representative point on each component of the contour in total time $O(n)$, provided that, for each $s_j \in R$, the subcell of the cell $V(s_i)$ of $Vor(L)$ containing w_j is already known.

To obtain this final item of information, we can use a simple plane-sweeping algorithm, similar to those described by Bentley and Ottman (1979), Nievergelt and Preparata (1982), and Shamos (1978). The algorithm sweeps a vertical line through the plane from left to right and maintains a vertical "front" $T(a)$ consisting of the segments into which the line $x = a$ is divided by the points of intersection of this line with the edges and spokes of $Vor(L)$, and by the boundaries of the objects in L. The structure of $T(a)$ will change only at points a that are either abscissae of Voronoi vertices of $Vor(L)$, or abscissae of endpoints of spokes in $Vor(L)$, or points for which the line $x = a$ is tangent to some Voronoi edge of $Vor(L)$ or to some object in L. The number of such "transition points" is plainly $O(n)$, and the total number of segments of $T(a)$ is also $O(n)$ for any real a. To start the algorithm, sort the set A, consisting of all transition points of $Vor(L)$ and all the leftmost points of the objects in R by their x-coordinates, and initialize the list $T(a)$ as a 2–3 tree for some sufficiently small negative real a. Both these tasks can be done in time $O(n \log n)$. Then scan A from left to right. For each $a \in A$, if a is a transition point of $Vor(L)$, update the list T by an appropriate combination of deletions, insertions, and merge operations applied to segments in T; this can be done in time $O(k_a \log n)$, where k_a is the number of segments that undergo these changes (note that the sum of these k_a's is $O(n)$). If a is the abscissa of a point w_i of some object in R, search T to find the segment in T containing w_i, from which the Voronoi subcell of $Vor(L)$ containing w_i is readily obtained. Proceeding in this way, we locate all the points w_i of objects in R in subcells of $Vor(L)$ in time $O(n \log n)$.

Together, the details just described yield the following algorithm for constructing $Vor(S)$:

1. Split S into two equal-size subsets L, R such that the leftmost point of each $s_i \in L$ lies to the left of the leftmost point of every $s_j \in R$ (we have assumed that no two leftmost points have the same abscissa).

2. Compute $Vor(L)$ and $Vor(R)$ recursively.

3. Apply the plane-sweeping procedure described above to locate the subcell $V(s_j)$ of $Vor(L)$ containing w_i for each leftmost point w_i of an object $s_i \in R$.

4. Construct the contour C as follows:

For each object $s_i \in R$ whose R-region (in $Vor(S)$) has not yet been identified

a. Connect its leftmost point w_i to the closest point y on the object $s_j \in L$ whose Voronoi cell $V(s_j)$ in $Vor(L)$ contains w_i.

b. Find the unique intersection z of the contour with the segment $e = w_i y$ by applying the tracing procedure described above. If that procedure detects an intersection of e with a cell $V(s_k)$ of $Vor(R)$ whose R-region M has already been found, it assigns M as the R-region of s_i and of all other objects of R whose cells in $Vor(R)$ have been crossed by e before $V(s_k)$ has been reached, and continues with the main loop of this phase.

c. Trace the whole contour component K containing z. An R-region indication is thereby assigned to all objects $s_k \in R$ whose cells in $Vor(R)$ are encountered.

5. Obtain the final diagram $Vor(S)$ by taking the union of $Vor(R)$ and of $Vor(L)$ with C, and then by discarding (portions of) edges of $Vor(R)$ or of $Vor(L)$ that are cut off from their cells by C.

The algorithm just sketched runs in time $O(n \log^2 n)$. Its costliest phase is step 3, that locates subcells of $Vor(L)$ containing the leftmost point w_i on each of the objects $s_i \in R$.

5.3. Bibliographical Notes on Voronoi Diagrams

Voronoi diagrams were introduced by Voronoi almost 80 years ago. They have been studied since then by many researchers, and have found applications in diverse areas such as crystallography (Brumberger, 1983), physics (Brostow, Dussault, & Fox, 1978), molecular chemistry (David & David, 1982), geology (Arnold & Milne, 1984), pattern analysis (Ahuja, 1982), image processing (Lee, 1982), statistics (Newman, Rinott, & Tversky, 1983), and so on. Various early algorithms have been proposed for the construction of Voronoi diagrams, as in Brostow et al. (1978), Finney (1979), Green and Sibson (1978), Tanemura, Ogawa, and Ogita (1983) and Bowyer (1981). However, these algorithms were generally less efficient than the ones described above, or did not involve any complexity analysis. (Some of these algorithms concern themselves with Voronoi diagrams in higher dimensional spaces, and are therefore doomed to be less efficient than those available for two-dimensional diagrams; see also Section 7 below). Shamos (1978) in his doctoral thesis was the first to analyze these diagrams from a complexity-theoretic point of view; Shamos and Hoey (1975) have developed efficient techniques for constructing the diagram of a set of n

points in the plane in time $O(n \log n)$, and have thereby obtained many efficient algorithms in computational geometry that use Voronoi diagrams (like those described in Section 4). Since then Voronoi diagrams have been recognized as an important tool in computational geometry and have been studied extensively. Kirkpatrick (1979) has generalized Shamos's algorithm for constructing the diagram for a set of points, using minimum spanning trees as a tool for locating points on the components of the contour. He has also outlined there a possible generalization of the minimum spanning tree technique for efficient construction of generalized diagrams for a set of points and line segments, but his arguments concerning this generalization are not clear (see, for example Leven & Sharir, 1985a, for some of the problems in this generalization). Lee and Drysdale (1981) have also studied generalized diagrams for sets of line segments and points or sets of circular disks and have obtained an $O(n \log^2 n)$ algorithm for constructing these diagrams, using a method similar to that outlined in Section 5.2 above. Related work by these authors on such generalized diagrams is found in Drysdale (1979), Drysdale and Lee (1978). Gowda, Kirkpatrick, Lee, and Naamad (1983) have studied the complexity of updating the Voronoi diagram of a point set S by insertion and deletion of points from S, obtaining $O(n)$ bounds on the complexity of these operations. Brown (1979) has shown how to obtain Voronoi diagrams from the convex hull of an appropriate embedding of the given set S in a higher-dimensional space, and this approach has been further extended by Seidel (1981). There are many additional ways in which Voronoi diagrams can be generalized. For example, Lee (1980), Lee & Wong (1980) have considered Voronoi diagrams under general l_p metrics instead of the standard Euclidean (l_2) metric. As another example, instead of partitioning the plane into cells, each containing points nearest to a single object $s \in S$, we could consider similar partitionings in which each cell consists of all points that are nearer to all objects in some subset of S than to any of the other objects in S. If the cell defining subsets of S are all subsets of S of some fixed size k, we obtain the k-nearest neighbor Voronoi diagram of S. In particular, if $k = |S| - 1$, we obtain the farthest-point Voronoi diagram. Properties of such generalized diagrams, and efficient algorithms for their construction given in Lee (1982a), Shamos (1978), Bhattacharya (1983), Dehne (1982), Chazelle & Edelsbrunner (1985). Another possible generalization is that in which distances from the test point to the objects in S are given multiplicative or additive weights (see Aurenhammer & Edelsbrunner, 1984, Boots, 1979). A special case of weighted diagrams is that which arises in connection with the problem of computing shortest paths from a given source point in two-dimensional space bounded by polygonal obstacles (Franklin, Akman & Verrilli, 1983). Further generalizations for sets of intersecting discs have been studied by Sharir (1985) who presented an $O(n \log^2 n)$ algorithm for constructing such diagrams (see also Section 4 above). Another recent generalization involves Voronoi diagrams for a set of discs (or balls in higher-dimensional space) under the Laguerre geometry (or the power distance), in which the distance of a point from such a disc is taken as the length of the tangent to the disc from the point. Such diagrams turn out to be polygonal (or polyhedral), and they can be calculated in $O(n \log n)$ time in the planar case (see Aurenhammer, 1983, Aurenhammer, 1985, Imai, Iri, & Murota, 1983). Another generalization considers Voronoi diagrams for a set of object under a convex distance function. Such a function is "almost" a metric, except that it is generally not symmetric. Nevertheless, Voronoi diagrams under such a distance function can still be defined in much the same way as above, and retain most of the properties of standard diagrams. Recent works on these diagrams yield $O(n \log n)$ algorithms for their calculation, and apply them to finding the largest empty homothetic copy of a convex polygon amidst a set of points (Chew & Drysdale, 1985), testing for polygon containment by translations (Fortune, 1985a), and planning a purely translational collision-free motion of a convex polygon amidst polygonal obstacles (Leven & Sharir 1985b). Voronoi diagrams for sets of objects in three or more dimensions have been studied by Avis & Bhattacharya (1983), by Klee (1980), and by Seidel (1981, 1982). Algorithms for their construction have been proposed in Brostow et al. (1978), Finney (1979), Tanemura et al. (1983), and Watson (1981) (See also Section 7 below). Voronoi diagrams in standard and generalized forms also proved useful in the calculation of Euclidean shortest paths in 3-dimensional space amidst polyhedral obstacles. Such generalizations and applications are discussed by Mount (1984, 1985) for a single convex polyhedron, and by Baltsan & Sharir (1985) for the case involving two convex polyhedral obstacles. A connection between Voronoi diagrams and arrangements of (hyper)surfaces in a higher-dimensional space is discussed by Edelsbrunner & Seidel (1985); this connection provides an alternative systematic and unified framework for defining and studying Voronoi diagrams. Perhaps the most complex generalization of Voronoi diagrams so far has been to configuration spaces of mechanical systems moving amidst stationary obstacles, where the diagram partitions such a space into cells, each of which contains all configurations in which the system lies nearest to one obstacle than to any of the others. These generalized diagrams are useful for motion planning of robot systems. Simple versions of these diagrams are described by Rowat (1979) and by Ó'Dúnlaing and Yap (1985), and more complex versions are analyzed by Ó'Dúnlaing, Sharir, and Yap (1983, 1984a, 1984b). As mentioned earlier, the problem of computing generalized Voronoi diagrams for sets of planar objects of general shapes in time $O(n \log n)$ is still open. Very recently Yap (1984) has obtained such an algorithm for constructing the diagram of a set of points and line segments, or a set of certain simple curves. Using a different approach, the authors have obtained an $O(n \log n)$ algorithm for constructing the diagram of sets of disjoint circular discs. There have also been several works on improving the pragmatic implementation of the algorithms described in this article; see for example Guibas and Stolfi (1983). Another very recent development is a new technique due to Fortune (1985b), which computes planar Voronoi diagrams using a line-sweeping algorithm. Finally, other expositions on Voronoi diagrams can be found in two recent books by Mehlhorn (1984) and by Preparata and Shamos (1985).

6. DYNAMIC INTERSECTION AND PROXIMITY PROBLEMS

In this section we briefly consider the dynamic versions of our intersection and proximity detection problems, as stated in Problem V in the introduction. That is, we suppose that we are given a collection S of objects $s_1, \ldots, s_n$, some of which may be moving along specified trajectories, and our aim is to determine whether at any time two of the objects intersect, and if not, what is the smallest distance ever attained between any two of the objects. We assume that the motion of these objects is simple enough so that, given a pair s_i, s_j of objects in S we can determine in constant time whether they ever intersect one another and, if not, what is the smallest distance ever attained between them. This implies however that by simply considering all pairs of objects in S, we can solve the dynamic problems posed above in time $O(n^2)$.

While we do not know of any more efficient "global" algorithm for these dynamic problems, the following alternative approach, which in fact is the one used implicitly so far, is also possible. Choose a sufficiently small time step τ, and partition the motion of the objects in S into discrete instances, τ time apart from one another. For each such instance, calculate the positions of the objects in S at that instance, and apply the Voronoi diagram based techniques described above to solve the problem for that time instance. Given an estimate on the velocities of the moving objects, and a safety distance Δ, we can choose τ sufficiently small so that if at time t no two objects in S intersect, and if the smallest distance between any two such objects exceeds Δ, no intersection will occur after τ more time. In this manner we can solve all the static proximity detection problems in total time $O(Kn \log n)$, where $K = O(1/\tau)$ is the number of motion instances that need be analyzed. Which of these two methods is better depends in general on the speed of the moving objects. If they move sufficiently slowly, K will be small and the second technique will be faster. If, on the other hand, the objects move too fast, many static proximity detections will have to be performed to guarantee safety of control, in which case a single global $O(n^2)$ algorithm may be faster.

7. THE THREE-DIMENSIONAL CASE

In the preceding sections we have presented techniques for efficient solutions of the two-dimensional versions of most of the intersection and proximity problems that concern us, using Voronoi diagrams. However, in the more realistic three-dimensional case, Voronoi diagrams are no longer adequate for efficient solution of these problems. It is straightforward to generalize the notion of Voronoi diagram to the three-dimensional case: Let S be a collection of n convex closed and disjoint objects in R^3 (having relatively simple shape). For each object $s \in S$ let $V(s)$, the *Voroni cell* of s, denote the set of all points $x \in R^3$ such that $d(x, s) \le d(x, s')$ for all $s' \in S$, and let $Vor(S)$, the *Voronoi diagram* of the set S, be defined as the set of all points $x \in R^3$ belonging to at least two distinct Voronoi cells. It is easy to see that $Vor(S)$ is a complex of two-dimensional *Voronoi surfaces*, meeting one another along one-dimensional *Voronoi edges*, that themselves end at *Voronoi vertices*. If S consists of points, then $Vor(S)$ is a polyhedral complex. Most of the properties of two-dimensional Voronoi diagrams extend to the three-dimensional case, with essentially the same proofs as given in Section 3. In particular, each Voronoi cell $V(s)$ is star-shaped with respect to s (or even convex if S consists of points). However, the *size* of the diagram, i.e. the total number of surface patches, edges and vertices in $Vor(S)$, can in general be as high as $\Omega(n^2)$. (Note that the size of $Vor(S)$ is always at most $O(n^2)$. To see this, one shows that the boundary of each Voronoi cell can be regarded as a planar graph with $O(n)$ faces, edges, and vertices.)

Indeed, a simple example of a "large" three-dimensional Voronoi diagram can be obtained as follows: Let S consist of the $2n$ points

$$(i, 0, 0), (0, i, 1), i = 1, \ldots ,n.$$

For each $1 \le i, j \le n$ consider the vertical line $L = \{(i, j, t): t \in R^1\}$. Each point on L is nearer to $(i, 0, 0)$ than to any of the points $(i', 0, 0)$, $i' \ne i$, and is nearer to $(0, j, 1)$ than to any of the points $(0, j', 1)$, $j' \ne j$. Hence L is included in $V((i, 0, 0)) \cup V((0, j, 1))$. Furthermore, as $t \to +\infty$ the point (i, j, t) becomes nearer to $(0, j, 1)$ than to $(i, 0, 0)$, and as $t \to -\infty$ the point (i, j, t) becomes nearer to $(i, 0, 0)$ than to $(0, j, 1)$. Hence, for each i, j the cells $V((i, 0, 0))$ and $V((0, j, 1))$ meet along a common boundary Voronoi surface, thus showing that $Vor(S)$ has at least n^2 surfaces in this case. In general, the size of the 3-D Voronoi Diagram need not be quadratic in the number of points. We do not know, however, of algorithms for the construction of such diagrams whose complexity tightly depends on the actual size K of the diagram (e.g., complexity such as $O(K\log n)$ etc.). On the other hand, some of the earlier algorithms (such as Brostow et al., 1978; Finney, 1979; and Watson, 1981) report quite satisfactory performance in practice.

Even though for our particular applications 3-dimensional Voronoi diagrams are too expensive a tool, they have been studied extensively for many other applications, and various algorithms for their construction have been recently devised. Among these algorithms let us mention the ones by Finney (1979), by Brostow et al. (1978), and by Tanemura et al. (1983). Generalizations of Voronoi diagrams for arbitrary d-dimensional spaces have also been studied. Algorithms for their construction are given by Watson (1981), by Avis and Bhattacharya (1983), and by Seidel (1981), and some properties of them are presented in Seidel (1981, 1982) and by Klee (1980).

Since, as we have remarked earlier, any of our intersection or proximity problems can be solved trivially in time $O(n^2)$, Voronoi diagrams are thus too expensive a tool for the solution of the three-dimensional versions of our problems, and other, less powerful techniques need be brought to bear. In this section we present two such techniques, both based on *space sweeping*, that solve efficiently only the simplest of our problems, i.e., the problem of detecting intersection in a set of n objects, and that only for objects of very simple and special shapes.

7.1. Intersection Detection Among Rectilinear Boxes

Consider first the case in which the objects in question are n *rectilinear boxes*, i.e. boxes whose faces are parallel to the standard coordinate planes. Each such box s_i can be parametrized by six parameters $x_i^- < x_i^+$, $y_i^- < y_i^+$, and $z_i^- < z_i^+$, so that

$$s_i = \{(x, y, z): x_i^- \le x \le x_i^+, y_i^- \le y \le y_i^+, z_i^- \le z \le z_i^+\}.$$

To tackle the three-dimensional intersection detection problem, we let $H = H(z)$ be a horizontal plane at height z, and let $r_i = r_i(z)$ denote the intersection of s_i with H, for $i = 1, \ldots, n$ (some of these rectangles can of course be empty). Suppose that $H(z)$ is swept upwards from $z = -\infty$ to $z = +\infty$. Then the collection $C(z)$ of rectangles $r_i(z)$ remains unchanged as long as z does not cross one of the "critical" heights z_i^-, z_i^+. At each height z_i^- the rectangle $r_i(z)$ is added to $C(z)$, whereas at each height z_i^+ the rectangle $r_i(z)$ has to be removed from $C(z)$. It is easy to see that there exists an intersection between some pair of boxes s_i, s_j if and only if the two rectangles $r_i(z)$, $r_j(z)$ intersect one another for $z = \max \{z_i^-, z_j^-\}$. Hence, to detect an intersection among the given boxes, it suffices to check, each time a new rectangle $r_i(z)$ is added to $C(z)$ during the sweeping process, whether $r_i(z)$ intersects another rectangle already present in $C(z)$ (see Fig. 5.8).

The problem therefore reduces to that of maintaining an appropriate data structure that stores a collection C of rectilinear rectangles in the plane, and that enables each of the following operations to be performed efficiently:

1. *Insertion* of a new rectangle into C.
2. *Deletion* of a rectangle from C.

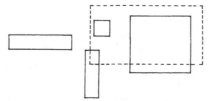

FIG. 5.8.   An instance of the 2-D rectangle intersection query.

3. *Intersection query* - given a rectangle r, find whether it intersects any rectangle in C (in particular we also look for cases in which r is wholly contained in, or contains a rectangle in C).

Various data structures of this kind have been recently devised (see e.g., Edelsbrunner, 1980a, 1980b, 1982; Edelsbrunner & Maurer, 1981). We sketch here the basic ingredients of such a data structure, but omit most of the more involved finer technical details, that can be found in the papers cited above, and also in Mehlhorn (1984).

To obtain the required data structure, we first observe that any rectangle intersection query as in (3) above can be decomposed into four separate subqueries: Given a rectangle r,

    (3.i)   Does r contain a vertex of some rectangle in C?

    (3.ii) Does there exist a rectangle in C that contains a vertex of r?

    (3.iii) Does there exist a vertical side of some rectangle in C that intersects a horizontal side of r?

    (3.iv) Does there exist a horizontal side of some rectangle in C that intersects a vertical side of r?

Note that neither of these subqueries can be dispensed with, as there can exist intersections of rectangles that manifest themselves in only one of the intersections or containments that these subqueries seek to detect.

Accordingly, we will maintain four separate data structures, one for each of the subqueries (3.i)–(3.iv). These data structures are composed of two basic one-dimensional substructures—the (one-dimensional) *range tree* and the (one-dimensional) *segment tree*.

A one-dimensional range tree represents a collection of points $x_1, \ldots, x_n$ on the real line. It is simply a (balanced) binary search tree T in which the points x_i are stored in its leaves.

A one-dimensional segment tree represents a collection A of (closed) intervals $I_1, \ldots, I_n$ on the real line. It is constructed as follows: Let $I_i = [a_i, b_i]$, $i = 1, \ldots, n$, and, assuming for simplicity of exposition that no two of the points a_i, b_j coincide, let $K_1, \ldots, K_{2n+1}$ be a partitioning of the real axis into intervals with disjoint interiors whose endpoints are the points a_i, b_i, $i = 1, \ldots, n$ (in particular each interval I_i is the union of a contiguous sequence of these subintervals). Let T be a binary search tree for the list of intervals $K_1, \ldots, K_{2n+1}$ (ordered so that K_i precedes K_j if K_i lies left on K_j), so that these subintervals are stored in the leaves of T. We regard each node v of T as representing an interval J_v that is the union of all intervals stored in the leaves of the subtree of T rooted at v. With each such node v we associate a list L_v of all intervals I in A that contain J_v, but not the interval $J_{v'}$ associated with the father v' of v. The desired segment tree is T augmented with these interval lists. Note that if T is a balanced binary tree, then each interval I in A appears in at most $O(\log n)$ lists L_v, so that the total size of the segment tree is $O(n \log n)$.

Note that if X is a given query point, we can find the leaf of T whose subinterval K_i contains X, by a simple tree search. Then the intervals I in A that contain X are precisely those that appear in one of the lists L_v along the path from the root to the leaf of K_i (note also that each $I \in A$ appears in at most one list along any such path in T).

To support our two-dimensional queries (3.i)–(3.iv), we will form the following four composite data structures:

(I) The *range-range tree*. Let $P_i = (x_i, y_i)$, $i = 1, \ldots, n$ be n points in the plane. The *primary structure* of a range-range tree T for these points is a one-dimensional range tree for their x-coordinates $x_1, \ldots, x_n$. We regard each primary node v as representing the set S_v of all points P_i whose x coordinates are stored in the leaves of the subtree of T rooted at v. With each such node v we associate a *secondary structure* T_v^* which is a one-dimensional range tree for the y-coordinates of the points in S_v. If each of the primary and secondary trees is balanced, then each point P_i appears in $O(\log n)$ secondary trees, all arranged along the path in the primary tree to the leaf containing (the x-coordinate of) P_i. Thus the size of T is $O(n \log n)$.

To see how range-range trees support queries of type (3.i), let T be a range-range tree representing the vertices of all the rectangles presently in C, and let $r = [a, b] \times [c, d]$ be a query rectangle. We first search in the primary tree of T for the two vertices $P = (x, y)$, $P = (x', y')$ for which x (resp. x') is the smallest (resp. largest) x-coordinate of a vertex which is $\geq a$ (resp. $\leq d$). Using the two paths in the primary tree to P and P' respectively we can find $O(\log n)$ subtrees representing disjoint sets of vertices whose union is the set of all vertices lying in the vertial strip $a \leq x \leq b$. Then we search each of the secondary trees corresponding to these $O(\log n)$ subtrees to see if any of them contains a point whose y-value lies between c and d. If so, we have found a rectangle vertex contained in r. If all the primary and secondary trees are maintained as balanced trees then the total search time is $O(\log^2 n)$.

(II) The *range-segment tree*. Let $K_i = \{x_i\} \times [c_i, d_i]$, $i = 1, \ldots, n$ be n vertical segments in the plane. The *primary structure* of a range-segment tree T for these intervals is a one-dimensional range tree for their x-coordinates $x_1, \ldots, x_n$. We regard each primary node v as representing the set S_v of all intervals K_i whose x coordinates are stored in the leaves of the subtree of T rooted at v. With each such node v we associate a secondary structure T_v^* which is a one-dimensional segment tree for the y segments $[c_i, d_i]$ of the intervals in S_v. If each of the primary and secondary trees is balanced, then each segment K_i appears in $O(\log n)$ secondary trees, all arranged along the path in the primary tree to the leaf containing (the x-coordinate of) K_i. Thus the size of T is $O(n \log^2 n)$.

To see how range-segment trees support queries of type (3.iii), let T be a range-segment tree representing the vertical sides of all the rectangles presently in C, and let $r = [a, b] \times [c, d]$ be a query rectangle. We will perform two searches on T, one to detect an intersection of some vertical side of a rectangle in C with the lower side of r, and another to detect such an intersection with the upper side of r. Since both these searches are identical, we consider here only the first case. We begin by searching in the primary structure of T as above, to obtain $O(\log n)$ subtrees representing disjoint sets of vertical sides whose union is the set of all such sides lying in the vertial strip $a \le x \le b$. Then we search each of the secondary segment trees corresponding to these $O(\log n)$ subtrees to see if any of the intervals appearing in their node lists contains the point c. If so, we have detected an intersection between the lower side of r an some vertical side of a rectangle in C. If all the primary and secondary trees are maintained as balanced trees then the total search time is plainly $O(\log^2 n)$.

(III) The *segment-range tree*. This is defined in complete symmetry to the definition of range-segment trees, and is used to represent a collection of horizontal segments, and to support detection of intersection between a vertical query segment and a segment in the given collection in time $O(\log^2 n)$ in a manner symmetric to that described above; in particular we can use this structure to support queries of type (3.iv). (Of course, by reversing the roles of the x and y coordinates, we could also use range-segment trees instead of segment-range trees for this purpose.)

(IV) The *segment-segment tree*. The final structure to be considered here is needed to support queries of type (3.ii). Here we represent the rectangles $r_i = [a_i, b_i] \times [c_i, d_i]$, $i = 1, \ldots, n$ themselves. The *primary structure* of a segment-segment tree T for these rectangles is a one-dimensional segment tree for their x-projections $[a_i, b_i]$. We regard each primary node v as representing the set S_v of all rectangles r_i whose x projection contains the interval J_v but not the interval $J_{v'}$ of the father v' of v. With each such node v we associate a *secondary structure* T_v^* that is a one-dimensional segment tree for the y projections $[c_i, d_i]$ of the rectangles in S_v. If each of the primary and secondary trees is balanced, then each rectangle r_i appears in $O(\log n)$ secondary trees, because at each level of the primary structure r_i can appear in at most two sets S_v, as is easily checked. Thus the size of T is $O(n \log^2 n)$.

To see how segment-segment trees support queries of type (3.ii), let T be a segment-segment tree representing the rectangles presently in C, and let $r = [a, b] \times [c, d]$ be a query rectangle. We will perform four searches on T, one for each of the vertices of r, to detect whether that vertex is contained in one of the rectangles in C. Consider for example the search for the vertex (a, c). We first search through the primary structure of T for the (horizontal) leaf-subsegment containing a. This gives us $O(\log n)$ subtrees, whose roots lie on the path to the leaf containing a, and we search each of their associated secondary segment trees to see if c is contained in any of the vertical sides that they represent, thereby detecting a containment of (a, c) in one of the rectangles in C. As before the total cost of these searches is easily seen to be $O(\log^2 n)$.

So far we have ignored the problem of dynamic maintenance of these four two-dimensional search trees, required by the insertion and deletion of rectangles from C (operations (i) and (ii) above). Such "dynamizations" of these data structures are possible, at the cost of $O(\log^2 n)$ per each update of C, but the details are somewhat involved, and we omit them here; they can be found (together with additional earlier references) e.g., in Edelsbrunner (1980a, 1980b, 1982), Edelsbrunner and Maurer (1981), Mehlhorn (1984).

Overall, we conclude the following

Theorem 7.1: The existence of an intersection among n arbitrary rectlinear boxes in 3-space can be tested by an algorithm requiring $O(n \log^2 n)$ space and $O(n \log^2 n)$ time.

Remark: The data structures described above, and some extensions and generalizations thereof, can be used to solve many additional related problems, involving multi-dimensional searching. See for example Edelsbrunner (1982) for a description of some of these applications. It should also be noted that there have recently been many improvements of these basic data structures that reduce their preprocessing time, space, or query time in several special cases. For example, the space complexity of our data structures can be improved to $O(n \log n)$, without changing the time complexity (Mehlhorn, 1984). In the following subsection we will use such an improved data structure for the detection of intersection among spheres.

7.2. Intersection Detection Among Spheres

We next consider a slightly more complicated case, in which the objects in question are n spheres $s_1, s_2, \ldots, s_n$ (this problem arises e.g., in computer simulation of large molecules). The algorithm presented below for detecting intersection among these spheres is due to Hopcroft, Schwartz, and Sharir (1983), and runs in time $O(n \log^2 n)$.

Before going into geometric details, we first describe the algorithm informally. Essentially, it uses a sweeping technique similar to that used in the case of rectilinear boxes. The swept plane H intersects each of the given spheres in a (possibly empty) circle. A natural generalization of the approach used above would be to maintain this collection C of circles dynamically as H is swept upwards, and to check with each change in C whether an intersection between two circles has occurred. There are however two problems with this approach. One problem is that the radii of the circles is C vary continuously with H. The other one is that it is more difficult to maintain a collection of circles, rather than rectangles, in a manner that will facilitate fast queries of the sort used in the preceding subsection.

We therefore use the following modified procedure: We sweep a plane not just in one direction, but in several directions that collectively form a sufficiently dense net of orientations on the unit sphere (nevertheless the number of sweeping directions is constant independent of n). For each such sweeping we maintain the collection C of spheres that intersect the current sweeping plane H, but represent each of these spheres s_i by the great circle on s_i parallel to H. Since such a representation is quite distorted, we will use a more complicated mechanism for detecting intersection among the spheres represented in C. This mechanism may miss some of the intersections, but if the spheres do intersect then it is guaranteed to find such an intersection of at least one pair of spheres, in at least one of the sweeping directions.

Details are as follows. We begin with the following geometric lemma.

Lemma 7.1: Let s_1, s_2 be two intersecting spheres having centers c_1, c_2 and radii r_1, r_2 respectively, and suppose that $r_1 \geq r_2$. Let K be a cube inscribed in s_1 and having faces parallel to the coordinate planes. Then if the angle ϑ between $c_1 c_2$ and the line l parallel to the z-axis and passing through c_2 is $\leq \alpha$, where $\sin\alpha = 1/2\sqrt{3}$ then l intersects K.

Proof: a standard exercise in stereometry.∎

Let Ψ be a finite set of orientations having the property that for each orientation ϑ there exists another orientation $\psi \in \Psi$ such that the angle between ϑ and ψ is $\leq \alpha$. Let M be the size of Ψ. For each $\vartheta_0 \in \Psi$ we apply the following space-sweeping technique. Assume without loss of generality that ϑ_0 points in the direction of the positive z-axis. Sweep a plane H parallel to the x, y plane from $z = -\infty$ to $z = +\infty$. As H is swept upwards we maintain a data structure T (to be described below) containing some of those spheres in S whose centers lie below H. Each time H passes through the center u of a sphere $s \in S$, we perform the following steps:

> (a) Let r be the radius of s. Perform a range query that enumerates all spheres σ in T whose radius ρ is not larger than r and whose center projects into a point of H that belongs to the square of side $2r/\sqrt{3}$ centered at u. Check each such sphere σ for intersection with s. If an intersection is found the algorithm reports this intersection and halts. Otherwise delete the sphere σ from T.
>
> (b) Add the sphere s to T.

It is clear that the sweeping procedure just described performs n range queries and n insertions into T. Moreover, each sphere $\sigma \in S$ can appear in at most one range query, since it is deleted from T immediately after any such query. Also the number of sweeps performed by the algorithm is a constant independent of n. We will show that T can be maintained in $O(n \log n)$ space, in such a way that individual insertions and deletions from T require time $O(\log^2 n)$, while each range query needed can be carried out in time $O(\log^2 n + t)$, t being the number of spheres satisfying the range condition of the query. The preceding remarks show that use of such a structure T will lead to an algorithm requiring $O(n \log n)$ space and $O(n \log^2 n)$ time.

The data structure T that we use represents a collection $s_1, \ldots, s_k$ of spheres. For each such sphere s_i let x_i, y_i be the x and y coordinates of its center and r_i its radius. T is simply a balanced binary range tree with a certain amount of auxiliary data stored at its nodes. The leaves of T correspond to the spheres $s_1, \ldots, s_k$, which are sorted by the x coordinates of their centers. Each internal node ξ of T represents an interval $[\xi_{low}, \xi_{high}]$ of x-values, and to each such ξ we attach a *priority search tree* Q_ξ (as defined by McCreight, 1982) that represents the set of all spheres whose centers have x coordinates lying in this range; each such sphere is represented in Q_ξ by the pair $[y, r]$ of the y coordinate of its center and its radius. As shown by McCreight, each such priority search tree Q has size $O(m)$ and supports query operations of the form $E(y_{min}, y_{max}, r_{max})$, that enumerate all pairs $[y, r]$ in Q for which $y_{min} \le y \le y_{max}$ and $r \le r_{max}$. Each such query requires time $O(\log m + t)$, where m is the number of pairs in Q, and where t is the number of pairs satisfying the query conditions to handle each query. Insertions and deletions of elements in such a tree require time $O(\log m)$ each. Note that the space occupied by the range tree T to whose nodes these priority search trees are attached is $O(n \log n)$.

It is easy to apply the operations we require to T. To insert a sphere s (represented by a triple $[x, y, r]$ giving the x and y coordinates of its center and its radius) into T we insert it into each of $O(\log n)$ priority search trees Q_ξ for ξ lying on the path from the root of T to an appropriate leaf of T, inserting the pair $[y, r]$ into each of these Q_ξ. Each of these updating operation can be accomplished in $O(\log n)$ time, so that the overall cost of inserting a sphere into T is $O(\log^2 n)$. Deletions are performed in a symmetric and similar fashion. Each of the queries that we apply has the form $EN(x_1, x_2, y_1, y_2, r_0)$, and calls for the enumeration of all spheres (i.e. triples $[x, y, r]$) in T satisfying $x_1 \le x \le x_2$, $y_1 \le y \le y_2$, and $r \le r_0$. To apply such a query, we find $O(\log n)$ nodes in T representing disjoint intervals that collectively span the interval between x_1 and x_2. For each such node ξ we can apply the query $E(y_1, y_2, r_0)$ to Q_ξ in time $O(\log n + t_\xi)$, thus enumerating all the t_ξ spheres satisfying the conditions of the original query whose centers have x-coordinates lying in $[\xi_{low}, \xi_{high}]$. Overall the query $EN(x_1, x_2, y_1, y_2, r_0)$ is handled in time $O(\log^2 n + t)$, where t is the total number of spheres satisfying the query conditions.

This description of the data structure T completes the definition of our algorithm and the analysis of its complexity. We now turn to prove the correctness of the algorithm. It is plain that the algorithm never reports a spurious intersection. Hence we only need to show that if any two spheres in S intersect each other, then at least one such intersection will be detected in at least one sweep performed by the algorithm. In fact we will show that the intersection between the two spheres s_1, s_2 for which $\min(r_1, r_2)$ is maximal must be detected in at least one sweep of the algorithm.

Suppose that these two spheres s_1, $s_2 \in S$ intersect each other, and let their centers be c_1, c_2 and their radii be r_1, r_2 respectively, so that $r_1 \le r_2$, and that no two spheres both having radii greater than r_1 intersect one another (in what follows we assume for simplicity that the radii of the spheres in S are all distinct). Let $\vartheta_0 \in \Psi$ be a direction that forms an angle $\vartheta \le \alpha$ with the directed segment $c_1 c_2$, and consider the sweep in the ϑ_0-direction that the algorithm performs. Rotating axes if necessary, assume that ϑ_0 points along the positive z-axis, and that c_1 is lower than c_2. By Lemma 7.1, the line l parallel to the z-axis and passing through c_1 intersects the cube K inscribed in s_2 with faces parallel to the coordinate planes, i.e. the projection c_1' of c_1 onto the x, y plane falls inside the projection of K onto that plane. Hence it suffices to show that by the time the swept plane P passes through c_2, the sphere s_1 is still in T, because then the set produced in response to the range query performed in step (a) of the algorithm at c_2 will include s_1, so that the intersection between s_1 and s_2 will be detected by the algorithm.

Suppose to the contrary that s_1 has been deleted from T prior to the processing of c_2. Then there must exist a third sphere $s \in S$ during whose processing s_1 was deleted from T. It follows that the radius r of s is $> r_1$, that the center c of s lies above c_1 but below c_2, and that c_1' also lies in the projection onto the x, y plane of the cube K' associated with the sphere s. Moreover, we can also assume that s and s_1 do not intersect each other, for otherwise this intersection would have been detected by the algorithm at the time c was processsed. Also by assumption the spheres s and s_2 do not intersect each other because both their radii are greater than r_1. However, such a configuration of spheres is impossible. Indeed, let J, J_1, J_2 be the intervals in which l intersects the spheres s, s_1, s_2 respectively. These intervals are assumed to be disjoint, but a straightforward estimation of their lengths will show this to be impossible (we omit here the calculation, which can be found in Hopcroft et al. (1983). This contradiction implies that the intersection between s_1 and s_2 will be detected, proving the correctness of our algorithm, thus justifying the following summary theorem.

Theorem 7.2: The existence of an intersection among n arbitrary spheres in 3-space can be tested by an algorithm requiring $O(n \log n)$ space and $O(n \log^2 n)$ time.

Remark: There are many related problems of detecting intersections in 3-D space, e.g. the problem of detecting intersections between ellisoids, or, in the extreme case, of lines. It is not clear, however, how to generalize our results.

The main difficulty is that the method of performing many sweeps does not extend easily to cases involving shapes which are not invariant to the direction of the sweep, as are spheres.

REFERENCES

Aho, A., Hopcroft, J., & Ullman J. (1974). *The Design and Analysis of Computer Algorithms.* Reading, MA: Addison-Wesley.

Ahuja, N. (1982). Dot pattern processing using Voronoi neighborhoods. *IEEE Trans. Patt. Anal. Mach. Int.* PAMI-4, 336–343.

Arnold, D. B., & Milne, W. J. (1984). The use of Voronoi tesselations in processing soil survey results. *IEEE Comp. Graphics Appl., 4,* No. 3, 22–30.

Aurenhammer, F. (1983). *Power diagrams: properties, algorithms, and applications.* (Tech. Report 120). Institut fur Informationverarbeitung, Technical University of Graz, Austria.

Aurenhammer, F. (1985). *Improved algorithms for discs and balls using power diagrams.* (Tech. Report 209). Institut fur Informationverarbeitung, Technical University of Graz, Austria.

Aurenhammer, F., & Edelsbrunner, H. (1984). An optimal algorithm for constructing the weighted Voronoi diagram in the plane. *Pattern Recognition, 17*(2), 251–257.

Avis, D., & Bhattacharya, B. (1983). Algorithms for computing d-dimensional Voronoi diagrams and their duals. in *Advances in computing research. Vol. 1: Computational geometry* (pp. 159–180). Greenwich, CT: JAI Press.

Baltsan, A., & Sharir, M. (1985). *On shortest paths between two convex polyhedra.* (Tech. Report 180). Computer Science Dept., Courant Institute of Mathematical Sciences.

Bentley, J. L. (1980). Multidimensional divide-and-conquer. *Comm. ACM, 23,* 214–229.

Bentley, J. L., Weide, B., & Yao, A. C. (1982). Optimal expected time algorithms for closest-point problems. *ACM Trans. Math. Software, 6,* 563–579.

Bhattacharya, B. (1983). *An algorithm for computing order k Voronoi diagrams in the plane.* (Tech. Report 83-9). Computer Science Dept., Simon Fraser University.

Boots, B. N. (1979). Weighting Thiessen polygons. *Econ. Geog.* 248–259.

Bowyer, A. (1981). Computing Dirichlet tesselations. *Computer J., 24,* 162–166.

Bentley, J. L., & Ottman, T. A. (1979). Algorithms for reporting and counting geometric intersections. *IEEE Trans. Computers* C-28, 643–647.

Brostow, W., Dussault, J. P., & Fox, B. L. (1978). Construction of Voronoi polehydra. *J. Comp. Physics, 29,* 81–92.

Brown, K. Q. (1979). Voronoi diagrams from convex hulls. *Inf. Proc. Letters, 9,* 223–228.

Brumberger, H. (1983). Voronoi cells - An interesting and potentially useful cell model for interpreting the small angle scattering of catalysts. *J. Appl. Crys., 16,* 83–88.

Chazelle, B. & Edelsbrunner, H. (1985). An improved algorithm for constructing k-th order Voronoi diagrams. *Proc. ACM Symp. on Computational Geometry,* pp. 228–234.

Cheriton, D., & Tarjan, R. E. (1976). Finding minimum spanning trees. *SIAM J. Computing, 5,* 724–741.

Chew, L. P., & Drysdale III, R. L. (1985). Voronoi diagrams based on convex distance functions. *Proc. ACM Symp. on Computational Geometry,* pp. 235–244.

Clarkson, K. L. (1983). Fast algorithms for the all nearest neighbors problem. *Proc. 24th IEEE Symp. on Foundation of Computer Science,* pp. 226–232.

David, E. E., & David, C. W. (1982). Voronoi polehydra as a tool for studying solvation structure. *J. Chem. Phys., 76,* 4511–4614.

Dehne, F. (1982). An optimal algorithm to construct all Voronoi diagrams for k nearest neighbor searching in the Euclidean plane. *Proc. 20th Allerton Conf. on Communication, Control, and Computing.*

Drysdale III, R. L. (1979). *Generalized voronoi diagrams and geometric searching*. Doctoral thesis, Dept. of Computer Science, STAN-CS-79-705, Stanford University.

Drysdale III, R. L. & Lee, D. T. (1978). Generalized Voronoi diagrams in the plane. *Proc. 16th Annual Allerton Conf. on Communications, Control and Computing*. pp. 833–842.

Edelsbrunner, H. (1980a). *Dynamic data structures for orthogonal intersection queries*. (Tech. Report F-59). Institut fur Informationverarbeitung, Technical University of Graz, Austria.

Edelsbrunner, H. (1980b). *On the intersection of orthogonal objects*. (Tech. Report F-60). Institut fur Informationverarbeitung, Technical University of Graz, Austria.

Edelsbrunner, H. (1982). *Intersection problems in computational geometry*. (Tech. Report F-93). Institut fur Informationverarbeitung, Technical University of Graz, Austria.

Edelsbrunner, H., & Maurer, H. A. (1981). On the intersection of orthogonal objects. *Inf. Proc. Letters, 13*, 177–181.

Edelsbrunner, H., & Seidel, R. (1985). Voronoi diagrams and arrangements. *Proc. ACM Symp. on Computational Geometry*, pp. 251–262.

Finney, J. L. (1979). Procedure for the construction of Voronoi polehydra, Note. *J. Comp. Physics, 32*, 137–143.

Fortune, S. (1985a). Fast algorithms for polygon containment. *Proc. 12th Int. Colloq. Automata, Languages, Programming*, Lecture Notes in Computer Science 194, Springer Verlag, New York, pp. 189–198.

Fortune, S. (1985b). *A sweepline algorithm for Voronoi diagrams*. (Computer Science Research Center Manuscript). AT&T Bell Laboratories.

Franklin, W. R., Akman, V., & Verrilli, C. (1983, November). *Voronoi diagrams with barriers and on polyhedra*. (Tech. Report). Electrical, Computer and Systems Engineering Dept., Rensselaer Polytechnic Institute, Troy, N.Y.

Gowda, I. G., Kirkpatrick, D. G., Lee, D. T., & Naamad, A. (1983). Dynamic Voronoi diagrams. *IEEE Trans. Info. Theory* IT-29, 724–731.

Green, P. J., & Sibson, R. (1978). Computing Dirichlet tesselations in the plane. *Computer J., 21*, 168–173.

Guibas, L., & Stolfi, J. (1983). Primitives for the manipulation of general subdivisions and the computation of Voronoi diagrams. *Proc. 15th ACM Symp. on Theory of Computing*, pp. 221–234.

Hopcroft, J. E., Schwartz, J. T., & Sharir, M. (1983). Efficient detection of intersection among spheres. *Int. J. Robotics Research*, 2(4), 77–80.

Imai, H., Iri, M., & Murota, K. (1983). *Voronoi diagram in the Laguerre geometry and its applications*. (Tech. Report RMI 83-02). Dept. of Mathematical Engineering and Instrumentation Physics, University of Tokyo.

Kirkpatrick, D. G. (1979). Efficient computation of continuous skeletons. *Proc. 20th Symp. on Foundations of Computer Science*, pp. 18–27.

Kirkpatrick, D. G. (1983). Optimal search in planar subdivisions. *SIAM J. Computing, 12*, 28–35.

Klee, V. (1980). On the complexity of d-dimensional Voronoi diagrams. *Arch. Math., 34*, 75–80.

Lee, D. T. (1980). Two-dimensional Voronoi diagrams in the L_p-metric. *J. Assoc. Comput. Mach., 27*, 604–618.

Lee, D. T. (1982a). On k-nearest neighbor Voronoi diagrams in the plane. *IEEE Trans. Computers*, C-31, 478–487.

Lee, D. T. (1982b). Medial axis transformation of a planar shape. *IEEE Trans. Patt. Anal. Mach. Int.* PAMI-4, 363–369.

Lee, D. T., & Drysdale III, R. L. (1981). Generalizations of Voronoi diagrams in the plane. *SIAM J. Computing, 10*, 73–87.

Lee, D. T., & Wong, C. K. (1980). Voronoi diagrams in L_1 (L_∞) metrics with 2-dimensional storage applications. *SIAM J. Computing, 9*, 200–211.

Leven, D. (in progress). *On problems in computational geometry and robotics*. Doctoral thesis.

Leven, D., & Sharir, M. (1985a). *On Voronoi diagrams for a set of discs*. (Tech. Report 28/85). Dept. of Computer Science, Tel-Aviv University, Israel.

Leven, D., & Sharir, M. (1985b). Planning a purely translational motion for a convex object in two-dimensional space using generalized Voronoi diagrams. (Tech. Report 34/85). Dept. of Computer Science, Tel-Aviv University,

McCreight, E. M. (1982, January). Priority search trees. (Tech. Report CSL-81-5). Xerox Corp., Palo Alto.

Mehlhorn, K. (1984). *Data structures and algorithms 3: Multi-dimensional searching and computational geometry*. Springer Verlag, Berlin.

Mount, D. M. (1984). *On finding shortest paths on convex polyhedra*. (Tech. Report 120). Center for Automation Research, University of Maryland.

Mount, D. M. (1985). *Voronoi diagrams on the surface of a polyhedron*. (Tech. Report 121). Center for Automation Research, University of Maryland.

Nievergelt, J., & Preparata, F. P. (1982). Plane-sweeping algorithms for intersecting geometric figures. *Comm. Assoc. Comput. Mach.*, 25, 739–747.

Newman, C. M., Rinott, Y., & Tversky, A. (1983). Nearest neighbor and Voronoi regions in certain point processes. *Adv. Appl. Prob.*, 15, 726–751.

Ó'Dúnlaing, C., & Yap, C. (1985). A 'retraction' method for planning the motion of a disc. *J. Algorithms*, 6, 104–111.

Ó'Dúnlaing, C., Sharir, M., & Yap, C. (1983). Retraction - a new approach to motion planning. *Proc. 15th ACM Symp. on Theory of Computing*. pp. 207–220.

Ó'Dúnlaing, C., Sharir, M., & Yap, C. (1984a). *Generalized Voronoi diagrams for a ladder: I. Topological considerations*. (Tech. Report 139). Computer Science Dept., Courant Institute of Mathematical Sciences.

Ó'Dúnlaing, C., Sharir, M., & Yap, C. (1984b). *Generalized Voronoi diagrams for a ladder: II. Efficient construction of the diagram*. (Tech. Report 140). Computer Science Dept., Courant Institute of Mathematical Sciences.

Preparata, F. P., & Shamos, M. I. (1985). *Computational geometry*. New York: Springer Verlag.

Rowat, P. F. (1979). *Representing spatial experience and solving spatial problems in a simulated robot environment*. Doctoral thesis, University of British Columbia.

Seidel, R. (1981). *A convex hull algorithm optimal for point sets in even dimensions*. (Tech. Report 81-14). Dept. of Computer Science, University of British Columbia.

Seidel, R. (1982). *On the size of closest point Voronoi diagrams*. (Tech. Report 94). Institut fur Informationverarbeitung, Technical University of Graz, Austria.

Shamos, M. I. (1978). *Computational geometry*. Doctoral thesis, Yale University.

Sharir, M. (1985). Intersection and closest-pair problems for a set of planar discs. *SIAM J. Computing*, 14, 448–468.

Shamos, M. I., & Hoey, D. (1975). Closest point problems. *Proc. 16th IEEE Symp. on Foundations of Computer Science*, pp. 151–162.

Tanemura, M., Ogawa, T., & Ogita, N. (1983). A new algorithm for 3-dimensional Voronoi tesselation. *J. Comp. Physics*, 51, 191–207.

Watson, D. F. (1981). Computing the n-dimensional Delaunay tesselation with applications to Voronoi polytopes. *Computer J.*, 24, 167–172.

Yap, C. K. (1984). *An O(n log n) algorithm for the Voronoi diagram of a set of simple curve segments*. (Tech. Report 161). Computer Science Dept., Courant Institute of Mathematical Sciences.

6 Fleshing Out Wire Frames: Reconstruction of Objects, Part I*

George Markowsky
Dept. of Computer Science, University of Maine, Orono, Maine

Michael A. Wesley
Manufacturing Research, IBM Thomas J. Watson Research Centre, Yorktown Heights, N.Y.

ABSTRACT

We present algorithms for reconstructing solid polyhedral objects. In this chapter we present an algorithm which discovers all objects with a given wire frame. This algorithm, which has a number of applications to mechanical design besides being of mathematical interest, has been implemented and has performed well on complex objects.

1. INTRODUCTION

The application of computers to problems in mechanical design was first recognized over 20 years ago (Sutherland, 1963). Since that time much work has been done on the development of production systems both for the entry of a design into a computer data base and for the use of a mechanical design data base in design analysis and in manufacturing.

In the field of data base entry, computer drafting systems allow a designer to interact with a display or tablet to produce drawings of objects, generally in the classic manner of two-dimensional projections of the edges of the three-dimensional object. Some systems also provide the capability of representing data in three dimensions; for example, depth coordinates may be added to the elements of a two-dimensional view, corresponding features in each of several views may

Reprinted from *IBM Journal of Research and Development*, Vol. 24, No. 5, September, 1980. Copyright 1980 by International Business Machines Corporation. Reprinted by permission.

be related, and isometric views may be constructed. These computer drafting systems have been engineered to very high levels of performance and can greatly enhance the productivity of a designer. As well as producing drawings of the edges of objects, computer drafting systems exist which allow the description of the surfaces of objects as smooth curves or patches splined together at their boundaries; in general these surfaces are represented by means of a discrete mesh superimposed on the surface.

Computer-based systems are also used in the analysis of designs and in the manufacture and assembly of objects. For example, parts can be checked for interference (Boyse, 1979; Wesley et al., 1980), finite element methods may be used for analysis of, for example, heat flow (Brown, 1977), the constraints between objects can be derived and mechanisms can be simulated (Taylor, 1976), numerically controlled machine tool tapes can be generated to allow manufacture of a part (Woo, 1975), and robot motions to assemble parts can be generated (Lozano-Perez & Wesley, 1979; Udupa, 1977). In general the full automation of these applications of the design data base requires three-dimensional volumetric information about an object, rather than just a description in terms of edges and surface mesh facets. At present the volumetric form of the data base is considered to be rather difficult and expensive to acquire, and the analysis data are generated in a computer-assisted manner. For example, in the case of numerically controlled machine tools, the path of the cutter may be entered over a drawing at a graphics terminal.

This paper presents an algorithm for automatically bridging the gap between these two fields of computer geometry, that is, from an object described in terms of its edges (a *wire frame*) to a volumetric description in terms of solid material, empty space, and the topology of surfaces and edges. In its present form, the algorithm is restricted to objects whose edges are straight lines and whose faces are planar; since the algorithm is a topological algorithm, it could be adapted to nonplanar surfaces.

Quite apart from its practical applications, the problem is of some theoretical interest. An edge description does not necessarily represent a unique object, and an algorithm should be able to detect ambiguities, enumerate solutions, and accept user decisions as to which solution is required. As with many geometrical problems, the simple cases are straightforward and the complex cases are extremely difficult; for example, many pathological cases can exist—vertices and edges contacting faces, and coplanar opposing faces meeting with edge contact.

Although the literature on geometric modeling is extensive (Baer, Eastman, & Henrion, 1979) and growing rapidly, few authors have chosen to represent objects formally. They are therefore generally unable to prove the correctness of their methods, to handle the full range of pathological cases and ambiguities that occur in practice, or even to describe objects precisely. However, the PADL project (Requicha & Tilove, 1978) is based on point set topology and its architects are able to prove the correctness of algorithms for computing the set operations of union, intersection, and difference between polyhedra. Other workers have used Euler operators (Baer et al., 1979) to ensure correctness of topology as an object is constructed. Idesawa (Idesawa, 1973; Idesawa, Soma, Goto, & Shibata, 1975) describes a wire frame reconstruction scheme as part of the general problem of constructing solids from many 2-D projections. The method used is based on finding sets of coplanar edges and fitting them together to form solid objects; however, the reconstruction method is not based on a formal description of objects and does not handle ambiguities or many of the pathological cases. Experience in modeling has taught us that even though pathological relationships among faces, edges, and vertices may not be physically realizable, they do occur frequently in the stylized world of geometric modeling, and a general-purpose modeling system should be able to handle them. Lafue (1976) also describes a program for generating solids from 2-D projections, but requires that objects be described in terms of faces rather than edges; further, faces are described in a stylized manner with extra edges to permit description of holes.

Other authors have considered the machine vision problem of recognizing polyhedral objects from incomplete edge descriptions (Clowes, 1971; Huffman, 1971; & Waltz, 1975). In this situation local ambiguities can exist and are resolved, if possible, by global propagation. The propagation is performed by labeling areas, whereas the algorithm described in this paper handles ambiguities in terms of volume regions. The use of volumes rather than areas leads to a much simpler handling of the process of labeling.

This paper is divided into four sections. Section 2 gives formal definitions of the concepts needed in order to be able to explain the algorithm and describes some of their consequences. Some standard topological notation is discussed in the appendix. Hocking and Young (1961) may be used as a reference for these terms. Section 3 describes the stages of the algorithm, which has been coded and has performed well even on complex objects. Section 4 gives a number of examples which illustrate the performance of the algorithm.

2. BASIC CONCEPTS

The concepts defined in this section are based on some fundamental topological ideas which are described in detail in Hocking and Young (1961) and to a lesser extent in the appendix. Throughout the paper the standard topology in $\mathbb{R}^3$ and the induced topology on subsets of $\mathbb{R}^3$ are assumed. Vertices refer to points in $\mathbb{R}^3$ and edges refer to line segments defined by two points in $\mathbb{R}^3$. The approach used in this section is to define faces, objects, and wire frames, and then describe the consequences of these definitions.

Definition 1

A *face*, f, is the closure of a nonempty, bounded, connected, coplanar, open (in the relative topology) subset of $\mathbb{R}^3$ whose boundary (denoted by ∂f) is the union of a finite number of line segments. P_f is used to denote the unique plane which contains f.■

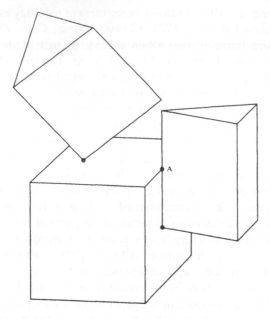

FIG. 6.1. An object exhibiting various kinds of intersections.

Definition 2

An *object*, $\mathbb{O}$, is the closure of a nonempty, bounded, open subset of IR^3 whose boundary (denoted by $\partial\mathbb{O}$) is the union of a finite number of faces.∎

The wire frame algorithm uses many geometric facts about objects. However, rather than define an object as being a set of points satisfying a long list of properties, we have preferred to offer a very simple definition of an object and then prove that it has all the desired properties. Thus from the definitions above it is easy to see that the "cube," $\{x, y, z \in \mathrm{IR}^3 | 0 \leq x \leq 1, 0 \leq y \leq 1, 0 \leq z \leq 1\}$ is an object and that $\{(1, y, z) \in \mathrm{IR}^3 | 0 \leq y \leq 1, 0 \leq z \leq 1\}$ is one of its "square" faces. Starting off with open sets means that faces and objects have nontrivial interiors.

Notice that it is not assumed that an object is the closure of a connected set. This allows objects that consist of disjoint "solids" or even objects which intersect in edges, etc. One can argue that this last case, illustrated in Fig. 6.1. does not represent a "real" object, but in practice all sorts of strange objects can appear. Thus, we decided to handle the most general case possible. Furthermore, this generality does not exact any penalty other than creating a larger number of solutions.

Another point worth noticing is that Definitions 1 and 2 allow many different representations of the boundaries of faces and objects by line segments and faces (respectively). However, canonical representations of the boundaries can be defined which correspond to one's intuitive notions about such things. To get to these representations it is necessary to introduce several additional concepts.

Definition 3

(a) Let f be a face. The *vertices* of f, V(f), are defined to be the set of all points for which two noncollinear line segments, contained in ∂f, can be found whose intersection is the given point.

(b) Let f be a face. The *edges* of f, E(f), are defined to be the set of all line segments e, contained in ∂f, satisfying the following conditions:

1. The endpoints of e belong to V(f);
2. No interior point of e belongs to V(f).

(c) Let $\mathcal{O}$ be an object. The *vertices* of $\mathcal{O}$, V($\mathcal{O}$), are defined to be the set of all points p for which faces $f_1, f_2, f_3 \subseteq \partial\mathcal{O}$ can be found such that $f_1 \cap f_2 \cap f_3$ and $P_{f_1} \cap P_{f_2} \cap P_{f_3}$ are both exactly the single point p.

(d) Let $\mathcal{O}$ be an object. The *edges* of $\mathcal{O}$, E($\mathcal{O}$), are defined to be the set of all line segments e, contained in $\partial\mathcal{O}$, satisfying the following conditions:

1. The endpoints of e belong to V($\mathcal{O}$);
2. No interior point of e belongs to V($\mathcal{O}$);
3. For every point p of e, two noncoplanar faces can be found, $f_1, f_2 \subseteq \partial\mathcal{O}$ such that $p \in f_1 \cap f_2$.

(e) Let $\mathcal{O}$ be an object. The *wire frame* of $\mathcal{O}$, WF($\mathcal{O}$), is defined to be the ordered pair [V($\mathcal{O}$), E($\mathcal{O}$)]. □

The concepts have been defined, but some work is required to show that things fit together as expected. For example, it is not clear that V($\mathcal{O}$) is finite. The reason for keeping the definitions so general is that it is fairly easy to check whether the concepts here include a particular class of entities.

It can be shown that V(f), E(f), V($\mathcal{O}$), and E($\mathcal{O}$) are all finite. V(f) and E(f) yield the intuitive representation of f which will be described more fully below. V($\mathcal{O}$) and E($\mathcal{O}$) do not quite represent $\mathcal{O}$, since the faces of $\mathcal{O}$, which have not been defined so far, are needed. Before getting into the definition of the faces of $\mathcal{O}$, the nature of faces is first described in somewhat greater detail. To do this an additional concept is needed.

Definition 4

A *1-cycle* is a collection of coplanar line segments $\{e_1, \ldots, e_k\}$ in $\mathbb{R}^3$ having the following properties:

1. The intersection of two distinct elements e_i and e_j is either the empty set or a point which is an endpoint of both line segments;
2. Every point of $\mathbb{R}^3$ is the endpoint of a *nonnegative, even* number (in most cases 0) of the e_j. □

In order to be able to give a complete definition of a face, it is necessary only to describe what is meant by the inside and outside of a 1-cycle. A 1-cycle has an inside (outside) if there is a bounded (unbounded), connected, open set whose boundary is the given 1-cycle. A 1-cycle may lack an inside or an outside, or may have them both. Some examples are discussed following Theorem 5. Whenever a point is said to be inside (outside) a 1-cycle, $\mathscr{C}$, it is meant that $\mathscr{C}$ has an inside (outside) and that the point in question belongs to some bounded (unbounded), connected, open set whose boundary is $\mathscr{C}$. Actually, if a 1-cycle, $\mathscr{C}$, has an inside (outside), there is a component of the complement of $\mathscr{C}$ whose boundary is $\mathscr{C}$ and which contains all other bounded (unbounded), connected, open sets whose boundary is $\mathscr{C}$.

Theorem 5

Let f be a face. Then 1-cycles, $\mathscr{C}_0$, $\mathscr{C}_1$, ..., $\mathscr{C}_k$ ($k \geq 0$), contained in ∂f can be found such that

1. $\partial f = \mathscr{C}_0 \cup \mathscr{C}_1 \cup \ldots \cup \mathscr{C}_k$;
2. Face f consists of all points inside $\mathscr{C}_0$ and outside

$$\mathscr{C}_i \ (i \geq 1), \text{ and the points of } \bigcup_{i=0}^{k} \mathscr{C}_i;$$

3. The 1-cycles are all disjoint. □

A typical face is pictured in Fig. 6.2. Note that the boundary can intersect itself at various points such as v_1, v_{16}, and v_{18}. The face in Fig. 6.2 can be defined in terms of three 1-cycles: $\mathscr{C}_0$ (traced out by following the sequence of vertices $v_1 v_2 \ldots v_9 v_{10} v_1 v_{11} v_{12} v_{13} v_1$), $\mathscr{C}_1$ (traced out by $v_{14} v_{15} v_{16} v_{17} v_{18}$ $v_{19} v_{20} v_{18} v_{16} v_{14}$), and $\mathscr{C}_2$ (traced out by $v_{21} v_{22} \ldots v_{28} v_{21}$). Thus the face in Fig. 6.2 consists of all points in the inside of $\mathscr{C}_0$ and in the outsides of $\mathscr{C}_1$ and $\mathscr{C}_2$, plus the points actually belonging to $\mathscr{C}_0$, $\mathscr{C}_1$, and $\mathscr{C}_2$. Note that the points v_2, v_1, and v_{11} are collinear, but by the definition of E($\mathcal{O}$), $v_2 v_1$ and $v_1 v_{11}$ actually belong to E($\mathcal{O}$), but $v_2 v_{11}$ does not. Note also that not every 1-cycle has an inside ($\mathscr{C}_1$, for example, fails the connected set requirement). Similarly, not every 1-cycle has an outside ($\mathscr{C}_0$, for example, again fails the connected set requirement).

The faces of an object are now defined. Again, the definition turns out to be rather straightforward, and it can be shown that the faces defined in this way actually turn out to be the things one would like to call faces anyway. From the following definition it is clear that the faces of an object are really determined by the object rather than by any particular representation of the object or its boundary.

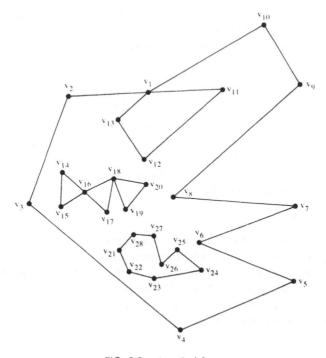

FIG. 6.2.   A typical face.

Definition 6

Let $\mathcal{O}$ be an object. The *faces* of $\mathcal{O}$, $F(\mathcal{O})$, are defined to be the closures of the connected components of $\partial\mathcal{O} - \cup\, E(\mathcal{O})$. □

A number of the results hold true for the faces of an object. Some of the important relationships are summarized in the following theorem.

Theorem 7

Let $\mathcal{O}$ be an object and $F(\mathcal{O}) = \{f_1, \ldots, f_m\}$. Then

1. $\partial\mathcal{O} = \displaystyle\bigcup_{i=1}^{m} f_i;$

2. $\displaystyle\bigcup_{i=1}^{m} E(f_i) \subseteq E(\mathcal{O}) \cup \Gamma$, where Γ is the set consisting of all line segments which are unions of elements of $E(\mathcal{O})$;

3. $\displaystyle\bigcup_{i=1}^{m} V(f_i) \subseteq V(\mathcal{O});$

4. Any face $f \subseteq \partial\mathcal{O}$ for which $E(f) \subseteq E(\mathcal{O}) \cup \Gamma$, where Γ is as in (2) above, is the union of elements of $F(\mathcal{O})$;

5. The intersection of two distinct elements of $F(\mathcal{O})$ is a union of elements of $E(\mathcal{O})$ and subsets of $V(\mathcal{O})$. $\square$

A few brief remarks about (2) and (3) above are in order. Vertices of an object need not be vertices of any face; e.g., in Fig. 6.1 point A is in $V(\mathcal{O})$ but is not a vertex of any face of $\mathcal{O}$. Thus the corresponding edges must be broken up into smaller line segments when considered as edges of $\mathcal{O}$, but this division does not occur if any face is considered separately.

To give the reader an idea of what objects look like in this model, two additional concepts are needed.

Definition 8

A *primitive object* is an object whose interior is connected. $\square$

The key point of Definition 8 is to prevent problems caused by the peculiar types of intersections illustrated in Fig. 6.1. In that figure, there are three primitive objects: a cube and two triangular prisms. In general, an object can be decomposed into primitive objects which do not have any 2-dimensional intersections between any two of them.

Definition 9

A 2-cycle is a collection of faces $\{f_1, \ldots, f_m\}$ in $\mathbb{R}^3$ having the following properties:

1. The intersection of two distinct elements f_i and f_j is the union of the elements of $E(f_i) \cap E(f_j)$ and a finite number of points;

2. Every element of $\displaystyle\bigcup_{i=1}^{m} E(f_i)$ belongs to an even number of distinct faces. $\square$

With all of these concepts a description of primitive objects can be given in terms of 2-cycles. From this one can extrapolate to objects in general. Note how similar the description is to the one given for faces. The definitions of inside and outside are similar to those defined for 1-cycles.

Theorem 10

Let $\mathcal{O}$ be a primitive object. Then 2-cycles $\mathcal{C}_0, \mathcal{C}_1, \ldots, \mathcal{C}_k$ ($k \geq 0$) contained in $\partial\mathcal{O}$ can be found such that

1. $\partial\mathcal{O} = \mathcal{C}_0 \cup \mathcal{C}_1 \cup \ldots \cup \mathcal{C}_k$;

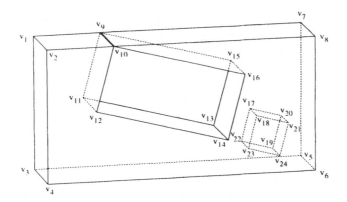

FIG. 6.3. A typical primitive object.

2. $\mathcal{O}$ consists of all points inside $\mathscr{C}_0$ and outside $\mathscr{C}_i$ $(i \geq 1)$ and the points in $\displaystyle\bigcup_{i=0}^{k} \mathscr{C}$;

3. The 2-cycles are all disjoint. □

The term cycle is used below in situations where it is clear from the context whether a 1-cycle or a 2-cycle is intended. Furthermore, the comment made to the effect that not all 1-cycles have a well-defined inside or outside applies as well to 2-cycles, where inside and outside are defined in a similar manner. It will be seen in the next section that to recover $\mathcal{O}$ from WF($\mathcal{O}$) will be necessary to decompose primitive objects further.

Figure 6.3 illustrates a typical object which is represented by two cycles $\mathscr{C}_0$ and $\mathscr{C}_1$. The exterior cycle, $\mathscr{C}_0$, consists of 11 faces, while the interior cycle, $\mathscr{C}_1$, has 6. The identification of the faces is left to the reader.

3. THE WIRE FRAME ALGORITHM

The goal of the wire frame algorithm is to construct all objects which have a given wire frame. It is a fairly elaborate algorithm with quite a few distinct stages. The key stages of the algorithm are outlined first below, followed by a more detailed description.

Stages of the Algorithm

1. Checking Input Data.   The input data are assumed to be a valid wire frame, that is, the ordered pair of vertices and edges [V($\mathcal{O}$), E($\mathcal{O}$)] (Definition

3(e), above). In this stage the input data may be checked for various kinds of errors, such as nondistinct vertices and edges. The choice of actual tests performed is based on the source of the input data and the expected types of errors.

2. Finding Planar Graphs.   All planes are found which contain at least two intersecting edges. For each distinct plane a canonical normal is defined and a graph of coplanar edges formed. For each vertex lying in a plane, a circular list of edges meeting that vertex is created and ordered counterclockwise with respect to the canonical normal.

3. Calculation of 1-Cycles and Virtual Faces.   In each planar graph the set of partitioning cycles is uncovered (bridges are ignored). The nesting relationships among these cycles are then determined, and all candidates for faces found. These candidates are called *virtual faces.*

4. Checking for Illegal Intersections Between Virtual Faces.   Two virtual faces can intersect illegally, i.e., so that both cannot be faces of the real object, in only two ways. These intersections are detected in this stage and appropriate action taken:

> A type I intersection occurs when an interior point of an edge of one pierces an interior point of the other. The latter virtual face is deleted.
> A type II intersection occurs when there is no type I intersection, yet a vertex of one is in the plane of the other and there exists a point that is interior to both. A decision on the faces cannot be made at this stage, and temporary additional edges called *cutting edges* are introduced. These cutting edges cut some of the virtual faces discovered in Stage 3 into new, smaller, virtual faces.

5. Calculation of 2-Cycles and Virtual Blocks.   For each edge a circular list of the virtual faces containing that edge is created. This list is ordered radially around the edge. These lists are used to find all partitioning cycles of the virtual face graph; the nesting relationships among these cycles are found and used to uncover all candidates for solid regions. These candidates are called *virtual blocks.* Virtual blocks are bounded by virtual faces and partition $\mathbb{R}^3$. Any virtual face which does not belong to two different virtual blocks is dropped.

6. Constructing All Solutions for the Wire Frame.   A decision tree, based on virtual blocks and using a few basic tests, assigns solid or hole state to all virtual blocks and thereby constructs all possible objects having a given wire frame. In this decision process, edges and cutting edges are treated separately; cutting edges are subsequently removed. □

The reader should keep in mind that the above description and the one below are designed for easy comprehension. As a result descriptions of each of the stages are given without describing every detail of the data structures and algorithms used. A more detailed description of the various stages follows.

Stage 1: Checking Input Data

The input to the wire frame algorithm must be a valid wire frame, that is, the ordered pair of vertices and edges $[V(\mathbb{O}), E(\mathbb{O})]$ (Definition 3(e), above). This input is assumed to be in the form of a list of vertices with their 3-dimensional coordinates and a list of pairs of vertices to represent edges. The wire frame algorithm described in the following sections requires that the input data represent a valid wire frame, that is, a wire frame that satisfies the definitions of edges and vertices given in Section 2. In this stage tests are performed to check the validity of the input data and to obtain information to be used in later stages. The exact choice of which tests to include depends on the characteristics of the input data and performance trade-offs between the cost of performing a test first, the usefulness of information generated for later stages, and the desirability of reporting errors before incurring the cost of executing the algorithm. These issues are not considered further here.

Two fairly straightforward tests check that vertices and edges are distinct and correctly defined. Furthermore, throughout the rest of the paper it is assumed that each vertex and edge has a unique index.

Another test ensures that every vertex belongs to at least three edges (this is a consequence of the definitions). This test is organized so that a table is generated showing which edges belong to which vertex. This table is important and will be used below.

A test which might also be performed at this point consists of checking that edges intersect only at endpoints, i.e., in elements of $V(\mathbb{O})$. Since two line segments can intersect only if they are coplanar, this test can also partition line segments into coplanar sets. Furthermore, it can even produce a list of edges which intersect a given plane. A test designed to work on the idea just put forth could be fairly expensive in terms of computer time (worst case $O[E(\mathbb{O})^2]$). Alternate tests are possible which are quicker but yield less information.

Depending on the operating environment, one can omit any of the above tests or substitute others if necessary.

Stage 2: Finding Planar Graphs

In this stage all planes which contain at least two intersecting edges are found, and for each plane a graph is constructed of the edges and vertices in that plane. For each vertex in $WF(\mathbb{O})$, a list is formed of the edges for which the vertex is an endpoint. For each noncollinear pair of edges in the list, the plane containing the edge pair is computed and a list formed of distinct planes at the vertex (each plane in $\mathbb{R}^3$ is specified uniquely once a normal and a distance from the origin along that normal are given). For each distinct plane at a vertex, a list is formed of edges in the plane for which the vertex is an endpoint, and the edges are sorted around the normal in a counter-clockwise direction. It is now straightforward to match up planes at vertices and, for each globally distinct plane, to form graphs of the edges and vertices contained in the plane. In practice, the number of edges at a vertex is quite small so the above procedure works quickly.

Thus, the output of this stage is a list of plane equations and, for each plane, graphs of the edges and vertices in the plane.

Stage 3: Calculation of 1-Cycles and Virtual Faces

In this stage each planar edge and vertex graph is processed to find all subgraphs that could represent faces in accordance with Definitions 1 and 3. These subgraphs are candidates for faces of the object and are called virtual faces.

From the discussion in Section 2, it is clear that virtual faces can be located by finding 1-cycles and determining the various nesting relationships among these 1-cycles. To make the discussion clearer, assume a plane P and a graph formed from the edges and vertices of $\mathcal{O}$ which lie in P. The edges of the graph are of two types: bridges and nonbridges. An edge is a nonbridge if and only if it lies on some cycle. In principle, bridges must be removed. The remaining edges can then be divided up into 1-cycles which partition the plane into regions so that any face of $\mathcal{O}$ lying in P is one of the regions. In practice 1-cycles are found and bridges removed in the same operation.

The algorithm proceeds by uncovering the cycle structure of the edges in P. The methods used are now described. It can be shown that the complement of the edges in P, Γ, is an open set with a finite number of open connected components. The number of connected components is the same if the bridges are removed. Every edge which is not a bridge belongs to the closures, in fact boundaries, of two distinct components of Γ. Since the edges are to be used to form 1-cycles to bound the various components of Γ, some conventions are needed for connecting edges and the components they belong to. For the time being bridges are ignored.

Let $e = v_i v_j$ be an edge. There are two ways to traverse e: either from v_i to v_j or from v_j to v_i. Suppose $i < j$, and write $+e$ to denote e traversed from v_i to v_j and $-e$ to denote e traversed from v_j to v_i. Since P has a normal defined on it, whenever an edge is traversed in some direction, left and right sides of the edge can be defined as if one were walking in the same direction above the plane in the positive normal direction.

Let Γ_1 and Γ_2 be the two components of Γ whose boundaries (denoted by $\partial\Gamma_1$ and $\partial\Gamma_2$) contain e. Γ_1 is defined to be on the left, traversing e from v_i to v_j if $+e \in \partial\Gamma_1$. In this case $-e \in \partial\Gamma_2$. Similarly, if $+e \in \partial\Gamma_2$, then $-e \in \partial\Gamma_1$. This is the notation of algebraic topology.

At this point it is probably helpful to illustrate some of these ideas. Figure 6.4 shows a typical graph in a plane. This particular graph consists of 19 vertices and 23 edges. The only bridges are e_6, e_{22}, and e_{23}. Note that the bridges are in the

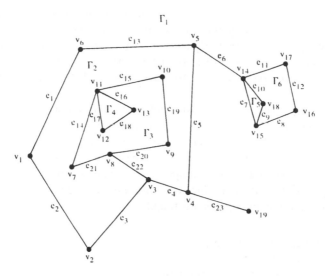

FIG. 6.4. A typical graph in a plane.

closure of exactly one component and are thus not in the boundary of any component. In this case Γ has 6 components and the following relations hold:

$$\partial\Gamma_1 = +e_1 -e_{13} -e_5 -e_4 -e_3 -e_2 +e_{11} -e_{12} -e_8 -e_7;$$
$$\partial\Gamma_2 = -e_1 +e_2 +e_3 +e_4 +e_5 +e_{13} -e_{21} +e_{14} -e_{15} -e_{19} -e_{20};$$
$$\partial\Gamma_3 = +e_{21} +e_{20} +e_{19} +e_{15} +e_{16} -e_{18} -e_{17} -e_{14};$$
$$\partial\Gamma_4 = +e_{17} +e_{18} -e_{16};$$
$$\partial\Gamma_5 = +e_7 +e_9 -e_{10};$$
$$\partial\Gamma_6 = +e_8 +e_{12} -e_{11} +e_{10} -e_9.$$

Now consider the various algorithmic steps needed to determine the information above. First pick an ordered edge, i.e., an edge and a direction, say $+e_1$; now attempt to discover the component, Γ_i, for which $+e_1 \in \partial\Gamma_i$, i.e., try to complete a cycle starting from v_1. Start at v_1 and move to v_6, pick the "next" (in a clockwise direction) edge at v_6, which is $-e_{13}$. At v_5 pick $+e_6$, then $+e_{11}$, then $-e_{12}$, then $-e_8$, then $-e_7$, then $-e_6$. Edges are checked off as they are added to a cycle; if an edge occurs twice in the same cycle, then a bridge exists (in this case e_6).

Whenever a bridge is found, there is a cycle between its two occurrences (in this case $+e_{11} -e_{12} -e_8 -e_7$). This cycle ($\mathscr{C}_1$) is set aside and the search resumed at v_5, ignoring e_6 which is removed from the graph. The sequence is now $+e_1 -e_{13} -e_5 +e_{23} -e_{23}$. The bridge detector now spots e_{23} as a bridge and removes it. The cycle between the two occurrences is the empty cycle, so the search is resumed at v_4. Another cycle ($\mathscr{C}_2$) is found as $+e_1 -e_{13} -e_5 -e_4 -e_3 -e_2$.

Each cycle is now examined to see whether the component it bounds is inside or outside of it, i.e., every point of the component is inside or outside the cycle. In this case Γ_1 is outside both $\mathscr{C}_1$ and $\mathscr{C}_2$. This information is recorded in the cycle tree described below. Also, for each vertex encountered, the cycles to which it belongs are recorded, and for each edge used, the sense in which it is used.

Now pick any other edge which has not been traversed in both directions and start all over again. Suppose $-e_{22}$ is picked now, giving the sequence $-e_{22}$ $+e_4$ $+e_5$ (e_{23} has already been eliminated) $+e_{13}$ $-e_1$ $+e_2$ $+e_3$ $+e_{22}$, at which point the bridge e_{22} is eliminated, leaving the cycle $+e_4$ $+e_5$ $+e_{13}$ $-e_1$ $+e_2$ $+e_3$, which bounds a bounded (inside) component. Data are recorded as before and the process repeated with another edge which has not yet been used in both senses. In this way the following cycles are found:

$$\mathscr{C}_1(\text{out}) = +e_{11} -e_{12} -e_8 -e_7;$$
$$\mathscr{C}_2(\text{out}) = +e_1 -e_{13} -e_5 -e_4 -e_3 -e_2;$$
$$\mathscr{C}_3(\text{in}) = +e_4 +e_5 +e_{13} -e_1 +e_2 +e_3;$$
$$\mathscr{C}_4(\text{out}) = +e_{14} -e_{15} -e_{19} -e_{20} -e_{21};$$
$$\mathscr{C}_5(\text{in}) = -e_{14} +e_{21} +e_{20} +e_{19} +e_{15} +e_{16} -e_{18} -e_{17};$$
$$\mathscr{C}_6(\text{in}) = -e_{16} +e_{17} +e_{18};$$
$$\mathscr{C}_7(\text{in}) = +e_7 +e_9 -e_{10};$$
$$\mathscr{C}_8(\text{in}) = -e_9 +e_8 +e_{12} -e_{11} +e_{10}.$$

The notation ''(in)'' above shows that when the cycle is traversed in the direction indicated, the unique connected component of the complement in P of the cycle, which always lies to the left, is bounded. Similarly, ''(out)'' denotes the case when the cycle is unbounded.

The amount of checking that must be done may be reduced. Suppose that a bridge or generalized bridge (i.e., a connected sequence of bridges) runs between two different cycles $\mathscr{C}$ and $\mathscr{C}'$ (the sense of the bridge must agree with that of the cycles). Then at least one of them is an (out) cycle. Also, if $\mathscr{C}$ is an (out) cycle and $\mathscr{C}'$ is a distinct cycle which intersects $\mathscr{C}$ (i.e., has at least a vertex in common), then it must be an (in) cycle.

Thus in the above example, once $\mathscr{C}_1$ and $\mathscr{C}_2$ are both found to be (out) cycles, the senses of the other cycles are determined if they are derived in the sequence shown. In particular, $\mathscr{C}_3$ is an (in) cycle because it intersects $\mathscr{C}_2$. $\mathscr{C}_4$ is an (out) because it is joined by a bridge (e_{22}) to the (in) cycle $\mathscr{C}_3$. $\mathscr{C}_5$ and $\mathscr{C}_6$ are both (in) cycles because they intersect the (out) cycle $\mathscr{C}_4$, while $\mathscr{C}_7$ and $\mathscr{C}_8$ are both (in) cycles because they intersect the (out) cycle $\mathscr{C}_1$.

At this stage the cycles in P have been found, and will be used to find candidates for faces, i.e., virtual faces. The description of a face given in Section 2 and the concepts introduced here show that a face is given by its outer boundary which is an (in) cycle, $\mathscr{C}_0$, and some finite number of disjoint (out) cycles, $\mathscr{C}_1$,

$\mathscr{C}_2, \ldots, \mathscr{C}_k$ which are contained in the inside of $\mathscr{C}_0$ and have the additional property that if any of them is contained in the inside of any other (in) cycle $\mathscr{C}'$, then $\mathscr{C}_0$ is contained in the inside of $\mathscr{C}'$ as well. This leads to consideration of the following tree structure.

The root is labeled by P. A cycle is a descendant of an (in) cycle, $\mathscr{C}$, if and only if it is contained in the inside of $\mathscr{C}$. A cycle is a descendant of an (out) cycle, $\mathscr{C}$, if and only if it is contained in the complement of the outside of $\mathscr{C}$. The tree structure for the cycles derived from Fig. 6.4 is given in Fig. 6.5.

A few observations aid in the construction of the tree. Any cycle which intersects an (out) cycle is automatically an (in) cycle and a son of the given cycle in the tree. Furthermore, at the finish (in) and (out) cycles must alternate. From the tree it is easy to determine that there are exactly five virtual faces at this stage: the regions bounded by $\mathscr{C}_5$, $\mathscr{C}_6$, $\mathscr{C}_7$, and $\mathscr{C}_8$; the region inside of $\mathscr{C}_3$ and outside of $\mathscr{C}_4$.

There is another point which is appropriate to bring up here. The wire frame algorithm has the property that if it fails to find an object having a given wire frame, then no such object exists. In practice, one works with wire frames of objects that exist. Thus if the final results of the algorithm indicate that no such objects exist, it is probable that some error was made in the input wire frame. Thus at the various stages there are a number of simple checks which can be performed to determine whether or not the wire frame is valid. At the end of Stage 3 one can check to see whether each edge belongs to at least two non-coplanar faces and that each vertex belongs to at least three faces which lie in planes whose intersection is exactly the vertex. Failure to meet any of these conditions would indicate the existence of an error at this point.

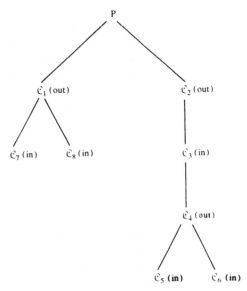

FIG. 6.5.   Tree of 1-cycles.

Stage 4: Checking for Illegal Intersections Between Virtual Faces

The description of objects in Section 2 is based on 2-cycles, which have the property that the faces belonging to them intersect only at boundary points of the faces. Two virtual faces intersect illegally when there exists a point in the intersection that is internal to both. In this case it is not possible for both to be real faces of the object. Illegal intersections can occur in either of two ways:

I. An interior point of an edge of one contains an interior point of the other;

II. The above type of intersection does not occur, yet a vertex of one is in the plane of the other, and there exists a point that is interior to both faces.

These illegal intersections, which are known as type I and type II intersections, respectively, are detected in this stage, and appropriate action taken.

A type I intersection occurs when any inside point of any virtual face is an inside point of any element of $E(\mathcal{O})$. If such a condition is found, the virtual face is dropped from the list of virtual faces because it is impossible for it ever to be a face. To see this, note that the edges of $\mathcal{O}$ belong to actual faces. If a virtual face intersects edges as described above, it would have to intersect the corresponding faces. Such an intersection would produce at least one edge emanating from an *inside* point of a face, which would be impossible. (A type I intersection is shown later in Table 6.1(d).)

Two ways are proposed to handle the second case. The second method, which is the preferred method, also suggests a quick means for checking for type II intersections.

The first method of handling type II intersections is to pick maximal subsets of virtual faces which lack a type II intersection and to proceed through the remaining stages of the wire frame algorithm to uncover all possible solutions under those assumptions. In some cases one can use any solutions found to resolve the true nature of type II intersections. In other cases it might be necessary to go through the remaining stages of the algorithm with several different maximal subsets lacking a type II intersection. For many practical objects, type II intersections are relatively rare (they arise from high degrees of symmetry), so this solution is quite a practical one. It also has the advantage that it simplifies the decision procedure in Stage 6, since there is only one kind of edge to consider.

The second method is based on the observation that a type II intersection consists of a finite number of line segments, the endpoints of which are elements of $V(\mathcal{O})$. To see this, let f_1 and f_2 be faces that have type II intersection. Let $l = P_{f_1} \cap P_{f_2}$. Let $p \in f_1 \cap f_2$ be an interior point of both f_1 and f_2. Let p_1 and p_2 be the points of l which give the maximal line segment containing p and contained in $f_1 \cap f_2$. Since f_1 and f_2 are compact, i.e., closed and bounded, p_1 and p_2 belong to $\partial f_1 \cup \partial f_2$. Since no boundary point of f_1 is an inside point of f_2 and

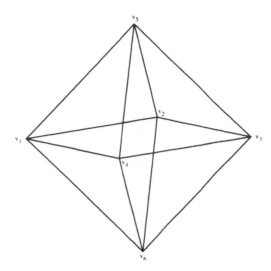

FIG. 6.6.   A regular octahedron ex-
hibits many type II intersections.

vice versa, $p_1, p_2 \in \partial f_1 \cap \partial f_2$. If the edges of f_1 and f_2 which contain p_1 (p_2) are collinear, then f_1 and f_2 must be coplanar and must overlap in nontrivial ways. This is impossible in view of the tests performed in Stage 3. Thus p_1 and p_2 belong to two noncollinear edges which can only intersect in an element of $V(\mathcal{O})$. To help visualize the preceding argument, look at Fig. 6.6. Here f_1 is given by v_1 v_2 v_3 v_4 v_1 and f_2 by v_2 v_6 v_4 v_5 v_2, p_1 is v_2, and p_2 is v_4. This gives a quick test for type II intersections: visit each vertex in turn and see if any of the virtual faces containing that vertex intersect.

Suppose that f_1 and f_2 are two virtual faces having a type II intersection; introduce the line segments of intersection as new edges, called *cutting edges.* Also introduce all the necessary points of intersection. The new vertices and edges are marked to distinguish them from the original vertices and edges. In general, these new vertices and edges will partition some of the virtual faces into smaller virtual faces. Using the algorithms described earlier, all those cutting edges which are bridges in a particular virtual face having type II intersections are identified. All virtual faces which induce these bridges are dropped, since they cannot possibly separate solid matter from empty space. Of course, after dropping some virtual faces, some of the type II intersections may disappear.

Since type II intersections are mostly the result of symmetry, we consider one of the most symmetrical cases possible, that shown in Fig. 6.6. After the regular octahedron of Fig. 6.6 passes through Stage 3, 11 faces will have been found: the usual 8 faces plus the 3 given by v_1 v_2 v_3 v_4 v_1, v_2 v_5 v_4 v_6 v_2, and v_1 v_5 v_3 v_6 v_1. The last 3 virtual faces (pairwise) have type II intersections. Each of the last 3 faces partitions each of the others into smaller virtual faces, which are all kept, ending up with 7 vertices, 18 edges, and 20 faces. The new wire frame is illustrated in Fig. 6.7.

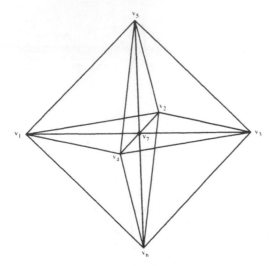

FIG. 6.7.   A regular octahedrom after Stage 4 with cutting edges inserted.

Stage 5: Calculation of 2-Cycles and Virtual Blocks

In this stage virtual faces are fitted together to form candidate objects called virtual blocks. From the definition and discussion in Section 2, it is clear that objects can be found by calculating all 2-cycles and finding the nesting relationships among them. This 3-D process is a close analog of the 2-D process of fitting edges together to form virtual faces. However, the definition of a 2-cycle is in terms of F($\mathcal{O}$), and at this stage of the algorithm only the virtual faces VF($\mathcal{O}$) are available, where F($\mathcal{O}$) $\subseteq$ VF($\mathcal{O}$). Thus, VF($\mathcal{O}$) can contain elements which are not faces of $\mathcal{O}$ and are known as *pseudo-faces*. Pseudo-faces arise through chance alignments of edges and may occur in two forms:

I. The interior of the virtual face is empty space;
II. The interior of the virtual face is interior to solid material.

It will be seen that type I pseudo-faces are always rejected and that type II may either be rejected or be used to partition a primitive object into smaller subobjects.

Some tests that detect pseudo-faces have been seen in Stage 4. An intersection of type I shows that the virtual face involved is really a pseudo-face. Similarly, an intersection of type II indicates that at least one of the virtual faces involved is a pseudo-face. Note that not every pseudo-face is involved in an illegal intersection of one of these two types. Another kind of pseudo-face that is detected in this stage is the 2-bridge, i.e., a virtual face which does not belong to any 2-cycle. After detecting and handling all of these pseudo-faces, the remaining virtual faces naturally break up into 2-cycles. These 2-cycles partition all of $\mathbb{R}^3$ into connected components in much the same way that the 1-cycles partition the planes. In fact, the remainder of this stage is very similar to the virtual face creation algorithm of Stage 3. However, since this is in 3-dimensional space, no new types of intersections can occur and no new tests are necessary. As in Stage 3 some conventions are needed for describing the relationship beween virtual faces and the components of an object which they bound.

Let B_1 and B_2 be the two components of an object B^* whose boundaries (denoted by ∂B_1 and ∂B_2) both contain given a virtual face f. If the canonical normal (introduced in Stage 2) erected at any interior point of f points away from (into) B_1, then $+f \in \partial B_1$ ($-f \in \partial B_1$). Clearly, $+f \in \partial B_1$ ($-f \in \partial B_1$) iff $-f \in \partial B_2$ ($+f \in \partial B_2$). The goal is to find the various components of B^* because the original object can be built out of them.

Before proceeding further, consider a simple example. An object can have 1-cycles which result accidentally; in Fig. 6.8, the virtual face, $v_5\, v_6\, v_7\, v_8\, v_5$, is a pseudo-face, because it is not an actual boundary between empty space and solid material. However, this cannot be detected until the object is considered globally, i.e., when virtual faces are being found in the various planes, there is no way of distinguishing between faces and pseudo-faces. Only when the construction of the complete object in Fig. 6.8 is attempted is $v_5\, v_6\, v_7\, v_8\, v_5$ seen to be a pseudo-face.

This problem of pseudo-faces is handled by working with virtual blocks, i.e., 2-cycles which do not contain any nonbridge, virtual faces in their interior. Thus,

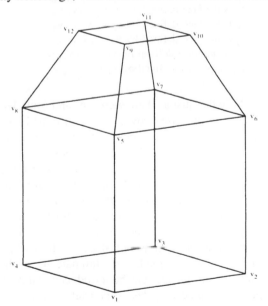

FIG. 6.8.   An example of a pseudo-face.

virtual blocks are the primitive building blocks for dissection of an object by pseudo-faces. The object in Fig. 6.8 has three virtual blocks associated with it:

1. The closure of the unbounded component of B^*;
2. The closure of the bounded component of B^* lying above v_5 v_6 v_7 v_8 v_5;
3. The closure of the bounded component of B^* lying below v_5 v_6 v_7 v_8 v_5.

To describe the boundaries in terms of the notation introduced above, assume that in Fig. 8 the origin is in the middle of the cube defined by v_1, v_2, v_3, v_4, v_5, v_6, v_7, and v_8, and that all plane positive normals radiate outward, giving

$$\partial B_1 = -f_1 -f_2 -f_3 -f_4 -f_5 -f_6 -f_7 -f_8 -f_9 -f_{11};$$
$$\partial B_2 = f_1 +f_2 +f_3 +f_4 +f_9 -f_{10};$$
$$\partial B_3 = f_5 +f_6 +f_7 +f_8 +f_{10} +f_{11};$$

where

f_1 is the face defined by v_5 v_6 v_{10} v_9 v_5;
f_2 is the face defined by v_6 v_7 v_{11} v_{10} v_6;
f_3 is the face defined by v_7 v_8 v_{12} v_{11} v_7;
f_4 is the face defined by v_8 v_5 v_9 v_{12} v_8;
f_5 is the face defined by v_1 v_2 v_6 v_5 v_1;
f_6 is the face defined by v_2 v_3 v_7 v_6 v_2;
f_7 is the face defined by v_3 v_4 v_8 v_7 v_3;
f_8 is the face defined by v_4 v_1 v_5 v_8 v_4;
f_9 is the face defined by v_9 v_{10} v_{11} v_{12} v_9;
f_{10} is the face defined by v_5 v_6 v_7 v_8 v_5;
f_{11} is the face defined by v_1 v_2 v_3 v_4 v_1.

Note that just as there were (in) and (out) 1-cycles, there are (in) and (out) 2-cycles. In the case above, ∂B_1 is an (out) 2-cycle, while ∂B_2 and ∂B_3 are both (in) cycles. As the reader probably suspects at this point, a tree of 2-cycles, similar to the one for 1-cycles, is constructed for an object. In fact, all the rules given for constructing a 1-cycle tree hold for 2-cycle trees. Virtual blocks are derived from this tree in the same way that virtual faces were derived from the 1-cycle tree. In this case the tree is represented in Fig. 6.9.

Before describing the procedure for finding 2-cycles, consider the case of the regular octahedron (Figs. 8.6 and 8.7). Because the octahedron is so symmetrical, it has three pseudo-faces each of which intersects the other two. Since there is no a priori method to eliminate any of them, either all possibilities can be tried or cutting edges can be introduced. Thus the octahedron of Fig. 6.7 decomposes into nine 2-cycles—one (out) and eight (in) 2-cycles.

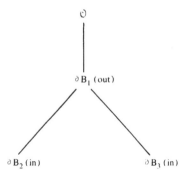

FIG. 6.9.   Tree of 2-cycles.

The algorithmic steps to discover all 2-cycles and 2-bridges are now described. For each edge e of $\mathcal{O}$, a circular list is formed of all virtual faces which have e contained in their boundary. The faces are ordered in the same way as the corresponding edges were ordered in Stage 2, that is they are ordered radially around the edge. The search for 2-cycles now proceeds very much like the search for 1-cycles. Pick a virtual face with an orientation, i.e., $+f$ or $-f$, and attempt to find a virtual block containing it. Process edges one at a time by adding the appropriate face with the correct orientation, and maintain information on the number of times the edge is used and the sense of each use. Choosing an orientation for a virtual face is equivalent to assuming that solid material lies on a particular side of the virtual face. Thus an edge is processed by seeing which oriented faces contain it and picking those oriented faces which are neighbors through the solid material. Figure 6.10 illustrates this point by giving an edge-on view of the process. Suppose that the virtual faces f_1, f_3, f_4, and f_6 have been selected to be in 2-cycles with the orientations suggested in the figure by the normals and the shading. Since f_3 and f_4 are neighbors through solid material, they can both be dropped from further consideration. To find a virtual block both f_2 and f_5 would need to be added to the proposed 2-cycle with the indicated orientation. Note that it would be impossible for f_7 to belong to the 2-cycle because each edge can only belong to an even number of faces. If a virtual face is found which would need to be incorporated into the same 2-cycle twice (it will turn out that it is with opposite orientations), then that virtual face is a bridge and is deleted from the list of virtual faces. The partial results are saved and the process continued until the 2-cycle is completed. At the end of this process all bridges have been eliminated and every remaining virtual face belongs to exactly two distinct virtual blocks. Furthermore, the interiors of the virtual blocks are exactly the components of the complement of the remaining virtual faces. The original object must be a union of some of these virtual blocks, thus showing that in principle the problem has been reduced to a problem which involves only a finite number of possibilities. The next stage handles this last problem efficiently.

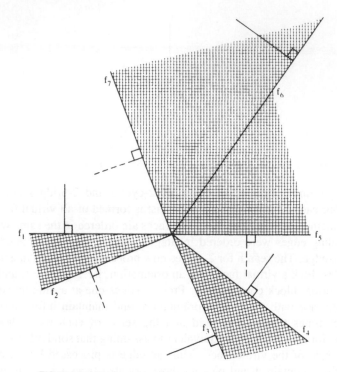

FIG. 6.10.   Finding virtual blocks—an edge's perspective.

It only remains to mention one complication which can arise. In some cases, several edges of $\mathbb{O}$ are collinear and can be combined into a single line segment. In this case it is possible for one face to have as an edge a line segment which contains edges from other faces as subedges. In this case, there are a number of straightforward modifications which must be made to the 2-cycle finding algorithm.

Stage 6: Constructing all Solutions for the Wire Frame

In this stage virtual blocks are fitted together to generate all objects with a given wire frame. Basically each virtual block may have *solid* or *hole* state and, when a state assignment has been made to each virtual block, an object is obtained. However, not all assignments of solid and hole yield the desired wire frame. An assignment of solid or hole to the virtual blocks yields an object with the correct wire frame if

1. *Every* element $e \in E(\mathbb{O})$ belongs to two noncoplanar virtual faces f_1 and f_2 each of which belongs to one virtual block assigned solid state and one assigned hole state;

2. *No* cutting edge belongs to two noncoplanar virtual faces f_1 and f_2 each of which belongs to one virtual block assigned solid state and one assigned hole state, i.e., every cutting edge must be inside material.

A decision tree is constructed by growing those edges having the smallest number of unassigned virtual blocks containing them. The unique infinite virtual block is always assigned the hole state. Condition (1) is not always used to make choices between states; the necessary condition that every edge belong to a solid block and to a hole block is also used. However, conditions (1) and (2) are the ones that must be satisfied. To illustrate this process consider the regular octahedron of Fig. 6.7.

There are nine blocks:

B_1—the infinite virtual block;
B_2—the virtual block determined by $v_2\ v_1\ v_7\ v_5$;
B_3—the virtual block determined by $v_1\ v_4\ v_7\ v_5$;
B_4—the virtual block determined by $v_4\ v_3\ v_7\ v_5$;
B_5—the virtual block determined by $v_3\ v_2\ v_7\ v_5$;
B_6—the virtual block determined by $v_2\ v_1\ v_7\ v_6$;
B_7—the virtual block determined by $v_1\ v_4\ v_7\ v_6$;
B_8—the virtual block determined by $v_4\ v_3\ v_7\ v_6$;
B_9—the virtual block determined by $v_3\ v_2\ v_7\ v_6$.

Each edge of $\mathcal{O}$ now belongs to two virtual blocks of undetermined status while each cutting edge belongs to four virtual blocks of undetermined status. The state hole is assigned to B_1. An edge is picked, say $e = \overline{v_1\ v_5}$, and the decision tree begun.

Note that e already belongs to a block with hole state, so solid state must be assigned to some block. Figure 6.11 shows the decision tree in this case. Notice that each time the state of one of the B_i $(i \geq 2)$ is set to hole a contradiction is quickly found. If B_2 has hole state, then B_3 and B_5 must be given solid state because $\overline{v_1\ v_5}$ and $\overline{v_2\ v_5}$ must belong to at least one solid block and there is only one candidate for this. However, if B_2 has hole state, and B_3 and B_5 solid state, the faces $v_1\ v_5\ v_7\ v_1$ and $v_2\ v_5\ v_7\ v_2$ contradict condition (2) for edge $\overline{v_5\ v_7}$. Similar contradictions arise whenever any B_i $(i \geq 2)$ is treated as being empty. Notice that after a few assignments the subsequent choices are determined and exponential growth of the tree is avoided.

In some cases, there is an exponential number of different objects having the same wire frame, so exponential growth cannot be entirely avoided. However, if the tree is grown for depth, some object can be found having the given wire frame. In practice, this stage is completed fairly quickly since the geometry generally takes over once several assignments have been made. In complex

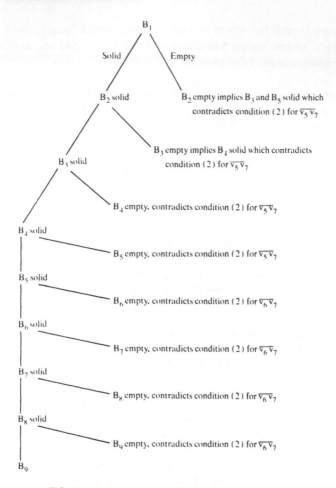

FIG. 6.11.   A decision tree for the regular octahedron.

objects it is often the case that many edges on the outer boundary belong to exactly one virtual block which can be marked solid. In particular, any vertex belonging to exactly three elements of $E(\mathcal{O})$ belongs to exactly two virtual blocks. Thus if one of them is empty, the other one must be solid.

Stage 6 feeds into an output module which puts the output together in forms which can be understood by the user of the system. The following section shows a number of examples in detail.

4.   EXAMPLES

In this section are described a number of examples chosen to illustrate particular features of the algorithm. The examples are illustrated in Table 6.1.

Table 6.1(a) shows a double tetrahedron. Seven triangular virtual faces are found—the six outside faces and the internal area bounded by the waist of the figure. Three virtual blocks are found; the decision process assigns solid state to (1) and (2); block (3) is the unbounded virtual block; (1) and (2) are combined to produce the output object.

Table 6.1(b) shows an object with 1-D bridges on the faces containing abcd and kmnp. The plane graphs contain three bridges ef, kl, and op, none of which appear in the virtual faces for the planes shown. Two virtual blocks are found, one the output object and one the unbounded virtual block.

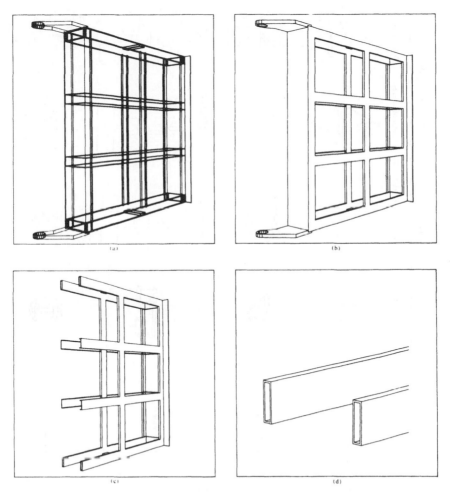

FIG. 6.12. (a) Wire frame with 1256 edges and 836 vertices; (b) volumetric representation of wire frame; (c) cross section of volumetric representation; and (d) close-up view of tubular members.

TABLE 6.1
Examples of the Wire Frame Algorithm

	Wire frame	Plane graphs of interest	Virtual faces of interest	Virtual blocks	Virtual block decision states (s = solid, h = hole)	Object
(a) Double tetrahedron					1 = s 2 = s 3 = h	
(b) 1-D bridges					1 = s 2 = h	
(c) 2-D bridges					1 = s 2 = s 3 = s 4 = s 5 = h 6 = h 7 = h 8 = h	
(d) Type I intersections			abcd and efgh have type I intersections and are not virtual faces		1 = s 2 = s 3 = s 4 = s 5 = h 6 = h	

254

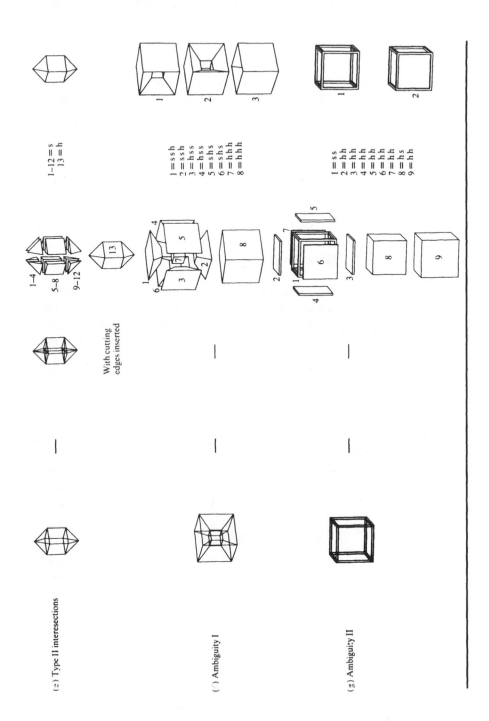

1–12 = s
13 = h

1 = s s h
2 = s s h
3 = h s s
4 = h s s
5 = s h s
6 = s h s
7 = h h h
8 = h h h

1 = s s
2 = h h
3 = h h
4 = h h
5 = h h
6 = h h
7 = h h
8 = h s
9 = h h

1–4
5–8
9–12

13

With cutting
edges inserted

(e) Type II intersections

(f) Ambiguity I

(g) Ambiguity II

255

Table 6.1(c) shows four cubes positioned on two levels with four shared vertices enclosing a rectangular area abcd; abcd is found to be a virtual face, but in the virtual block building process is detected to be a 2-D bridge (i.e.,it is assigned opposite directions in the same virtual block to become a zero thickness sheet) and is not used in the output objects.

Table 6.1(d) contains an octahedron extended by a cube and pierced by a vertical square prism. The two plane graphs containing abcd and efgh have type I intersections with the vertical sides of the hole and therefore are not virtual faces. Six virtual blocks are found and assigned states as shown.

Table 6.1(e) shows the object of Table 6.1(d) without the piercing hole. Four face graphs with type II intersections occur and are shown as virtual faces with cutting edges inserted. Thirteen virtual blocks are found and assigned states as shown.

Table 6.1(f) shows a well known ambiguous wire frame (Voelker & Requicha, 1977); eight virtual blocks are found, and the decision process enumerates three valid solutions: one pair of opposing blocks (Sutherland, 1979; Boyse, 1979), (Wesley et al., 1980; Brown, 1977), or (Taylor, 1976; Woo, 1975) must have hole state, the center block (7) always has hole state.

Table 6.1(g) shows another ambiguous wire frame that could well occur in practice. Nine virtual blocks are formed; the decision process finds that block (8) can have hole or solid state.

Figure 6.12(a) shows a more complicated wire frame with 1256 edges and 836 vertices. In the course of the reconstruction process the wire frame algorithm finds 93 virtual blocks, most of them being window holes and enclosed volumes inside tubular members of the structure, and generates the volumetric representation shown in Fig. 6.12(b). Figures 6.12(c) and (d) show a cross section of the reconstruction with the nested interiors of tubular members correctly represented.

ACKNOWLEDGMENTS

Our thanks are due to D. D. Grossman for encouraging us to tackle the wire frame reconstruction problem and for providing the ambiguous example of Table 6.1(g), and to T. Lozano-Pérez for his contributions to early discussions on the problem.

At the time this work was done, the authors were located at the IBM Thomas J. Watson Research Center, Yorktown Heights, New York 10598. G. Markowsky is now at the University of Main, Orono, ME.

APPENDIX A: TOPOLOGICAL CONCEPTS

A brief introduction is given to those standard topological concepts used in this paper. For more details, see Hocking and Young (1961).

Definition A.1

Let $x \in \mathbb{R}^3$ and r be a positive number. $B_r(x)$ is used to denote the set of all points of $\mathbb{R}^3$ whose Euclidean distance from x is less than r. $B_r(x)$ is called the *open ball at x of radius r.* $\square$

Definition A.2

A subset $X \subseteq \mathbb{R}^3$ is said to be *open* if, for all $x \in X$, there exists $r > 0$ such that $B_r(x) \subseteq X$. A subset $Y \subseteq \mathbb{R}^3$ is said to be *closed* if $\mathbb{R}^3 - Y$ is open. Note that open balls are open and that $\varnothing$ and $\mathbb{R}^3$ are both open and closed. $\square$

Definition A.3

Let $X \subseteq Y \subseteq \mathbb{R}^3$. Then X is said to be *open in Y in the relative (induced) topology* [or open in the relative (induced) topology for short] if, for all $x \in X$, there exists $r > 0$ such that $B_r(x) \cap X = B_r(x) \cap Y$. X is *closed in the relative topology* if $Y - X$ is relatively open. $\square$

In the cases most of interest here, i.e., subsets of a plane in $\mathbb{R}^3$, being relatively open means containing open disks (the intersection of a plane and an open ball). The following definitions will be stated only for the standard topology of $\mathbb{R}^3$ (Definition A.2) and the reader should verify that they make sense for any relative topology.

Definition A.4

The *closure*, $\bar{X}$, of a subset X of $\mathbb{R}^3$ is the set $\{x \in \mathbb{R}^3 \mid$ for all $r > 0$, $B_r(x) \cap X \neq \varnothing\}$. In particular, $X \subseteq \bar{X}$. $\square$

It can be shown that $\bar{X}$ is a closed set and that a subset $Y \subseteq \mathbb{R}^3$ is closed if and only if $Y = \bar{Y}$.

Definition A.5

The *boundary*, ∂X, of a set $X \subseteq \mathbb{R}^3$ is the set $\bar{X} \cap \overline{(\mathbb{R}^3 - X)}$. $\sqcup$

Thus a point, x, is in ∂X if and only if there are points of both X and $\mathbb{R}^3 - X$ arbitrarily close to x.

Definition A.6

A subset X of $\mathbb{R}^3$ is said to be *connected* if two nonempty open subsets U_1, U_2 of $\mathbb{R}^3$ cannot be found such that $U_1 \cap U_2 = \varnothing$, $U_1 \cap X \neq \varnothing \neq U_2 \cap X$, and $X \subseteq U_1 \cup U_2$. $\square$

In the case of $\mathbb{R}^3$ and its subplanes, all connected open subsets have the property that any two points in a given subset can be connected by a path which lies entirely in the given set.

Definition A.7

Let $X \subseteq \mathbb{R}^3$. A *connected component* of X is a subset Y of X which is connected and such that for any other connected subset $Z \subseteq \mathbb{R}^3$, either $Z \subseteq Y$ or $Z \cap Y = \varnothing$. $\square$

Any set in $\mathbb{R}^3$ can be written as the disjoint union of its components.

Definition A.8

A subset X of IR^3 is said to be *bounded* (unbounded) if there exists for all $r >$ 0 a point $p \in IR^3$ such that $X \subseteq B_r(p)$ [$X \not\subseteq B_r(p)$].

REFERENCES

Baer, A., Eastman, C. & Henrion, M. (1979, September). Geometric Modeling: A Survey. *Computer Aided Design, 11*, 253–272.

Boyse, J. (1979, January). Interference detection among solids and surfaces. *Commun. ACM, 22*, 3–9.

Brown, B. E. (1977, June). *Modeling of solids for three-dimensional finite element analysis.* Doctoral dissertation, Department of Computer Sciences, University of Utah, Salt Lake City.

Clowes, M. B. (1971). On seeing things. *Artificial Intelligence, 2*, 79–116.

Hocking, J. G. & Young, G. S. (1961). *Topology.* Reading, MA: Addition-Wesley.

Huffman, D. A. (1971). Impossible objects as nonsense sentences. In B. Meltzer & D. Michie, (Eds.), *Machine intelligence* (Vol. 6, pp. 295–324). Edinburgh University Press, Scotland.

Ikesawa, M. (1973, February). A system to generate a solid figure from a three view. *Bull JSME, 16*, 216–225.

Idesawa, M., Soma, T., Goto, E., & Shibata, S. (1975). Automatic input of line drawing and generation of solid figure from three-view data. *Proceedings of the International Joint Computer Symposium*, pp. 304–311.

Lafue, G. (1976, July). Recognition of three-dimensional objects from orthographic views. *Proceedings 3rd Annual Conference on Computer Graphics, Interactive Techniques, and Image Processing*, ACM/SIGGRAPH, pp. 103–108.

Lozano-Pérez, T., & Wesley, M. A. (1979, October). An algorithm for planning collision-free paths among polyhedral objects. *Commun. ACM, 22*, 560–570.

Requicha, A. G., & Tilove, R. B. (1978, March). *Mathematical foundations of constructive solid geometry: General topology of closed regular sets.* (Tech. Memo. No. 27). Production Automation Project. University of Rochester, New York.

Sutherland, I. E. (1963). SKETCHPAD: A man-machine graphical communication system. *Proc. SJCC, 23*, pp. 329.

Taylor, R. H. (1976, July). A synthesis on manipulation control programs from task level specifications. *Report No. STAN-CS-76-560* Stanford Artificial Intelligence Laboratory). Computer Science Department, Stanford University, Palo Alto, CA.

Udupa, S. (1977). *Collision detection and avoidance in computer controlled manipulators.* Doctoral thesis, California Institute of Technology, Pasadena.

Voelker, H. P., & Requicha, A. A. G. (1977, December). Geometric modeling of mechanical parts and processes. *Computer, 10*, 48–57.

Waltz, D. (1975). Understanding line drawings of scenes with shadows. In P. H. Winston (Ed.), *The psychology of computer vision* (pp. 19–91). New York: McGraw-Hill.

Wesley, M. A., Lozano-Pérez, T., Lieberman, L. I., Lavin, M. A., & Grossman, D. D. (1980, January). A geometric modeling system for automated mechanical assembly. *IBM J. Res. Develop., 24*, 64–74.

Woo, T. C.-H. (1977). *Computer understanding of design.* Doctoral thesis, University of Illinois at Urbana-Champaign.

7

Fleshing Out Projections: Reconstruction of Objects, Part II*

Michael A. Wesley
Manufacturing Research, IBM Thomas J. Watson Research Centre, Yorktown Heights, N.Y.

George Markowsky
Department of Computer Science, University of Maine, Orono, Maine

ABSTRACT

We now extend the Wire Frame algorithm described in Chapter 6 to find all solid polyhedral objects with a given set of two dimensional projections. These projections may contain depth information in the form of dashed and solid lines, may represent cross sections, and may be overall or detail views. The choice of labeling conventions in the projections determines the difficulty of the problem. It is shown that with certain conventions and projections the problem of fleshing out projections essentially reduces to the problem of fleshing out wire frames. Even if no labeling is used, the Projections algorithm presented here finds all solutions even though it is possible to construct simple examples with a very large number of solutions. Such examples have a large amount of symmetry and various accidental coincidences which typically do not occur in objects of practical interest. Because of its generality, the algorithm can handle pathological cases if they arise. This Projections algorithm, which has applications in the conversion of engineering drawings in a Computer Aided Design, Computer Aided Manufacturing (CAD-CAM) system, has been implemented. The algorithm has successfully found solutions to problems that are rather complex in terms of either the number of possible solutions or the inherent complexity of projections of objects of engineering interest.

1. INTRODUCTION

The Wire Frame algorithm described in the previous chapter was based on the concepts of algebraic topology and rigorous definitions of the geometric entities

Reprinted from *IBM Journal of Research and Development*, Vol. 25, No. 6, November, 1981. Copyright 1981 by International Business Machines Corporation. Reprinted by permission.

involved. It recognized that many solid objects may have the same wire frame and was able to find all possible solutions efficiently.

We now extend the Wire Frame algorithm to polyhedral objects described by a set of two dimensional projections such as might be seen on an engineering drawing. The projection process may introduce another level of ambiguity into reconstruction problems and increases the possibility of there being many objects with the same set of projections. The Projections algorithm presented here can work with very little information, for example, only two projections, and find all possible objects matching the data. However, it is seen that the number of solutions may be very large and that it may be reasonable to provide more information in the form of three or more projections, by labeling corresponding features in divers views, and by providing depth information. The Projections algorithm is able to make use of this extra information and can also accept other forms of advice, such as whether given points are inside material.

Quite apart from its mathematical interest, the algorithm has practical applications in the automatic conversion of digitized engineering drawings into solid volumetric representations of the geometry of objects. These solid volumetric representations become the basis for the simulation and synthesis of large parts of the design validation, analysis, manufacture, inspection, and documentation process (Wesley, 1980; Wesley et al., 1980).

The subject of reconstruction of solid polyhedral objects from their projections has been studied over a period of years. Early work (Idesawa, 1973; Idesawa, Soma, Goto, & Shibata, 1975; Sutherland, 1974) was largely based on labeling corresponding information in different views and requiring the user to conform to constraints on the manner of description of features such as faces. The historical trend has been to free the user of as many constraints as possible (Lafue, 1976). However, the relaxation of constraints has led to the possibility of multiple solutions to a given problem, and workers have tended to concentrate on heuristic approaches to find a probable solution. Wesley and Markowsky report such a heuristic approach that allows complete freedom of input and has been implemented (see Chapter 6 in this volume); another paper (Woo & Hammer, 1977) outlines an approach that would allow certain views of cylindrical surfaces but does not include an implementation. None of this work appears to be based on formal geometric definitions and the concepts of algebraic topology. A closely related development path has been followed by workers in the fields of Computer Vision and Scene Analysis. This path has been based on vertex and edge configurations in a single view (Clowes, 1971; Huffman, 1971; Waltz, 1975) and has generally been restricted to objects with trihedral vertices and views with no chance alignments; this approach has led to the Origami World (Kanade, 1978) and a linear programming approach (Sugihara, 1981).

This paper presents a very general and complete approach based on the authors' previously published Wire Frame algorithm. In addressing the problem of constructing a solid object from a number of two dimensional views, it is shown that, on the one hand, complete labeling of edges and vertices leads to the previously published Wire Frame algorithm. On the other hand, the Projections algorithm described here is capable of working with no further information than the lines and points of the two dimensional projections and is able to enumerate all possible solutions to a given set of projections, with a cost commensurate with the number of solutions. The techniques presented are applicable when two or more projections are available. Of course, the one projection case has, in general, infinitely many solutions and is not discussed further in this paper. The chief advantage gained from providing more projections is quite naturally to reduce the number of possible ambiguities.

The Projections algorithm constructs polyhedral objects from projections containing only straight lines. The logical component of this algorithm is topological in nature and is, in principle, independent of whether the components are linear or nonlinear. While extension to objects with curved surfaces and projections with curved lines appears to be feasible, the ease of actually carrying out such an extension would depend greatly on the family of allowable curves and surfaces, as well as the projection conventions used.

The paper is organized as follows: Section 2 reviews the definitions of objects, faces, edges, and vertices used in the paper describing the Wire Frame algorithm (see Chapter 6 in this volume) and then develops the basic results dealing with back projections and labeled projections. Section 3 outlines the original Wire Frame algorithm and describes the Basic Projections algorithm which handles the general case of unlabeled projections of wire frames of objects. Section 4 presents some extensions to the Basic Projections algorithm which enable it to make use of more general forms of input data. For example, various types of views (overall, detail, and cross section) and depth information distinguishing between visible and occulted lines are considered. In Section 5, some examples are given to clarify this discussion. These examples illustrate the execution of the algorithm in both the stylized world of geometric puzzles with multiple solutions and the practical world of engineering drawings. The engineering objects successfully constructed from their projections are sufficiently complicated that a human unfamiliar with the solid object generally has some difficulty envisioning it. Thus, the algorithm appears capable of handling real world problems.

2. BASIC CONCEPTS AND RESULTS

The basic concepts defined in this section are based on some fundamental topological ideas which are described in detail in (Hocking & Young, 1961). Throughout the paper the standard topology in $\mathbb{R}^3$ and the induced topology on subsets of $\mathbb{R}^3$ are assumed. Vertices refer to points in $\mathbb{R}^3$ and edges refer to line segments defined by two points in $\mathbb{R}^3$. The approach used in this section is to define faces, objects, wire frames, and projections, and then describe the consequences of these definitions.

Definition 1

A *face*, f, is the closure of a nonempty, bounded, connected, coplanar, open (in the relative topology) subset of $\mathbb{R}^3$ whose boundary (denoted by ∂f) is the union of a finite number of line segments. P_f is used to denote the unique plane which contains f. □

Definition 2

An *object*, $\mathbb{O}$, is the closure of a nonempty, bounded, open subset of $\mathbb{R}^3$ whose boundary (denoted by $\partial\mathbb{O}$) is the union of a finite number of faces. □

From the definitions above it is easy to see that the "cube," $\{x, y, z \in \mathbb{R}^3 \mid 0 \le x \le 1, 0 \le y \le 1, 0 \le z \le 1\}$ is an object and that $\{(1, y, z) \in \mathbb{R}^3 \mid 0 \le y \le 1, 0 \le z \le 1\}$ is one of its "square" faces. Starting off with open sets means that faces and objects have nontrivial interiors. Notice that it is not assumed that an object is the closure of a connected set. This allows objects that consist of disjoint "solids" or even objects which intersect only in edges, etc. One can argue that this last case does not represent a "real" object, but in practice all sorts of strange objects can appear. Thus, we decided to handle the most general case possible. Furthermore, this generality does not exact any penalty other than creating a large number of solutions.

Another point worth noticing is that Definitions 1 and 2 allow many different representations of the boundaries of faces and objects by line segments and faces (respectively). However, there are canonical representations of the boundaries which correspond to one's intuitive notions about such things. To get to these representations it is necessary to introduce several additional concepts.

Definition 3

(a) Let f be a face. The *vertices* of f, V(f), are defined to be the set of all points for which two noncolinear line segments, contained in ∂f, can be found whose intersection is the given point.

(b) Let f be a face. The *edges* of f, E(f), are defined to be the set of all line segments e, contained in ∂f, satisfying the following conditions:

1. The endpoints of e belong to V(f);
2. No interior point of e belongs to V(f).

(c) Let $\mathbb{O}$ be an object. The *vertices* of $\mathbb{O}$, V($\mathbb{O}$), are defined to be the set of all points p for which faces $f_1, f_2, f_3 \subseteq \partial\mathbb{O}$ can be found such that $\{p\} = f_1 \cap f_2 \cap f_3 = P_{f1} \cap P_{f2} \cap P_{f3}$.

(d) Let $\mathbb{O}$ be an object. The *edges* of $\mathbb{O}$, E($\mathbb{O}$), are defined to be the set of all line segments e, contained in $\partial\mathbb{O}$, satisfying the following conditions:

1. The endpoints of e belong to $V(\mathcal{O})$;
2. No interior point of e belongs to $V(\mathcal{O})$;
3. For every point p of e, two noncoplanar faces can be found, f_1, $f_2 \subseteq \partial\mathcal{O}$ such that $p \in f_1 \cap f_2$.

(e) Let $\mathcal{O}$ be an object. The *wire frame* of $\mathcal{O}$, WF($\mathcal{O}$), is defined to be the ordered pair $(V(\mathcal{O}), E(\mathcal{O}))$. □

It can be shown that the edges of an object can intersect only at vertices of the object, i.e., at their endpoints.

The Wire Frame algorithm detailed in this volume allows one to construct all possible objects which have a given wire frame. It happens to be true, but not immediately obvious from the definitions, that $V(f)$, $E(f)$, $V(\mathcal{O})$, and $E(\mathcal{O})$ are all finite and well-defined. These facts and others are discussed in greater detail on pp. 232–237.

The Wire Frame algorithm runs on any collection of points and line segments in $\mathbb{R}^3$ and either returns all objects having the given collection as their wire frame or shows that the given collection could not be a valid wire frame. In presenting the Projections algorithm the first things to consider are the projections of the wire frame of a valid object. At this point it is necessary to make clear exactly what is meant by a projection.

Definition 4

Let $\mathcal{O}$ be an object, $P \subset \mathbb{R}^3$ a plane, and $\pi_p : \mathbb{R}^3 \to P$ the perpendicular projection. By the P-*projection* of $\mathcal{O}$, denoted by $\mathcal{O} \mid P$, is meant the ordered pair, $(V(\mathcal{O}|P), E(\mathcal{O}|P))$, of P-*vertices* and P-*edges* of $\mathcal{O}$ defined by the following process. Let E* be the set of images under π_p of all edges of $\mathcal{O}$ which are not perpendicular to P. Then the P-vertices of $\mathcal{O}$ are those points of P which lie on at least two noncolinear line segments in E*. The P-edges of $\mathcal{O}$ are those line segments of P which have elements of $V(\mathcal{O}|P)$ as endpoints, have no points of $V(\mathcal{O}|P)$ as interior points, and are subsets of unions of elements of E*.

XY, YZ, and ZX are used to denote the planes $Z = 0$, $X = 0$, and $Y = 0$, respectively. □

Figure 7.1 shows some of the things that can happen as a result of projection. The vertex A disappears in the front and top views. Furthermore, the edges AB, AC, AD, and AE do not appear as such in these views. Rather a single edge appears which is the union of the projections of the four aforementioned line segments. However, in the side view the vertex A projects into a vertex, and the projections of AB, AC, AD, and AE form distinct line segments.

At this point it seems appropriate to discuss the situations in which vertices of an object project into vertices in a given projection. Note that if a vertex of a polyhedral object is the intersection of at least three noncoplanar line segments,

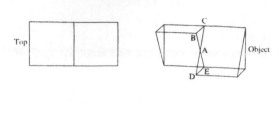

FIG. 7.1. Examples of projections.

the image of that vertex under any projection is the intersection of at least two noncolinear line segments and is thus a vertex in that projection. For convenience, vertices which are the intersections of at least three noncoplanar line segments are called *Class I vertices*. Thus, if two different projections of an object are given, the Class I vertices are a subset of the set of all intersections of all the perpendiculars erected at the vertices in each projection.

All vertices of an object which are not Class I are to be called *Class II vertices*. In Fig. 7.1, vertex A is Class II; all other vertices are Class I. In general one cannot expect to recover Class II vertices simply by erecting perpendiculars and computing their intersections.

There are a number of properties of Class I and Class II vertices which are useful in recovering an object from its projections. The key observation, which is formalized below, is that the wire frame of $\mathcal{O}$ can be recovered from the Class I vertices of $\mathcal{O}$ and certain line segments joining these vertices.

Definition 5

The *skeleton*, $S(\mathcal{O})$, of an object $\mathcal{O}$ is the ordered pair $(SV(\mathcal{O}), SE(\mathcal{O}))$ of *skeletal vertices* and *skeletal edges* where $SV(\mathcal{O})$ is the set of the Class I vertices of $\mathcal{O}$ and $SE(\mathcal{O})$ is a set of line segments joining the elements of $SV(\mathcal{O})$. For $v_1, v_2 \in SV(\mathcal{O})$, there exists $w \in SE(\mathcal{O})$, joining v_1, v_2 iff there exists an edge or colinear sequence of edges of $\mathcal{O}$ joining v_1 and v_2 and not containing any other Class I vertex. ∎

Theorem 6

Let $\mathcal{O}$ be an object. Then the wire frame of $\mathcal{O}$, $(V(\mathcal{O}), E(\mathcal{O}))$, can be recovered from the skeleton of $\mathcal{O}$, $(SV(\mathcal{O}), SE(\mathcal{O}))$, as follows. First, $V(\mathcal{O}) = V^*(\mathcal{O})$ where

$$V^*(\mathcal{O}) = SV(\mathcal{O}) \cup \{v | \{v\} = e_1 \cap e_2, e_1, e_2 \in SE(\mathcal{O})\}.$$

Thus, to get all vertices of $\mathcal{O}$ it is enough to add all intersection points of skeletal edges to the skeletal vertices. Second, $E(\mathcal{O})$ is simply the set of line segments which result from partitioning the skeletal edges using their points of intersection.

Proof: Observe that from Definition 5 it follows that every skeletal edge is the union of edges of $\mathcal{O}$. Thus, the intersection of two skeletal edges is a point of intersection of two edges. However, edges of $\mathcal{O}$ intersect only in vertices of $\mathcal{O}$. Thus, $V^*(\mathcal{O}) \subseteq V(\mathcal{O})$.

It remains to show that $V(\mathcal{O}) \subseteq V^*(\mathcal{O})$. In particular it must only be demonstrated that every Class II vertex of $\mathcal{O}$ belongs to $V^*(\mathcal{O})$. To see this it is necessary to consider briefly the nature of the edges of $\mathcal{O}$. Let $e \in E(\mathcal{O})$, $p \in e$, and l the infinite line through p containing e. Let X be the set of disjoint line segments formed by the intersection of l and the boundary of $\mathcal{O}$. Now either p is in the interior of X (i.e., there are points of X on both sides of p which are arbitrarily close to p) or p has arbitrarily close neighbors only to one side of it. Let f_1 and f_2 be the two noncoplanar faces whose intersection contains e. If p is not in the interior of X, then, since p is on the boundaries of f_1 and f_2 but is not in the interior of any of the edges, it must be a vertex of each of the faces, i.e., there must be edges, $e_1 \in f_1$, $e_2 \in f_2$, not colinear with e such that $\{p\} = e \cap e_1 = e \cap e_2$. But in this case there are three noncoplanar edges through p, namely, e, e_1, and e_2. Thus, p is a Class I vertex.

The point of the preceding paragraph is to show that either a point, p, of an edge, e, is a Class I vertex or the line through p containing e has boundary points of $\mathcal{O}$ arbitrarily close to p, i.e., there exists a line segment $s \supseteq e$ contained in $\partial\mathcal{O}$ for which p is an interior point. In particular, an edge, e, containing a Class II vertex p can be extended to a line segment s lying in $\partial\mathcal{O}$ containing e whose endpoints are Class I vertices, i.e., every edge of $\mathcal{O}$ is contained in some skeletal edge. Since every vertex of $\mathcal{O}$ must lie on at least three edges, every Class II vertex of $\mathcal{O}$ must lie on at least two skeletal edges and hence $V(\mathcal{O}) = V^*(\mathcal{O})$.

Since every edge of $\mathcal{O}$ lies in some skeletal edge and $V(\mathcal{O}) = V^*(\mathcal{O})$, it follows that the edges of $\mathcal{O}$ are exactly the pieces into which the skeletal edges are partitioned by the vertices of $\mathcal{O}$. ∎

Theorem 6 gives some insight into the working of the Projections algorithm. Back projection yields a *pseudo skeleton* consisting of a set of vertices which includes the Class I vertices and a set of edges. This pseudo skeleton is processed to produce a *pseudo wire frame*. In general, the pseudo skeleton and pseudo wire frame contain vertices and edges not in the skeleton and wire frame of the original object. However, they do contain all the vertices and a partition of the edges of the skeleton and wire frame of the original object. In fact, the additional complexity of the Projections algorithm is based on the fact that back projection generally yields many vertices and edges not in the original object. The Projections algorithm thus proceeds along the lines laid down by the Wire Frame algorithm, but with suitable modifications made to deal with surplus information.

The discussion of Class II vertices in the proof of Theorem 6 shows that they have various properties, one of which appears as Theorem 7. Theorem 7 is very useful in showing that certain points which arise from back projection cannot be vertices of $\mathcal{O}$. Example 4 later in the paper illustrates the power of this observation.

Theorem 7

Let $\mathcal{O}$ be an object and v a Class II vertex of $\mathcal{O}$. Any plane, P, through v separates $\mathbb{R}^3$ into two components each of which contains interior points of $\mathcal{O}$ which are arbitrarily close to v.

Proof: From the proof of Theorem 6 it follows that v is the intersection of two noncolinear line segments, e_1 and e_2, which are unions of line segments of $\mathcal{O}$, are contained in the boundary of $\mathcal{O}$, and contain v as an interior point. Any plane through v not containing e_1 or e_2 is clearly going to contain interior points of $\mathcal{O}$ near v, and in this case this theorem is true. Also, there is only one plane, P, containing e_1 and e_2. If all of $\mathcal{O}$ were to one side of P, there would be a contradiction, since at least four noncoplanar faces would go through v, but all the edges containing v would be coplanar. ∎

The remainder of this section shows how much simpler things are when items are labeled or when special projections are used. The discussion of the unlabeled case is resumed in Section 3.

In mechanical drawing practice, one generally starts with $\mathcal{O}|XY$, $\mathcal{O}|YZ$, and $\mathcal{O}|ZX$, although it is always possible to use other planes. In fact, as will now be shown, for each object $\mathcal{O}$ it is always possible to find a plane P such that π_p distinguishes all the elements of $WF(\mathcal{O})$.

Proposition 8

Let $\mathcal{O}$ be an object. Then there exists a plane P containing the origin for which π_p projects each element of $V(\mathcal{O})$ into a distinct vertex of $\mathcal{O}|P$, elements of $E(\mathcal{O})$ project into distinct line segments which can intersect in at most one point, and no point in $V(\mathcal{O})$ projects into a projection of an element of $E(\mathcal{O})$ unless it is a member of it.

Proof: The set of all planes in $\mathbb{R}^3$ containing the origin can be identified with the unit sphere, S^2, in $\mathbb{R}^3$, where each unit vector corresponds to the plane for which it is a unit normal. Clearly, in this manner exactly two points of S^2 correspond to each plane through the origin. In order for a projection π_p to map each vertex of $\mathcal{O}$ to a distinct member of $V(\mathcal{O}|P)$, P cannot be perpendicular to any line which goes through at least two of the points of $V(\mathcal{O})$ and cannot be perpendicular to any plane containing all edges incident with a Class II vertex. Each of these restrictions rules out exactly one plane, i.e., two points on S^2. Thus, in order to get an injection on $V(\mathcal{O})$, at most

$$2\left[\left(\begin{array}{c}|V(\mathcal{O})|\\2\end{array}\right)+|V(\mathcal{O})|\right]$$

points on S^2 must be avoided.

Two elements of $E(\mathcal{O})$ can have an intersection of more than one point in some projection if and only if they are coplanar. Furthermore, they can project with a nontrivial overlap only into planes which are perpendicular to the plane containing both of the elements of $E(\mathcal{O})$. The set of all planes through the origin perpendicular to a given plane corresponds to a great circle of S^2. Thus, to get the desired behavior at most

$$\left(\begin{array}{c}|E(\mathcal{O})|\\2\end{array}\right)$$

great circles on S^2 must be avoided.

To keep a point from projecting onto a line segment not containing it there are two cases to consider. First, the point and line segment might be colinear. In this case, one must avoid the plane perpendicular to the given line. Again this means avoiding two points. Thus, at most $2|V(\mathcal{O})|\,|E(\mathcal{O})|$ points must be avoided. Second, the point and line segment are not colinear. In this case it is enough to avoid all planes perpendicular to a given plane as before. Thus, at most $|V(\mathcal{O})|\,|E(\mathcal{O})|$ great circles on S^2 must be avoided.

Since points and great circles are nowhere dense in S^2 and the number of sets which must be avoided is finite, it follows from the Baire Category Theorem (see Hocking & Young, 1961) that there must be points of S^2 which do not lie in any of the forbidden sets. Using any such point yields a plane with the desired properties. ■

Definition 9

Let $\mathcal{O}$ be an object, P a plane in 3-space and π_p the projection of 3-space onto P. Projection π_p is said to be a *distinguishing projection* for $\mathcal{O}$ if it has all the properties of Proposition 8. ■

Note that the proof of Proposition 8 shows that for a given object "most" projections are distinguishing projections since the nondistinguishing ones have a two dimensional measure of 0. The probability of picking a nondistinguishing projection at random is thus zero in an ideal model. However, in most practical situations there are only a finite number of choices for coordinates, and there is a nonzero probability of picking a nondistinguishing projection. Many objects of engineering interest have planar features aligned with the "natural" axes of the object, and the set of three standard views contains a maximum degree of concealment and self alignment.

At this point it is worthwhile to consider two cases. In the first case, the image of each vertex in each projection carries the labels of all the vertices of $\mathcal{O}$ that project into it, i.e., the P-projections are labeled. In the second case, there are no labels on the vertices of the P-projection.

In the first case there is, quite naturally, significantly less ambiguity than in the second case. The following theorem shows exactly how much information can be recovered from labeled P-projections.

Theorem 10

Let P_1 and P_2 be two nonparallel planes in $\mathbb{R}^3$, and let $\mathcal{O}$ be an object. Assume that the P_1 and P_2 projections of $\mathcal{O}$ are labeled. Then there is a unique set of points in $\mathbb{R}^3$ which can be V($\mathcal{O}$). Furthermore, if either of the projections is distinguishing or if all the edges in at least one P-projection are labeled with the pairs of vertices they connect, then WF($\mathcal{O}$) can be reconstructed uniquely. In this case, reconstructing objects from projections reduces to the problem of reconstructing objects from wire frames.

Proof: If P_1-vertices and P_2-vertices are labeled, to reconstruct a point x $\in$ SV($\mathcal{O}$), the images of x under the two projections are found and perpendiculars erected at those points. Since P_1 and P_2 are not parallel, these perpendiculars can meet in at most one point. Since they both go through x, x can be recovered as their unique intersection point. In this way SV($\mathcal{O}$) can be reconstructed uniquely, which, by Theorem 6, means that V($\mathcal{O}$) can also be reconstructed uniquely.

Clearly, if the edges of at least one P-projection are labeled as described above, E($\mathcal{O}$) can be uniquely reconstructed. If one of the projections is distinguishing, E($\mathcal{O}$) can be reconstructed by joining together two points of V($\mathcal{O}$) if and only if they are joined together in the distinguishing projection (or in both projections). ∎

Thus, given a fairly small amount of information on projections, one can quickly and easily reconstruct a unique wire frame. In many practical situations, where the emphasis is on getting things done and not on creating puzzles, it seems quite likely that there will be ample information for constructing the correct wire frame easily. Unfortunately, there will also be many situations with inadequate information. The techniques developed for handling the unlabeled case are of great importance in such situations.

To complete the development of the labeled case, the situation in which there are no distinguishing projections must be discussed. Since this problem is a subset of the unlabeled case, the unlabeled case is considered next.

In the unlabeled case, there can be a number of distinguishing projections and it may not be possible to recover a wire frame uniquely. The following example illustrates this in the case of three distinguishing projections.

Example 11

Let $\mathcal{O}_1$ be the tetrahedron with vertices $\{(1, 1, 1), (1, 2, 2), (2, 1, 2), (2, 2, 1)\}$ and $\mathcal{O}_2$ the tetrahedron with vertices $\{(1, 1, 2), (1, 2, 1), (2, 1, 1), (2, 2, 2)\}$. The projections of O_1 and O_2 into the XY, XZ, and YZ planes are all distinguishing and are identical in each plane, but do not allow construction of a unique wire frame. Actually, the projections into the various planes are all essentially the same, i.e., by ignoring the coordinate which is fixed at 0 in each case, one gets the points $\{(1, 1), (1, 2), (2, 1), (2, 2)\}$ and the six possible lines between them, i.e., each projection looks like a square with both of its diagonals drawn in. In Section 5, the problem of reconstructing all objects for which all three standard projections look like a square with its diagonals is discussed in more detail. As shall be seen, there are surprisingly many solutions to this problem. ■

The above discussion shows that labeling projections can be very useful in reducing the difficulty of reconstructing objects from projections. The truth of the preceding sentence becomes even more apparent after the discussion of the algorithm for reconstructing objects from unlabeled projections in Section 3 and the discussion of the examples in Section 5.

3.   FLESHING OUT UNLABELED PROJECTIONS

In order to aid in the comprehension of this rather complex algorithm, a basic form of the algorithm, which accepts only limited data, is presented here (Section 3). The basic algorithm constructs all polyhedral solid objects whose wire frames have a given set of projections (or *views*). The extension of the algorithm to a more general set of projections forms (i.e., overall, detail, and cross section), and to the use of depth information to distinguish between visible and occulted edges, is deferred until Section 4. Since the Projections algorithm is an extension to the Wire Frame algorithm, the basic concepts of the Wire Frame algorithm and its terminology are reviewed first.

In the Wire Frame algorithm the input data [a wire frame, Fig. 7.2(a)] are processed to find all graphs containing more than two noncolinear edges. For each such graph, minimum enclosed areas are found and nested in a tree hierarchy. From this hierarchy candidate faces with an exterior boundary and possibly interior boundaries (i.e., a face may have holes) are constructed—these are called *virtual faces* [Fig. 7.2(b)]. For each edge, a list of virtual faces is formed and ordered radially around the edge. Minimum enclosed volumes are found and nested, again in a tree hierarchy. From this hierarchy, candidate volume regions called *virtual blocks* are found [Fig. 7.2(c)]. A final decision process assigns state solid or hole to each virtual block [Fig. 7.2(d)], glues the solid blocks together, and finds all possible solid objects with the input wire frame. Note that one virtual block is always an infinite envelope block (i.e., it is inside out) and is always a hole.

The ability to handle all possible cases is embedded in the parts of the algorithm for finding enclosed regions (for example, bridges are ignored), for the handling of illegal intersections between virtual faces (Type I and Type II inter-

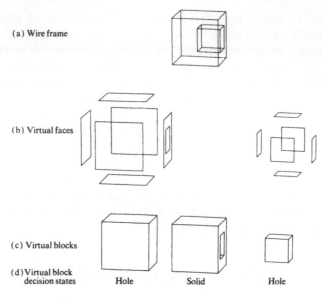

(a) Wire frame

(b) Virtual faces

(c) Virtual blocks

(d) Virtual block
     decision states   Hole   Solid   Hole

FIG. 7.2.   The Wire Frame algorithm in action.

sections, see below), and in the final decision process. The correctness of objects is derived from the use of directed edges and faces and from rules governing the number of times and directions with which edges and faces are used.

The several stages of the Projections algorithm are now described. Since many of these stages are quite similar to the corresponding stages of the Wire Frame algorithm, details are given about only those points which are different. The presentation is given in two parts: first, a brief outline of the stages, and second, a more detailed description of each stage.

The early stages (1, 2, and 3) of the Projections algorithm are concerned with converting, by means of a back projection process, a set of projections of an object to a pseudo skeleton and thence to a pseudo wire frame for the object. This pseudo wire frame contains supersets of the vertices of all objects with the given projections. Furthermore, the edges of this pseudo wire frame partition the edges of all objects with the given projections. The existence of various edges and vertices in objects may be known for certain or may be uncertain. All components of the pseudo wire frame are consistent with all the views.

The later stages (i.e., 4–7) apply an extended form of the Wire Frame algorithm to a pseudo wire frame to find all polyhedral solid objects with the given projections.

• Outline of the Basic Projections Algorithm

1. Check Input Data.   The input data to the basic algorithm are assumed to be a set of at least two distinct parallel projections of the wire frame of a polyhedral object. Extensions to handle more general forms of input data are presented in Section 4. The data are checked for validity and reduced to canonical form with edges and vertices distinct and with edges intersecting only in vertices.

2. Construct Pseudo Vertex Skeleton. The vertices in each view are back projected to find all Class I vertices (i.e., vertices formed by the intersection of noncoplanar edges) and some Class II vertices (i.e., vertices formed by the intersection of only coplanar edges); at this point it is not possible to distinguish between vertex classes. The vertices discovered here, and the remainder of any Class II vertices missed in this stage and found in Stage 3, are called *candidate vertices*. While not all vertices of $\mathbb{O}$ may be recovered at this stage, enough are recovered to enable the recovery of all vertices after passing through the next stage. Note also that candidate vertices may not be vertices or even points of $\mathbb{O}$.

3. Construct Pseudo Wire Frame. The vertices constructed in Stage 2 form a skeleton for the pseudo wire frame in the same sense that WF($\mathbb{O}$) derives from S($\mathbb{O}$). Edges are introduced based on the edges in the projections. These edges are checked for mutual internal intersections. Intersections are introduced as additional vertices and used to partition the edges. The remaining Class II vertices are constructed in this manner. The vertices constructed here and in Stage 2 are the set of candidate vertices (denoted CV($\mathbb{O}$)), and the final set of edges constructed in this stage is the set of candidate edges (denoted CE($\mathbb{O}$)). Together the candidate edges and vertices form the *pseudo wire frame*. The candidate vertices are a superset of V($\mathbb{O}$), and the candidate edges partition the elements of E($\mathbb{O}$). The edge connectivity of all vertices is examined and the candidate edge and vertex lists edited. The editing process may remove impossible items, simplify colinear edges, and update the classification of vertices as Class I or II. Candidate edges and vertices which are the only possible candidates for some edges and vertices appearing in one of the projections are labeled as certain and must appear in a solution object; all others are labeled uncertain and may or may not appear in solution objects. For both candidate edges and vertices, cross reference lists are maintained between view edges and vertices and pseudo wire frame edges and vertices and *vice versa*.

4. Construct Virtual Faces. Beginning with the pseudo wire frame generated in Stage 3, all virtual faces are found in a manner analogous to that used in the Wire Frame algorithm. All uncertain edges are checked for containment in at least two noncoplanar virtual faces. Any edges not meeting this criterion are deleted and the virtual faces updated. Any impossible virtual faces (e.g., a certain edge piercing the interior of a virtual face) are deleted. The consequences of deletions are propagated until a stable condition is reached.

5. Introduce Cutting Edges. Illegal intersections between two virtual faces such that both faces cannot exist in an object are handled by the introduction of a temporary *cutting edge* along their line of intersection. The cutting edge partitions the virtual face into smaller independent virtual faces and will be removed in the final stages. All the partitioning processes in the algorithm, be they of edges or faces, generate lists of siblings with common parent edge or face, and also lists of correlations between edges or faces which cannot co-exist in an object; these data structures are used in the final stages of the algorithm.

6. *Construct Virtual Blocks.*   Virtual faces are pieced together to form *virtual blocks* in exactly the same manner as in the Wire Frame algorithm.

7. *Make Decisions.*   A depth first decision process is used to assign solid or hole state to the virtual blocks and to find all objects with the given projections. The process ensures that all cutting edges disappear in solution objects (i.e., that they are either totally surrounded by space or by material or they separate coplanar surfaces). Efficiency in the search process is obtained by careful pruning of the decision tree, for example, by recognizing that decisions involving partitioned edges and virtual faces may be propagated to the whole original edge or virtual face.

• Detailed Description of the Basic Projections Algorithm

To make the description of the algorithm more comprehensible, the example based on Fig. 7.1 is used to illustrate the various stages, i.e., the problem is to recover the object in Fig. 7.1 from its three views. For brevity, this problem is referred to as the Two Wedges problem.

1. *Check Input Data.*   The input data to the basic algorithm are assumed to be a set of two dimensional views of the whole wire frame of a polyhedral object. The views may be at arbitrary projection directions, but must meet a minimum requirement of at least two distinct projections. Each view is an ordered pair of vertices and edges (Definition 4) expressed relative to a local two dimensional coordinate frame and accompanied by a transformation matrix between the coordinate frame of the three dimensional object and the two dimensional view.

In this and later stages, tests are performed on the data input to a stage of the algorithm, for detection of inconsistencies in the data, for reduction of the data to canonical form for the stage, and to obtain information to be used in later stages. The exact choice of which tests to include depends on the characteristics of the input data and performance trade-offs between the cost of performing a test first, the usefulness of information generated for later stages, and the desirability of reporting errors before incurring the cost of executing the algorithm. These issues are not considered further here. However, it will be seen that the combinatorial problems of the projections algorithm may be very severe, and there is therefore a need to minimize the quantity of surplus information generated in the early stages of the algorithm.

2. Construct Pseudo Vertex Skeleton.   As stated earlier, in this stage perpendiculars are erected at each vertex of each view. Then, only those vertices lying on at least two noncolinear perpendiculars and which are consistent with all other projections, i.e., their images are either vertices or interior points of edges, are selected. As noted after Definition 4, all Class I vertices and possibly some Class II vertices are recovered. In order for the projections to be consistent, it is necessary that every **P**-vertex have at least one element of $CV(\mathbb{O})$ in its inverse image. This check may be performed as part of this stage. In addition, if some **P**-vertex has a unique element of $CV(\mathbb{O})$ in its inverse image, then that element of $CV(\mathbb{O})$ must actually be an element of $V(\mathbb{O})$. Such a vertex is assigned type certain, and all other vertices are assigned type uncertain.

Each intersection is tested to see if it coincides with a previously found vertex and, if not, is introduced as a new vertex. Each vertex found is accompanied by a list of cross references to the view-vertex pairs from which it has been generated. Conversely, for each view vertex, a list is formed of the wire frame vertices into which it projects.

The pseudo vertex skeleton of the Two Wedges problem consists of 12 points: the 8 points corresponding to the vertices of a cuboid and 4 points corresponding to the mid-points of the 4 horizontal edges [see Fig. 7.3(a)].

3. Construct Pseudo Wire Frame.   In this stage all pseudo skeletal edges are constructed as a prelude to constructing the pseudo wire frame. To do this, simply join two vertices in the pseudo vertex skeleton by an edge iff in every projection the images of these two vertices coincide or are joined by an edge or colinear set of edges and no other vertex of the pseudo vertex skeleton would be an interior point of the edge.

In general, these pseudo skeletal edges may intersect in mutually interior points. To obtain the pseudo wire frame from this skeleton it is only necessary to duplicate the techniques of Theorem 6, i.e., to introduce edges in the obvious way so that all edges have vertices as endpoints, that two edges intersect only in a vertex, and that no vertex be an interior point of an edge.

Note that the proof of Theorem 6 shows that $V(\mathbb{O}) \subseteq CV(\mathbb{O})$ and that every edge of $\mathbb{O}$ can be written as the union of candidate edges.

Many of the checks of Stage 2 are used on the vertices produced in this stage. With modification these checks are used on candidate edges. Thus, it should be verified that every **P**-edge has some element of $CE(\mathbb{O})$ in its inverse image. In particular, if some **P**-edge has a unique inverse image, then that element of $CE(\mathbb{O})$ must be real, i.e., it must actually be an element of $E(\mathbb{O})$ and, like the rule for vertices above, is classified as type certain. At the end of this stage pruning operations are performed. All vertices with edge connectivity of degree ≤ 1 are

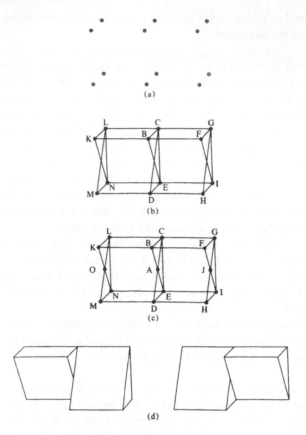

FIG. 7.3.   (a) The vertex pseudo skeleton of the Two Wedges problem. Edge recovery in the Two Wedges problem: (b) the pseudo skeleton and (c) the pseudo wire frame. (d) The two solutions to the Two Wedges problem.

removed, together with any incident edges. If the vertex has degree 2, the incident edges are checked for colinearity. If they are colinear, the vertex is removed and the two edges are merged into a single edge. If they are not colinear, they are removed together with the vertex. If a vertex of degree ≥ 3 has only coplanar edges, then any edges not having a colinear extension, and possibly also the vertex, are removed. Whenever edges are removed, the effects of the change are propagated until a stable configuration is achieved. In a similar manner to the vertices, cross reference lists are maintained from psuedo wire frame edges to view-edge pairs, and conversely, for each view edge, a cross reference list to the pseudo wire frame edges is formed.

Figures 7.3(b) and (c) show the results obtained during this stage in the case of the Two Wedges problem. Note that vertex A of the original figure appears in the pseudo wire frame exhibited in Fig. 7.3(c) but does not appear in the skeleton

[Fig. 7.3(b)]. Note also that by Theorem 7 vertices J and O are clearly spurious since all solid material lies to one side of the planes KLN and FGI. However, these conditions cannot be derived until a later stage of the algorithm.

The stages described above are fairly straightforward. Before describing the later stages of the Projections algorithm it will be helpful to understand exactly what has been produced so far. The pseudo wire frame (CVⓄ, CE(Ⓞ)) looks like a wire frame. Indeed, in many cases (CV(Ⓞ), CE(Ⓞ)) is exactly the wire frame of Ⓞ and feeding (CV(Ⓞ), CE(Ⓞ)) to the Wire Frame algorithm will yield the correct solutions directly. The important thing is to understand the way in which simply applying the Wire Frame algorithm to (CV(Ⓞ), CE(Ⓞ)) can fail to find all solutions. The chief problem is that the original Wire Frame algorithm treats vertices and edges as real entities, whereas the pseudo wire frame contains uncertain edges and vertices, any of which may or may not exist in a solution. Any solid object having a *subset* of (CV(Ⓞ), CE(Ⓞ)) as its wire frame and producing the correct projections is a solution of the projections problem. Thus, the Wire Frame algorithm approach may fail to find all solutions of the projections problem (it may in fact fail to find any). The assumption of reality of edges and vertices is crucial to two places in the Wire Frame algorithm:

- Dealing with illegal intersections between virtual faces, and
- Making decisions.

Whenever an edge pierces a virtual face (a Type I intersection) in a legitimate wire frame problem, it is safe to drop the virtual face since it is known that the edge is "real" and that "real" edges cannot pierce faces which separate solid material from space (these are the only important faces). In the present situation, it might very well be that the edge is not real and should itself be dropped instead. Of course, if it is known that a particular edge is real (i.e., certain), the algorithm can proceed as before.

In the Wire Frame algorithm the decision process was concerned with finding those combinations of virtual blocks which made every edge (except the cutting edges) an edge of a real object. In the case of the Projections algorithm it is necessary only to find combinations of virtual blocks with projections agreeing with the given projections. In general, this means that not every uncertain element of (CV(Ⓞ), CE(Ⓞ)) is actually a member of (V(Ⓞ), E(Ⓞ)). Thus, the decision procedure must be modified to check that every edge in each projection comes from a candidate edge which becomes a real edge in the corresponding solution.

Cutting edges were introduced to handle illegal intersections between virtual faces when no internal point of an edge from one face was contained in the interior of another face, but there were points common to the interior of both faces (a Type II intersection) (see Chapter 6, pp. 244–245). This situation was interpreted as one where the two faces could not co-exist in the solution, and temporary edges—cutting edges—were introduced along the line of intersection of the two faces. The cutting edges partitioned the faces into nonintersecting sub-faces, which could be used to build more, smaller, virtual blocks. The decision process ensured that cutting edges did not remain in the final solutions. Although introduced originally for Type II intersections, cutting edges are applicable also to Type I intersections, and are particularly relevant to the case of uncertain edges.

4. *Construct Virtual Faces.* This stage is essentially identical with Stage 4 of the Wire Frame algorithm. As noted earlier, each candidate edge is checked to see whether it lies in at least two noncoplanar virtual faces. Thus, in the Two Wedges problem, 19 virtual faces [KLO, LON, MON, ABC, ACE, ADE, FGJ, GIJ, HIJ, KLCB, NLCE, MNED, BCGF, ECGI, DEIH, MOLCAD, KONEAB, DACGJH, BAEIJG in Fig. 7.3(c)] are discovered.

5. *Introduce Cutting Edges.* This stage is very similar to its equivalent in the wire frame algorithm but has a minor modification to allow for uncertain edges. If an interior point of a certain edge is contained in the interior of a virtual face, then the virtual face cannot be a face of the object and is deleted. All other illegal intersections between virtual faces, i.e., both faces cannot exist in the object, are handled by the introduction of temporary cutting edges. Cutting edges separate virtual faces into independent regions so far as the illegal intersection was concerned and are removed in the final decision process in Stage 7. When a virtual face is partitioned into subfaces, mapping tables and correlation lists are generated in a manner similar to that described for partitioned edges.

Note that if records are kept in the correct manner all reprocessing of virtual faces is done with reference to a particular virtual face, rather than starting with a general wire frame problem. Furthermore, if, when reprocessing a virtual face, f, to determine the smaller virtual faces into which it is partitioned by the cutting edges, a cutting edge, e, is found which is not on the boundary of one of the smaller virtual faces, then it can be dropped together with any virtual face, g, whose intersection with f is e. Face g can be dropped since it is impossible for g to be a member of a virtual block. As usual, dropping a virtual face will in general have other repercussions which are exploited until a stable situation results. For brevity, virtual faces found in Stage 4 will be called *original virtual faces*. Those arising because of cutting edges will be called *new virtual faces*.

In the Two Wedges problem, two cutting edges [OA and AJ in Fig. 7.3(c)] are introduced. These two edges partition four virtual faces (CADHJG, BAEIJG, MOLCAD, KONEAB) into eight virtual faces (CAJG, ADHJ, BAJF, AEIJ, KOAB, ONEA, MOAD, OLCA).

6. *Construct Virtual Blocks.* This stage is identical with the corresponding stage in the Wire Frame algorithm. In the Two Wedges problem, six finite virtual blocks are uncovered:

B_1:(MONEAD),
B_2:(NOLCAE),
B_3:(LOKBAC),
B_4:(DAEIJH),
B_5:(EACGJI),
B_6:(BACGJF),

where the description of virtual blocks is in terms of the labeling of Fig. 7.3(c). The seventh virtual block, B_0, is the unique infinite empty block.

7. Make Decisions.   The set of virtual blocks is fed to a decision procedure, which is an extension of the decision procedure used in the Wire Frame algorithm. The differences between the two procedures revolve around the fact that the Projections algorithm is aware that not every vertex and edge must be real.

The chief difference consists of the fact that whenever the nature of a new virtual face is determined (i.e., whether or not it separates solid material and space), the same determination can be made for all other new virtual faces which are subdivisions of the same original virtual face. Furthermore, as soon as it is determined (or assumed) that an original virtual face, f, does separate solid material and space, all original virtual faces sharing a cutting edge with f are forced to be spurious. This means that any pair of virtual blocks using any part of any virtual face "cutting" f as a common boundary must both be assigned the same state. Similarly, if a virtual face is known to be spurious, all virtual blocks using any part of it as a boundary must have the same state.

These facts speed up the decision procedure considerably and offset the greater number of virtual blocks that have been introduced. Similar arguments apply to entire edges which have been partitioned in Stage 3. In the final solution, no cutting edge can be a real edge. Of course, all decisions respect the fact that the final outcome must be consistent with the original projections.

In this stage virtual blocks are fitted together to generate all objects with the given projections. Basically, each virtual block may have solid or hole state and, when a state assignment has been made to each virtual block, an object is obtained. However, not all assignments of solid and hole yield the desired projections. An assignment of solid or hole to the virtual blocks yields an object with the correct wire frame if, and only if

1. *Every* certain edge element $e \in E(\mathbb{O})$ belongs to two noncoplanar virtual faces f_1 and f_2 each of which belongs to one virtual block assigned solid state and one assigned hole state;

2. *No* cutting edge belongs to two noncoplanar virtual faces f_1 and f_2 each of which belongs to one virtual block assigned solid state and one assigned hole state.

3. *Every* uncertain edge element $e \in E(\mathbb{O})$ may be assigned either to state certain and obeys the rule for certain edges (1) above or to state not-visible and

obeys the rule for cutting edges (2) above, in a manner consistent with the input projections.

The decision process is performed by assigning states in a virtual block state vector, whose elements are ordered *a priori*. The first element of the state vector is the unique infinite virtual block, which is assigned the empty state. For each edge, a list is formed of the faces containing the edge and the blocks they bound; this list is sorted around the edge and allows the angular sequence of block state transitions to be discovered.

The decision process proceeds as a depth first search in the virtual block decision space tree. At any node in the tree, the current state vector is checked for consistency and consequential states are assigned. Thus, although the state vector may have dimension of many hundreds, the consistency check may be expected to prune large sections of the tree, while the propagation of consequential states may be expected to reduce substantially the number of decisions to be made.

The checks for consistency are essentially those listed above. The consequential state assignments are performed to meet the following criteria:

• A certain edge with all except one containing block assigned the same state forces the remaining block to be assigned the opposite state.

• An uncertain edge totally surrounded by either all material or by all space becomes nonvisible; an uncertain edge contained in blocks producing exactly two coplanar state transitions around the edge becomes nonvisible; an uncertain edge contained by blocks of both hole and solid states and with at least two noncoplanar state transitions around the edge becomes certain.

• An uncertain edge that is the only edge remaining to create a view edge becomes certain.

• A cutting edge whose surrounding blocks have the same state, i.e., both solid or both hole, spanning regions 180 degrees apart, allows the same state to be assigned to *all* blocks around the edge.

• A cutting edge whose surrounding blocks have the same state < 180 degrees apart around the edge allows any intermediate blocks to be assigned to the same state.

• A new virtual face which is a real face, i.e., it separates blocks of different states, and which is a subdivision of an original virtual face formed by cutting edges allows the same solid-hole relationship to be given to all blocks containing sibling faces from the original virtual face. Similar rules apply when the face is not real.

• An uncertain edge which becomes a certain edge and which is a subdivision of an original wire frame edge allows its sibling edges to be upgraded to certain state.

In some cases, particularly those where there are high degrees of symmetry and a limited number of views, giving rise to many highly correlated uncertain edges, there may be a very large number of objects producing the given projections. Thus, although the depth first search and also heuristic search approaches to this problem (Preiss, 1980) allow a solution to be found efficiently, an exhaustive search must ultimately be used, and efficient pruning of the decision tree is very important. It is evident that, in the case of problems with multiple solutions, the provision of rather small amounts of extra information by the user, for example, labeling of some uncertain edges, and assigning states to points in 3-space, can resolve the ambiguities completely. Thus, in a practical system, the user may be requested to assist with extra information when requested. The basis for the system requesting extra information in the early stages of the algorithm is the preponderance of uncertain edges, discovered in Stage 3, and self intersection of uncertain edges, discovered in Stages 4 and 5.

At this point it can be appreciated that the use of cutting edges has allowed construction of a set of virtual blocks having the property that every solution of the projection problem can be built out of the virtual blocks in this set.

Stage 7 feeds into an output module which puts the output together in forms which can be understood by the user of the system. In our implementation of the algorithm, the output is in the form of a polyhedron for the Geometric Design Processor system (Wesley et al., 1980).

In the case of the Two Wedges problem, this stage produces the two solutions shown in Fig 7.3(d). The decision procedure works as follows in this case. Suppose that the search in this case deals with the virtual blocks B_0, ..., B_6 in that order. B_0 is known to be empty. Thus, the first branch of the decision tree corresponds to determining the state of B_1.

If B_1 is assumed to be solid, MOAD is seen to separate solid from space. This means that the entire virtual face MOLCAD must separate solid from space. In particular, B_2 must be solid and B_3 empty. Thus, the next step is to decide whether B_4 is solid or empty. Assuming that B_4 is solid forces B_5 to be solid and B_6 to be empty. However, the object resulting from making B_1, B_2, B_4, B_5 solid and B_0, B_3, B_6 empty clearly fails to have the right projections. Thus, the decision procedure backs up to the B_4 decisions and assigns hole to B_4. This means that the new virtual face DAJH is spurious and that the original virtual face DACGJH is spurious. Thus, B_5 and B_6 must have the same state. If they are both assumed to be empty, the object that results is just a simple wedge, which clearly has the wrong projections. Thus, B_5 and B_6 must both be assumed to be solid. The object that results is a left-right transform of the original object in Fig. 7.3 and clearly has the correct projections.

On the other hand, if B_1 is assigned hole, B_2 and B_3 must both be assigned the same state. Clearly, if B_2 and B_3 are also empty, it is impossible to obtain the correct front and top views. Thus, B_2 and B_3 must be solid in this case. Furthermore, assuming B_4 to be solid forces B_5 to be solid and B_6 to be empty. This yields the original object. Assuming B_4 to be empty forces B_5 and B_6 to have the same state. The objects that result are both wedges of differing width and are clearly not solutions.

It is clear that keeping track of the number of objects remaining in the inverse image of a projected artifact can be helpful in the decision procedure, i.e., if assigning a particular state to a given virtual block removes the last vertex or edge in the back projection of some vertex or edge, then that assignment can be rejected and its consequences need not be explored further. ■

The following section describes ways in which additional information can be extracted from various drawing conventions. The final section contains examples which should clarify the discussion in this and the next section.

4. ADDITIONAL INFORMATION FROM DRAWING CONVENTIONS

Designers and draftsmen use a number of conventions and aids to clarify and help reduce ambiguity in engineering drawings. Extensions to the Basic Projections algorithm are presented in this section. These extensions cover two concepts: the generalization of the set of types of views to include overall, detail, and cross sectional, and the use of depth and detail information expressed by line types. The presentation is made within the context of the various stages of the algorithm presented previously.

• Stages of the Algorithm Reconsidered

1. Check Input Data.   In extending the basic algorithm to handle several different types of view (i.e., overall, detail, and cross sectional), the central problem is to be able to relate information from the different types of views. This is achieved here by classification of the edges of the object into two types: gross and detail. The gross edges describe the main structure of the object; the detail edges add more information in regions where there is fine structure in the object.

The edges of the views are labeled with edge types according to an agreed drawing standard. For example, visible edges are generally drawn with line type *solid* and occulted edges with type *dashed*, which provides depth information. Another possibility, namely the omission of occulted edges, is not permitted; the Projections algorithm is based on geometric concepts and the premise that all edges are shown in all projections. An algorithm that attempts to fill in missing information would have to be based on heuristic ideas of what a most likely object would be as well as on the concepts of geometry.

An overall view is a projection of the major features of the whole object onto a plane outside the object. The set of overall views of the object contain projections of only the gross edges. Thus, every gross edge of the object is represented as an edge or a vertex in every overall view. Similarly, every object vertex that is the intersection of gross edges appears as a vertex or a point in an edge in every overall view.

A detail view is a projection of a portion of the object. The view has a defined polyhedral boundary and two extents along the projection direction. The boundary and extents define a right prismatic region in 3-space. The detail view is a projection of all edges and vertices of the object contained in the region. A detail view contains projections of both the gross and detail edges, without distinction, contained within its defined region.

A cross sectional view may be either overall or detail. The view is a planar cross section normal to the projection direction. In this case the view transformation contains the location of the section plane in the coordinate frame of the object. Note that edges are shown at the cross section plane that may not be present in the object (they lie in surfaces of the object), and may not be shown in other views of the object.

2. Construct Pseudo Vertex Skeleton.   This stage proceeds in a manner similar to before. However, somewhat greater care must be taken to treat the various projections consistently. Intersections between back projections of vertices from appropriate pairs of different views are considered candidate vertices. Appropriate means noncolinear projection directions and the same type of view, i.e., both overall or both detail. In the case of pairs of detail views, the intersection point must lie within the intersection of their respective prismatic regions. In the case of a cross sectional view, the intersection point must lie in the halfspace defined by the section plane and projection direction. Also, a cross sectional view generates a set of vertices and edges in the plane of the view.

3. Construct Pseudo Wire Frame.   This stage is essentially unchanged from Stage 3 in Section 3. However, the following is a very useful observation: whenever a view shows two noncolinear solid (i.e., visible) lines intersecting internally in a point, p, then there must be some vertex of $\mathcal{O}$ visible in the appropriate direction which projects onto p and which has only visible edges incident with it corresponding to the solid lines incident with p. In particular, if in moving along the perpendicular from p one first encounters candidate vertices which are clearly not vertices of $\mathcal{O}$ (see discussion of Stage 3 in Section 3), then these vertices and all incident edges may be discarded. To appreciate the power of this observation see Example 4.

4. Construct Virtual Faces.   This stage is essentially the same as Stage 4 in Section 3. However, it is possible at this point to use line type depth information to edit out some type II vertices and uncertain candidate edges, as well as to extract additional information for use at a later time.

The cross reference lists from view edges to edges in the wire frame are concatenated with the list of original (i.e., before any partitioning) virtual faces and sorted by distance along the projection direction from the mid-point of the view edge. For any edge that is visible, i.e., not dashed, the nearest pseudo wire frame edge is identified. Any interposing virtual faces cannot exist and are deleted. For an edge to be dashed, there must be at least one occulting virtual face in the projection direction. If there is only one such face, then it must be a real face separating solid material from space, and since the projection is from outside, the directedness of the face is known. This information is fed forward to the decision process as initial certain states of blocks and faces. As before, the consequences must be fully propagated.

5 and 6. Introduce cutting edges and form virtual blocks. These stages are the same as in Section 3.

7. Make Decisions. This stage again is very similar to the corresponding stages described in Section 3. Clearly, however, the decision procedure must accommodate the drawing conventions in the correct manner. It is fairly apparent how this is to be done. Thus, for example, in the case that occulted edges are represented explicitly in views, each view edge must contain a visible edge in the view projection direction, and each nonvisible view edge must be occulted by an interposed face in the view projection direction. ■

The examples in the next section illustrate the points made above. As shall be seen, pathological features do not appear to be common in objects of practical interest.

5.   EXAMPLES

To clarify the discussion in Sections 3 and 4, several examples are presented in this section. The examples are chosen to illustrate particular features of the algorithm and some of the performance trade-offs involved in providing extra information.

• Example 1—Octahedron Projections

The octahedron illustrates a simple problem having many solutions, but for which the Projections algorithm does not need to introduce any cutting edges. Figure 7.4 shows three views of an octahedron. It is interesting to determine the set of all objects having the identical projections. The back projection process generates the 12 edges of the octahedron with type certain and the three intersecting diagonals with type uncertain. In a wire frame example of an octahedron (see p. 245, this volume) it was shown that the diagonal edges must be introduced as cutting edges for the Wire Frame algorithm to handle the mutually intersecting

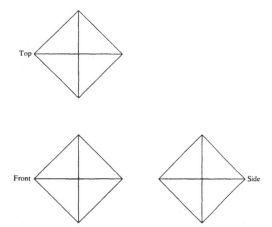

FIG. 7.4.   Three views of an object related to an octahedron.

interior virtual faces. In the Projections case, the algorithm proceeds with no need to generate further edges and enters the decision process with eight virtual blocks, one for each octant around the intersection point of the diagonals. Since the interior edges are of type uncertain and the exterior are all of type certain, any selection of octants such that no two hole octants share a face is a solution. The decision process finds 35 solutions:

   1 with all octants solid,
     8 with one octant a hole,
   16 with two octants holes,
     8 with three octants holes, and
     2 with four octants holes.

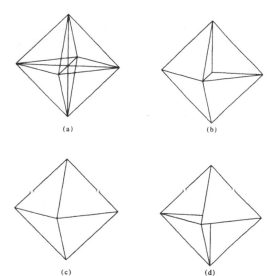

FIG. 7.5.   The solution to the problem of Fig. 7.4: (a) the pseudo wire frame; all external edges are of type certain, internal edges are of type uncertain; (b, c, d) some of the 35 solid objects with the views of Fig. 7.4.

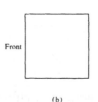

Top

(a)

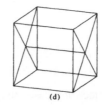

(c)

Front

(b)

(d)

FIG. 7.6. (a, b) Two views of an object related to a cube; (c) the pseudo wire frame; (d) the pseudo wire frame with a cutting edge inserted.

A sampling of these solutions is shown in Fig. 7.5. Note that in this case dashed lines do not reduce the amount of ambiguity. ∎

• Example 2—Cube Projections

The cube illustrates a simple use of cutting edges. Figure 7.6 shows two views, front and top, of a cube. Again the Projections algorithm determines the number of objects having the same two views. The back projection process finds the cube

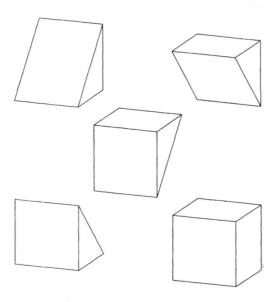

FIG. 7.7. Objects with the views shown in Fig. 7.6 (a, b).

FIG. 7.8. The Two Y's, a well known mechanical drawing puzzle: front and top views of an object.

edges, albeit as type uncertain. However, in the direction perpendicular to the two given views, the cube face diagonals are found without intersection. A cutting edge is inserted between the intersection points of the face diagonals, and five virtual blocks are found (the envelope and four quadrant blocks). Five solutions are found as shown in Fig. 7.7. Note that if all three views of a cube were furnished, there would be a unique solution to this projections problem. ■

• Example 3—Two Y's Problem

Figure 7.8 shows a well known mechanical drawing puzzle: find all objects having the top and front views shown. Because of the way edges line up in the two views, the back projection process finds the pseudo skeleton with 29 edges and 12 vertices shown in Fig. 7.9. Intersections of the edges yield three addi-

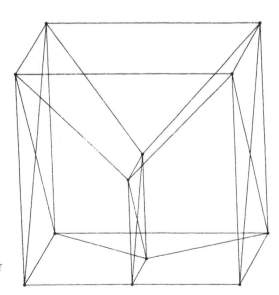

FIG. 7.9. The pseudo skeleton of the Two Y's problem.

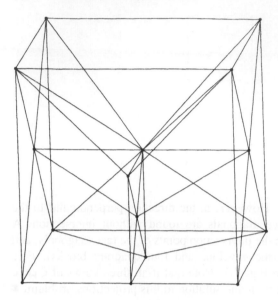

FIG. 7.10.   The pseudo wire frame with cutting edges added.

tional vertices where the diagonals intersect, and intersections of virtual faces yield eight cutting edges. The final pseudo wire frame is shown in Fig. 7.10. The 16 internal virtual blocks found in Stage 7 are shown in Fig. 7.11. Under the assumptions of Section 3 there are 55 solutions to this problem. Under the assumptions of Section 4 (i.e., all lines in the views are assumed to be solid, that is, visible) there are seven solutions. These are shown in Fig. 7.12.

FIG. 7.11.   Sixteen virtual blocks found from the two views of Fig. 7.8.

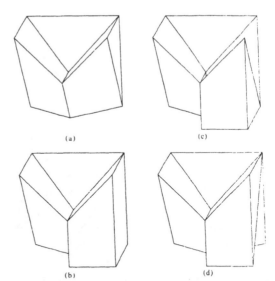

FIG. 7.12. Objects with the two views of Fig. 7.8: (a) is symmetric; (b.c.d) are asymmetric and each is typical of a pair of objects.

The Two Y's problem is very sensitive to numerical considerations. If the branch point of one of the Y's is moved from the center of its view, there are no solutions to the corresponding projections problem. ■

• Example 4—Three X's Problem

The Three X's problem illustrates vividly the savings that can result from the use of depth information. Figure 7.13 shows an apparently minor modification to the problem of Fig. 7.4; the object is now clearly contained within a cube. However, further investigation shows that the solution process becomes surprisingly com-

Top

Front

Side

FIG. 7.13. The Three X's problem: three views of an object whose extent is bounded by a cube.

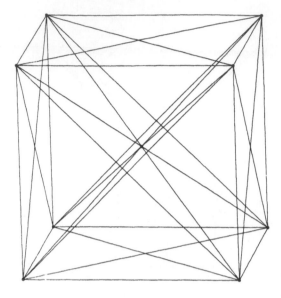

FIG. 7.14.   The pseudo skeleton for the Three X's problem.

plex. The back projection process produces nine vertices—the cube vertices (uncertain) and its midpoint (certain) and thirty-two edges, all uncertain—the twelve cube edges, twelve face diagonal edges (initially with type II intersections, but later changed to mutually exclusive intersections), and eight cube diagonals from the midpoint. Note that, in contrast to the situation with Fig. 7.4, none of the edges found are of type certain and that ambiguities can be expected to stem from this lack of definite information. The pseudo skeleton that is obtained by back projection is shown in Fig. 7.14.

In the case without depth information, i.e., all edges drawn regardless of occultation, the partitioning process, of intersecting edges to generate sub-edges and virtual faces to generate sub-virtual faces with cutting edges, divides space into many small regions. A total of 96 internal virtual blocks are found and the decision process uncovers 38 065 solutions. Clearly, searching a 96-level decision tree for 38 065 solutions is a complex process. The solution is made practicable by making heavy use of the mappings and correlations between original faces and edges and their partitioned forms. One solution, picked at random, is shown in Fig. 7.15. The object is hard to understand, even with a model in one's hand. It is a set of three tetrahedra, a pair with a common face and a third with edge contact only, i.e., it is decomposable into two disjoint objects. The solutions found could be filtered to reject unstable objects of this form, but this test has not been executed.

The analysis of the case with depth information shows the power of Theorem 7 when used in Stage 3. Each of the three views shows solid lines intersecting in the center point. Following a perpendicular from any of the center points of any

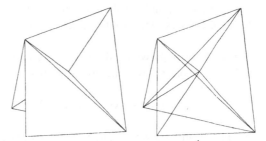

FIG. 7.15. One of 38,065 objects found with the views of Fig. 7.13 and assuming that all edges are shown in each projection. The object is based on three tetrahedra.

view leads first to a point in the center of a face of the cube containing the pseudo wire frame. This vertex cannot be a Class I vertex since all candidate edges incident with it are coplanar. It also cannot be a Class II vertex since all solid material lies to one side of the plane containing the face in question. Thus, the center points and all diagonal edges may be discarded from the front, top, and appropriate side faces of the cube. Furthermore, in Stage 4 corresponding faces of the cube are also discarded since they would obscure a vertex and lines in the interior. With these faces discarded, three of the leading edges of the cube must be discarded also since they no longer contain at least two noncoplanar virtual faces.

After these reductions, the algorithm goes on to find the ten solutions shown in Fig. 7.16. The solutions may be considered as being based on the union of three pyramids, as shown in Fig. 7.16(a). In all solutions, the view of the objects in the projection directions are the four triangular faces of the union of the three pyramids. The distinguishing features between the solutions are cavities in the

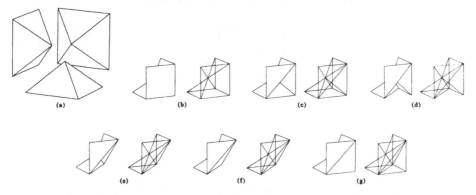

FIG. 7.16. All ten solutions to the Three X's problem in dashed line mode: (a) three pyramids forming the basic solutions; (b) three pyramid solutions; (c) one pyramid bisected; (d) two pyramids bisected; (e) all three pyramids bisected; (f) solution (b) cut by plane containing the diagonals of the square faces; (g) solution (b) with an internal tetrahedral cavity.

"rear"; the viewpoint for the solutions in Figs. 7.16(b)–(g) is chosen to illustrate these cavities. The solutions are grouped as follows:

- Fig. 7.16(b) shows all three pyramids complete,
- Fig. 7.16(c) shows one pyramid bisected and is one of a set of three solutions,
- Fig. 7.16(d) shows two pyramids bisected and is one of a set of three solutions,
- Fig. 7.16(e) shows all three pyramids bisected,
- Fig. 7.16(f) shows the object (b) above cut by the plane containing the diagonals of the three square faces,
- Fig. 7.16(g) shows the object (b) with an internal tetrahedral cavity just visible as diagonal edges of the square faces.

• Example 5—Two Ramps Problem

The Two Ramps problem illustrates the effectiveness of the pruning operations of Stage 3. Figure 7.17 shows this well-known two-view puzzle problem reputed to have twelve solutions. The back projection process produces an array of three by four, i.e., twelve, vertices on the left-hand face and an array of two by four for the right-hand face. Twelve edges are found linking the left and right sides. However, the number of possible edges in the end faces, i.e., in the direction normal to the two given views, is large; see Fig. 7.18. Fortunately many of those in the left-hand face are rejected by the edge and virtual face connectivity test at the end of Stage 3 (Fig. 7.19). Some 108 internal virtual blocks are found and 107 distinct solutions. Only 12 of the solutions, however, pass the stable object criterion. Some of the solutions are shown in Fig. 7.20. ■

Top

Front

FIG. 7.17.   The Two Ramps problem: Two views of an object.

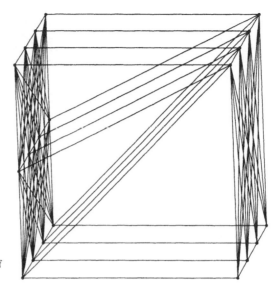

FIG. 7.18. The pseudo skeleton of
the Two Ramps problem.

• Example 6—Real Engineering Objects

After developing the algorithm to be as general as possible, and proving it with
problems chosen for their geometric difficulty and ambiguities, it is refreshing to
look at some real engineering objects and consider their reconstruction from their
three standard views. Figures 7.21 and 7.22, parts (a), (b), and (c), show two
examples of engineering objects. Even without using depth information, only

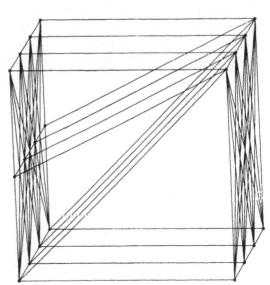

FIG. 7.19. Pseudo wire frame after
pruning in Stage 3; note the reduction
of edges in the left-hand face.

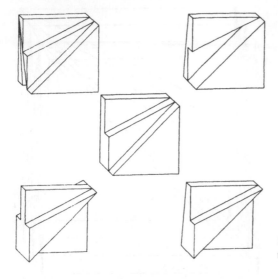

FIG. 7.20. Some of the 107 solutions to the Two Ramps problem.

one solution is found to each object, and the reconstructed objects are shown in Figs. 7.21(d) and 7.22(d). It is apparent from the views that the polyhedral approximations of the cylindrical holes in the objects greatly increase the number of vertices and edges to be handled and that the projections of these polyhedral features can lead to many small edges in the view, indicating potential for numerical problems. However, our implementation of the Projections algorithm does not have problems in these areas with these examples. Further, it is clear that objects of this complexity raise real problems in ensuring the validity of the input data. The three views used as input in this example were obtained from an existing model and were therefore guaranteed to be correct. A human generating these views directly would have some difficulty ensuring their correctness and self consistency. The Projections algorithm in its present form does not attempt to handle incorrect (or incomplete) data. ∎

6. SUMMARY

The Projections algorithm presented in this paper finds all polyhedral objects $\mathcal{O}$ with a given set of projections. It has been shown that, if the projections are labeled, the problem may be solved by the Wire Frame algorithm (see ch. 6); in the unlabeled case an extended form of the Wire Frame algorithm, the Projections algorithm, is needed.

It has been shown that an inverse projection process may be used to construct a superset of Class I vertices of $\mathcal{O}$ (vertices contained in at least three noncoplanar edges) together with a superset of unions of edges of $\mathcal{O}$. These edges and vertices constitute the skeleton of $\mathcal{O}$.

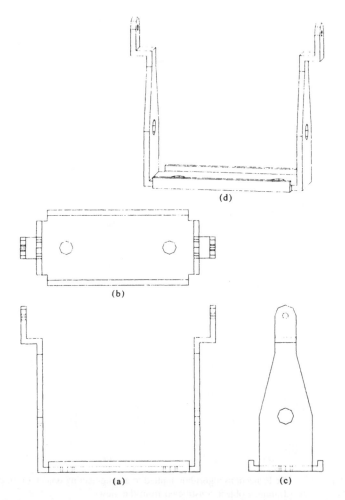

FIG. 7.21.   The Projections algorithm applied to an engineering object: (a, b, c) three views; (d) unique object constructed from the views.

It has also been shown that a superset of the Class II vertices of $\mathcal{O}$ (vertices contained only in a set of coplanar edges) may be found as intersections of skeleton edges. These updated vertices and edges constitute a pseudo wire frame.

A pseudo wire frame differs from a wire frame in that it contains supersets of the edges and vertices of the wire frame. Some of these elements have been identified uniquely and have type certain; the rest are of type uncertain. Any object whose wire frame is composed of the certain elements of the pseudo wire frame and any subset of the uncertain elements and produces the correct projections is a solution.

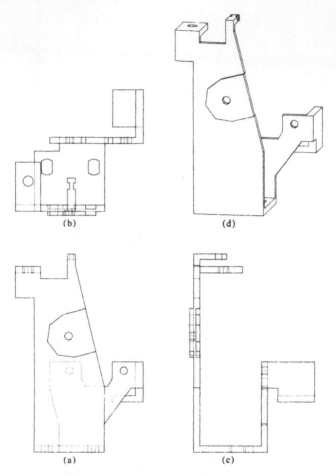

FIG. 7.22. The Projections algorithm applied to an engineering object: (a, b, c) three views; (d) unique object constructed from the views.

The pseudo wire frame is processed to find candidate faces (virtual faces). Virtual faces are connected to enclose volume regions (virtual blocks). A depth first decision process with heavy pruning is used to find all state assignments of hole or solid to virtual blocks that produce solid objects with the correct projections.

The Basic Projections algorithm accepts projections of the wire frame of O; extensions handle a more general set of projection types (detail, overall, and cross sectional) and projection conventions such as depth information obtained from occulted edges in a projection being shown as dashed.

The Projections algorithm has been implemented and its operation has been illustrated by a set of examples. These examples have shown that problems of a mechanical drawing puzzle nature, which typically have high degrees of symmetry leading to large numbers of uncertain elements in the pseudo wire frame, can have very large numbers of solutions. On the other hand, engineering objects, with projections sufficiently complex to require careful thought from a human, have been run and have produced unique solutions.

ACKNOWLEDGMENTS

Our thanks are due to D. D. Grossman for encouraging us to tackle the projection reconstruction problem. A. F. Nightingale suggested the Two Ramps problem, and V. J. Milenkovic implemented the stable object test.

REFERENCES

Clowes, M. B. (1971). On seeing things. *Artificial Intelligence 2*, 79–116.

Hocking, J. G. & Young, G. S. (1961). *Topology*. Reading, MA: Addison-Wesley.

Huffman, D. A. (1971). Impossible objects as nonsense sentences. In B. Meltzer & D. Michie (Eds.), *Machine Intelligence* (Vol. 6, pp. 295–324). Edinburgh, Scotland: Edinburgh University Press.

Idesawa, M. (1973, February). A system to generate a solid figure from a three view. *Bull. JSME, 16*, 216–225.

Idesawa, M., Soma, T., Goto, E., & Shibata, S. (1975). Automatic input of line drawing and generation of solid figure from three-view data. *Proceedings of the International Joint Computer Symposium*, pp. 304–311.

Kanade, T. (1978, September). A theory of origami world. *(Report No. CMU-CS-78-144)*. Department of Computer Science, Carnegie-Mellon University, Pittsburgh.

Lafue, G. (1976, July). Recognition of three-dimensional objects from orthographic views. *Proceedings 3rd Annual Conference on Computer Graphics, Interactive Techniques, and Image Processing*, ACM/SIGGRAPH, pp. 103–108.

Preiss, K. (1980, April). Constructing the 3-D representation of a plane-faced object from a digitized engineering drawing. *Proceedings of International Artificial Intelligence Conference*, Brighton, England.

Sugihara, K. (1981, May). *Mathematical structures of line drawings of polyhedra—towards man-machine communication by means of line drawings*. Third Laboratory of the Department of Information Science, Faculty of Engineering, Nagoya University, Nagoya Japan.

Sutherland, I. (1974, April). Three dimensional data input by tablet. *Proc. IEEE, 62*.

Waltz, D. (1975). Understanding line drawings of scenes with shadows. In P. H. Winston (Ed.), *The psychology of computer vision* (pp. 19–91). New York: McGraw-Hill.

Wesley, M. A. (1980). Construction and use of geometric models. Ch. 2, In J. Encarnacao (Ed.), *Computer aided design. Lecture notes in computer science No. 89*, New York: Springer-Verlag.

Wesley, M. A., Lozano-Pérez, T., Lieberman, L. I., Lavin, M. A., & Grossman, D. D. (1980, January). A geometric modeling system for automated mechanical assembly. *IBM J. Res. Develop., 24*, 64–74.

Woo, T. C. & Hammer, J. M. (1977). Reconstruction of three-dimensional designs from orthographic projections. *Proceedings of 9th CIRP Conference*, Cranfield Institute of Technology, Cranfield, England.

ACKNOWLEDGMENTS

REFERENCES

Author Index

Numbers in *italics* indicate pages with complete bibliographic information.

A

Aggarwal, A., 158, *184*, 165, *181*
Aho, A., 150, 153, 161, *181*
Ahuja, N., 166, *182*, 213, *226*
Akers, S., 97, *141*
Akman, V., 214, *227*
Ambler, A. P., 17, *42*
Anderson, D. C., 15, *42*
Ariel-Sheffi, E., 98, *143*
Arnold, D. B., 213, *226*
Arnon, D. S., 127, 130, *141*
Asano, T. (1), 169, *182*
Asano, T., (2), 169, 170, *182*
Aurenhammer, F., 214, 215, *226*
Avis, D., 154, 169, *184*, 215, 217, *226*

B

Baer, A., 230, *258*
Baker, B., 99, 138, *141*, 165, 166, *182*
Baldwin, M., 158, *184*
Ballard, D. H., 28, *40*
Baltsan, A., 215, *226*
Bates, J. L., 21, *40*
Bellos, I. M., 17, *42*

Ben-Or, M., 128, 137, *141*, 148, *182*
Bentley, J. L., 65, *91*, 149, 150, 154, 166, *182*, 192, 200, 212, *226*
Berman, L., 137, *141*
Bhattacharya, B. K., 154, *182*, 214, 215, 217, *226*
Binford, T. O., 16, 18, 20, 39, *41*, *42*
Birdwell, N., 166, *182*
Birrell, N. K., 18, *42*
Blinn, J. F., 25, *40*
Blum, M., 97, *141*
Boots, B. N., 214, *226*
Borning, A., 23, *40*
Bowyer, A., 213, *226*
Boyce, J. E., 164, *182*
Boyse, J., 230, 256, *258*
Boyer, C. B., 90, *91*
Brady, M., 95, 107, *141*
Brooks, R., 18, 20, 34, *41*, 99, *141*
Brostow, W., 213, 215, 217, *226*
Brown, C. M., 18, 28, *40*, *41*
Brown, K. Q., 65, 90, *91*, 214, *226*
Brown, B. E., 230, 256, *258*
Brumberger, H., 213, *226*
Burridge, R., 30, *41*

297

C

Chand, D. R., 154, *182*
Chang, J. S., 164, 165, *181, 182*
Chazelle, B., 65, 90, *91, 92*, 154, 165, 166, 167, 168, 169, 170, 180, *182*, 214
Cheriton, D., 202, *226*
Chew, L. P., 215, *226*
Chien, R. T., 166, *182*
Clarkson, K. L., 200, *226*
Clocksin, W. F., 23, *41*
Clowes, M. B., 231, *258*, 260, *295*
Cole, R., 106, 140, *141*
Collins, G. E., 15, 34, *41*, 125, *141*
Constable, R. L., 21, *41*
Cook, R. L., 25, *41*
Cooke, G. E., 120, 122, *141*

D

David, E. E., 213, *226*
David, C. W., 213, *226*
Dehne, F., 214, *226*
DePano, A., 165, *183*
Dijkstra, E. W., 21, *41*
Dobkin, D. P., 65, 77, 90, *91, 92*, 164, 168, 169, 170, 180, *182*
Drysdale, R. L. III, 110, *141*, 164, *182*, 189, 193, 194, 208, 214, 215, *227*
Dussault, J. P., 213, *226*

E

Eastman, C., 230, *258*
Eddy, W., 153, *183*
Edelsbrunner, H., 90, *92*, 190, 214, 215, 219, 222, *226, 227*
El Gindy, H., 154, *183*
Ernst, H. A., 98, *141*

F

Faust, G. M., 154, *182*
Feng, H., 169, *183*
Ferguson, J., 11, *41*
Ferrari, L., 170, *183*
Finney, R. L., 120, 122, *141*
Finney, J. L., 213, 215, 217, *227*
Fisher, W. B., 18, *42*
Fitzgerald, W., 18, 20, *41*
Floyd, R. W., 21, *41*

Foley, J. D., 11, 25, *41*
Forbus, K. D., 20, *41*
Fortune, S. J., 138, *141*, 165, *182*, 202, 215, 227
Fox, B. L., 213, *226*
Franklin, W. R., 214, *227*
Franzblau, D. S., 169, *183*
Freeman, P., 39, *41*

G

Garey, M. R., 65, *92*, 133, *141*, 166, *183*
Goldman, R. N., 15, *42*
Goto, E., 231, *258*, 260, *295*
Gowda, I. G., 214, *227*
Gracer, F., 18, 20, *41*
Graham, R. L., 77, *92*, 146, 154, *183*
Green, P. J., 213, *227*
Greene, D. H., 168, 169, *183*
Greiner, R., 18, 20, *41*
Grosse, E. H., 138, *141*, 166, *182*
Grossman, D. D., 16, 18, *42*, 230, *258*, 260, *295*
Guibas, L. J., 90, *92*, 154, 164, *182*, 215, 227

H

Hammer, J. M., 260, *295*
Hanrahan, P., 25, *41*
Hartquist, E. E., 18, *42*
Henrion, M., 230, *258*
Hertel, S., 167, 168, *183*
Hoare, C. A. R., 21, *41*
Hocking, J. G., 231, *258*, 261, *295*
Hoey, D., 44, 65, *93*, 110, *142*, 158, 159, *184*, 213, *228*
Hollerbach, J. M., 95, 107, *141*
Hong, S. J., 77, *93*, 150, 154, *184*
Hopcroft, J., 33, 35, 37, *41*, 98, 99, 129–132, 134, *141*, 150, 153, 161, *181*, 190, 222, *227*
Huffman, D. A., 231, *258*, 260, *295*

I

Idesawa, M., 231, *258*, 260, *295*
Imai, H., 169, 170, *182, 183*, 215, *227*
Incerpi, J., 65, *92*, 167, 168, *182*
Iri, M., 215, *227*

J

Jarvis, R. A., 148, *183*
Johnson, T., 95, 107, *141*
Johnson, D. S., 65, *92*, 133, *141*, 166, *183*
Joseph, D., 33, 35, *41*, 98, 134, *141*

K

Kahn, J., 168, *183*
Kanade, T., 260, *295*
Kantabutra, V., 99, *141*
Kapur, S. S., 154, *182*
Katz, B., 16, 39, *42*
Keil, J. M., 169, 170, *183*
Kirkpatrick, D. G., 65, 90, *92*, *93*, 110, *141*, 151, 160, *183*, 193, 194, 199, 202, 204, 214, *227*
Klawe, M., 168, *183*
Klee, V., 158, 164, *183*, 215, 217, *227*
Kleitman, D. J., 168, 169, *183*
Knuth, D. E., 65, *93*, 150, 153, 157, 161, 166, *183*
Kosaraju, S. R., 99, *141*
Kozen, D., 97, 128, 129, 130, 137, *141*

L

Lafue, G., 231, *258*, 260, *295*
Laskowski, M. C., 158, 164, *183*
Lavin, M. A., 16, 18, *42*, 230, *258*, 260, *295*
Lee, C., 97, *141*
Lee, D. T., 43, 90, *92*, *93*, 110, *141*, 154, 165, *182*, 189, 193, 194, 208, 213, 214, *227*
Leu, M. C., 20, *41*
Leven, D., 119, *141*, 215, *227*, *228*
Levin, J., 14, 25, *41*
Lewis, H. R., 136, *142*
Lieberman, L. I., 16, 18, *42*, 230, *258*, 260, *295*
Lingas, A., 169, 170, *184*
Lipski, W. Jr., 170, *184*
Lipton, R. J., 65, *92*, *93*
Lowry, M., 16, 39, *42*
Lozano-Perez, T., 12, 16, 18, 33, 34, 38, *41*, *42*, 95, 97, 98, 99, 107, *141*, *142*, 230, *258*, 260, *295*

M

Maddila, S., 158, *184*
Mahajan, R., 20, *41*
Mahaney, S. R., 165, *182*
Mason, M. T., 37, 38, *42*, 95, 107, *141*
Massey, W. S., 36, *42*
Maurer, H. A., 90, *92*, 190, 219, 222, *227*
McCallum, D., 154, *184*
McMahon, T. A., 32, *42*
Meagher, D. J., 11, *42*
Megiddo, N., 159, 160, 165, *184*
Mehlhorn, K., 167, 168, *183*, 216, 219, 222, *228*
Mellish, C. S., 23, *41*
Milne, W. J., 213, *226*
Moore, E. F., 97, *142*
Moravec, H., 97, 112, *142*
Mount, D. M., 215, *228*
Muller, D. E., 90, *93*
Munro, J. I., 77, *92*
Murota, K., 215, *227*

N

Naamad, A., 214, *227*
Nelson, G., 23, *42*
Newell, A., 39, *41*
Newman, W. M., 155, *184*
Newman, C. M., 213, *228*
Nievergelt, J., 212, *228*
Nilsson, N. J., 97, *142*

O

O'Donnell, M. J., 21, *41*
O'Dunlaing, C., 99, 108, 119, 121, 137, *142*, 215, *228*
O'Rourke, J., 90, *92*, 158, 165, 169, *184*
Ocken, S., 14, *42*
Ogawa, T., 213, *228*
Ogita, N., 213, *228*
Ottmann, T. A., 65, *91*, 166, *182*, 192, 212, *226*
Overmars, M. H., 77, *93*, 154, *184*

P

Papadimitriou, C. H., 90, *93*, 136, *142*
Paul, R., 98, *142*
Pavlidis, T., 169, *183*

Pieper, D., 98, *142*
Pinter, R., 170, *184*
Popplestone, R. J., 17, *42*
Post, M. J., 164, *184*
Preparata, F. P., 43, 65, 77, 90, *92, 93,* 150,
    154, 166, *182, 183, 184,* 212, 216, *228*

R

Rafferty, C. S., 166, *182*
Raibert, M. H., 32, *42*
Rajan, V. T., 30, *42,* 107, *142*
Ramshaw, L., 154, *183*
Reif, J. H., 98, 128, 138, *142*
Requicha, A. A. G., 11, 18, 22, *42,* 230,
    256, *258*
Rinott, Y., 213, *228*
Rivest, R., 170, *184*
Rowat, P. F., 99, *142,* 215, *228*

S

Sack, J. R., 168, 170, *184*
Sankar, P. V., 170, *183*
Saxe, J. B., 65, *91*
Schachter, B., 169, *184*
Schorr, A., 99, 112, *143*
Schwartz, J. T., 14, 30, 33, 34, 35, *41, 42,*
    98, 99, 113, 116, 122–130, 134, *141, 142,*
    190, 222, *227*
Sederberg, T. W., 15, *42*
Sedgewick, R., 148, 150, 153, 166, *184*
Seidel, R., 90, *92,* 151, 154, 160, *183, 184,*
    215, 217, *227, 228*
Shamir, A., 170, *184*
Shamos, M. I., 44, 65, *93,* 110, *142,* 149,
    150, 154, 158, 159, 166, *182, 184,* 189,
    190, 193, 201, 206, 212, 213, 216, *228*
Shannon, C. E., 97, *143*
Shapiro, J. E., 18, *42*
Sharir, M., 14, 33, 34, 35, *42,* 98, 99, 112,
    113, 116, 119, 122–130, 134, 137, 138,
    141, 142, 143, 189, 190, 193, 200, 208,
    214, 215, 222, *227,* 28
Shibata, S., 231, *258,* 260, *295*
Shimano, B. E., 98, *143*
Sibson, R., 213, *227*
Silverman, B. W., 164, *185*
Sklansky, J., 170, *183*
Soma, T., 231, *258,* 260, *295*
Sommerville, D. M. Y., 15, *42*

Spirakis, P., 99, 135, *143*
Sproull, R. F., 155, *184*
Steiglitz, K., 90, *93*
Stolfi, J., 154, *183,* 215, *227*
Sugihara, K., 260, *295*
Supowit, K. J., 165, 169, *184*
Sutherland, I. E., 97, *143,* 229, 256, *258*

T

Tanemura, M., 213, 215, 217, *228*
Tarjan, R. E., 65, *92,* 166, *183,* 202, *226*
Tarski, A., 124, *143*
Taylor, R. H., 230, 256, *258*
Tilove, R. B., 230, *258*
Titterington, D. M., 164, *185*
Torrance, K. E., 25, *41*
Toussaint, G. T., 155, 158, 168, 169, 170,
    184, 185
Tversky, A., 213, *228*

U

Udupa, S. M., 98, *143,* 230, *258*
Ullman, J. D., 150, 153, 161, *181*

V

Van Dam, A., 11, 25, *41*
Van Leeuwen, J., 77, *93,* 154, *184*
Van Wyk, C. J., 23, *42*
Verrill, C., 214, *227*
Vick, J. W., 36, *42*
Voelcker, H. G., 11, 18, *42,* 256, *258*

W

Walker, R. J., 14, *42*
Waltz, D., 231, *258,* 260, *295*
Watson, D. F., 215, 217, *228*
Weide, B., 200, *226*
Wesley, M. A., 11, 12, 16, 18, *42,* 97, 99,
    142, 230, 256, *258,* 260, *295*
Whitesides, S. H., 33, 35, *41,* 98, 134, *141*
Widdoes, C., 98, *143*
Wilfong, G., 3, *41,* 129–132, 134, *141*
Winston, P. H., 16, 39, *42*
Wolfe, R., 18, 20, *41*
Wong, C. K., 214, *227*
Woo, T. C.-H., 230, 256, *258,* 260, *295*

Y

Yao, A. C., 148, *185,* 200, *226*
Yao, F. F., 154, *183*
Yap, C. K., 99, 106, 108, 119, 121, 129, 130, 135, 137, 140, *142, 143,* 164, 165, *181, 182,* 193, 215, *228*
Yen, R., 166, *182*
Young, G. S., 231, *258,* 261, *295*

Subject Index

A

ACRONYM, 18, 20
adjacency graph, 118, 120
adjacency of cells, 120, 127, 129
algebraic surfaces, 11–16, 25
algebraic topology, 36
analysis of algorithms, 2
approximate algorithms, 154

B

bicubic patches, 15
binary search, 44

C

cell complex, 122, 129, 130
cell decomposition, 15, 120
circle enclosure, 158–163
clearance, 103, 109
collision-avoidance (*see* motion planning)
complete problems, 133ff.
complexity theory, 1, 5, 132–133
compliant motion, 22, 139
computational geometry, 2
computer aided assembly (*see* computer aided design)
computer aided design (CAD), 229–231, 260
computer drafting (*see* computer aided design)

computer graphics, 25
computer science, 1, 8
constraint programming, 23, 24
constraints, 101
contact motion, 130
convex hull, dynamic, 72–74, 154
convex hulls, 67–74, 146–158
corner, 109
couplings, 101
covering of polygons, 169
CSG (constructive solid geometry) (*see* geometric modeling)
cycles, 1- and 2-cycles, 233, 236
cylindrical decomposition (*see* cell decomposition)

D

decomposable problems, decomposable search, 62–65
decomposable problems, order decomposable, 75–77
decomposition of polygons, 168ff.
detection of intersection and proximity, 166, 187ff., 230
divide-and-conquer, 66, 149–150
duality (geometric) (*see* transformations, geometric)
dynamic polygon search, 60–62

dynamic intersection and proximity problems, 216
dynamic motion planning, (*see* motion planning)
dynamization, 222, (*see also* decomposable search problem)

E

enclosure problems, 158–165
environment, 102
error tolerance, 22
Euclidean group (*see* rigid transformations)
Euler operators, 231
Euler's relation, 58

F

feature, 16
findpath problem (*see* motion planning)
finite element method (FEM), 29, 32
friction, 30

G

GDP/GRIN, 18, 20
generic object (*see* object class)
geometric modeling, constructive solid geometry, 11–12, 26–28, 230
geometric transformations (*see* transformations)
gift-wrapping, 148
Gouraud shading, 25
Graham scan, 146
graphic display, 25
grasping, 138

H

Hausdorff distance, 109
homology theory, 36

I

interference of objects (*see* intersection of objects)
internal structure, functional dependency, 19
internal structure, object states, 19, 20
intersection, of colored objects, 188, 198
intersection, of rectilinear boxes, 218–222
intersection, of spheres, 222–
intractibility, 2 (*see* also tractability)

K

kinematic motion-planning, 107, 137

L

Laguerre diagram and geometry, 214–215
line-sweep, 166

M

marriage-before-conquest, 151–153
Mayer-Vietoris theorem, 36
mechanical design system (*see* computer aided design)
minimum spanning tree, Euclidean, 201
motion, 103
motion planning, 95ff
motion planning, coordination problems, 139
motion planning, decomposition approach, 120
motion planning, fine, 37
motion planning, for a disc, 108–112
motion planning, for a ladder, 112–119
motion planning, heuristics, 105–106, 145
motion planning, manipulator arms, 98
motion planning, mobile robots, 97
motion planning, offline and realtime, 106, 188–189
motion planning, the basic problem, 105, 137
motion planning, the decision problem, 106, 128
motion planning, the dynamic problem, 138
motion planning, the general problem, 103, 123
motion planning, the local problem, 105
movers' problem (*see* motion planning)

N

NP-hard, 133
numerically controlled machining (*see* computer aided design)

O

object, 232
object class, 18, 20, 22
object editing, 23
object, internal states, 19
object, primitive, 236
objects, ambiguities of reconstruction, 256, 282–292

objects, reconstruction from projections, 259ff.
objects, reconstruction from wire frames, 230ff.
obstacle-avoidance (*see* motion planning)
offline programming, 106, 188–189 (*see* also robot simulation)

P

PADL II, 18, 230
parameterized object (*see* object class)
partial differential equations, 29, 32
physical space, 99
PL/CV, 21
placements, 100
placements, admissible, 100
placements, free, 100
placements, semifree, 102
polygon search, 58–60
polygon-cutting theorem, 166
PRL, 21
program correctness, 21
program verification (*see* program correctness)
projection, 263
projection, unlabeled, 269
Prolog, 23

Q

quadric surface, 14

R

raytracing, 25
rectangle search, 48–58
reduction of problems, 133
representation, boundary, 13
representation, hierarchical, 17
representation, surfaces and solids, 11–16
retraction, 99, 111, 123, 129
rigid bodies, 100
rigid transformations, 99–100, 125
robot simulation (offline programming), 8, 20, 28–32
robotics, 1

robots, arms, 98
robots, mobile, 97
rotating calipers, 155

S

Sard's lemma, 116
scene analysis, 231, 260
semi-algebraic sets, 124
sensory feedback, 22, 140
shape complexity (sinuosity), 167
shortest paths, 99, 112
simulation of physical processes. (*see* robot simulation)
skeleton, 264
stereophenomenology, 10, 40
swept volume, 12

T

Tarski, formulae and language, 124, 137
topological concepts, 256–258
tractability, 128
trajectory planning, 107
transform, binary, 61–63
transform, reverse triangular, 64
transform, triangular, 63–64
transformations, geometric, 44, 77–90
transformations, point/line dualities, 78–87
transformations, point/point duality, 87–89
triangulation, 165, 200

V

vision, 231, 260
Voronoi diagrams, 44, 99, 110, 187ff, 192
Voronoi diagrams, applications, 213
Voronoi diagrams, for convex objects, 208
Voronoi diagrams, various generalizations, 213
Voronoi dual, 200

W

wall, 109
wire frame, 11, 229, 230, 233, 237ff.